Reading the Shape of Nature

Science and Its Conceptual Foundations

DAVID L. HULL, EDITOR

Reading the Shape of Nature

COMPARATIVE ZOOLOGY AT THE AGASSIZ MUSEUM

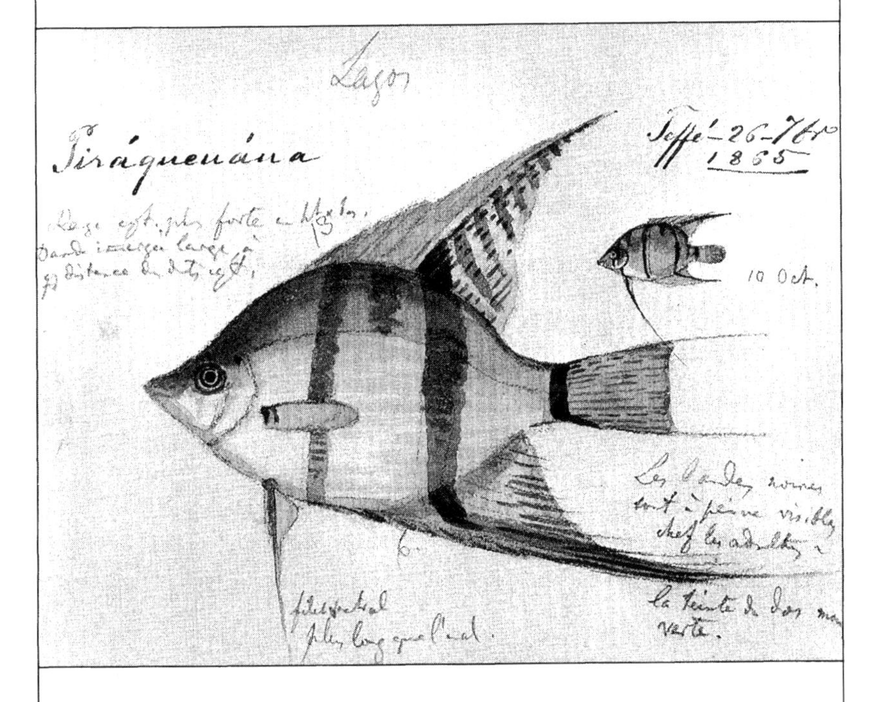

MARY P. WINSOR

The University of Chicago Press
CHICAGO AND LONDON

Mary P. Winsor is associate professor at the Institute for the History and Philosophy of Science and Technology at Victoria College, University of Toronto.

Title page illustration: Field sketch of Amazonian angelfish made by Louis Agassiz's artist Jacques Burkhardt in 1865, with notations by Agassiz. (By permission of the Museum of Comparative Zoology Archives, Harvard University)

The University of Chicago Press, Chicago 60637
The University of Chicago Press, Ltd., London

Printed in the United States of America

00 99 98 97 96 95 94 93 92 91 5 4 3 2 1

Library of Congress Cataloging-in-Publication Data

Winsor, Mary P.
Reading the shape of nature : comparative zoology at the Agassiz Museum / Mary P. Winsor.
p. cm.—(Science and its conceptual foundations)
Includes bibliographical references and index.
ISBN 0-226-90214-5 (cloth);
ISBN 0-226-90215-3 (pbk.)
1. Natural history—Classification. 2. Evolution—Philosophy.
3. Harvard University. Museum of Comparative Zoology—History.
I. Title. II. Series.
QH83.W56 1991
574′.012—dc20 91-8742
CIP

The paper used in this publication meets the minimum requirements of the American National Standard for Information Sciences—Permanence of Paper for Printed Library Materials, ANSI Z39.48-1984.

To Ruth Dixon Turner

Comparative Zoologist *par excellence*

Contents

Illustrations

Preface

It is recorded that God brought before Adam every living creature for naming,* which would have made him not only the first man but the first taxonomist; it is recorded, too, that Noah took aboard his ship "every beast after his kind, and all the cattle after their kind, and every creeping thing that creepeth upon the earth after his kind, and every fowl,"[1] which would have made him the greatest of natural history collectors. At first we feel both stories imaginable, because we review in our minds all the creatures we can think of, and we picture Adam and Noah doing this and more. However, a fuller acquaintance with the actual diversity of living species (leaving aside the millions extinguished in the vastness of time) makes it obvious that the biblical patriarchs had been given tasks beyond any person's grasp, short of miraculous assistance. The diversity of life on earth immeasurably exceeds our imagination. A definitive inventory was undertaken by Carl Linnaeus in the mid–eighteenth century and has been carried on ever since by thousands of naturalists, yet although they have been scouring the wilds, naming hundreds of thousands of beasts and creeping things, we may now have the end in sight only because of a terrible cheat: we are erasing from existence the wilds themselves, residents and all. A pale reflection of the wonderful variety in the world around us can be found in field guides and old-fashioned natural history museums, both of which are expressions of the human urge to copy Adam by trying to encompass this magical diversity and subject it to our dominion.

In paying systematic attention to living creatures in all their delightful variety, botanists and zoologists discover a further quite unexpected dimension: profound resemblances connect all the species to one another.

* "And out of the ground the Lord God formed every beast of the field, and every fowl of the air; and brought *them* unto Adam to see what he would call them; and whatsoever Adam called every living creature, that *was* the name thereof. And Adam gave names to all cattle, and to the fowl of the air, and to every beast of the field" (*Genesis* 2:18–19).

One could suppose that hundreds of thousands of species should imply hundreds of thousands of different forms, each unique, but instead we discover a symphony of themes and variations. Sets of alikenesses link them, layer by layer. As Charles Darwin said in 1859, "From the first dawn of life, all organic beings are found to resemble each other in descending degrees, so that they can be classed in groups under groups."[2] A God of limitless power and imagination could have peopled a planet, we may imagine, with any quantity of species, each one different, but, in fact, on this planet we find weevils and ladybirds and june bugs, each unique in some respects, but every one bearing the stamp of some great beetle factory. From familiarity we rarely pause to appreciate the nonchaotic nature of life's diversity, but it is distinctive and significant. As surely as a footprint or a belly button is evidence of a prior event, the descending degrees of similarities among organisms attest to their family history.

The science that studies living diversity is now called systematics, while the art of classifying, so central to systematics, is called taxonomy, though at various times in the past these terms have been used nearly interchangeably. The history of systematics and taxonomy is complex and still little known.[3] One of its great themes, the naturalness of taxonomic groups, is of enduring philosophical interest, for we know that classifying is a human act imposed upon the world by language and reason, yet the classes seem to be more than mere invention. Living things are composed in such a way as to reward the diligent taxonomist and invite belief in their connectedness. Another leading theme in history must be the relationship between taxonomy and evolution. Certainly botanists and zoologists were improving their recognition of kinds, and were casting species into higher groups, for some two hundred years previous to the publication of Darwin's *Origin of Species* in 1859. Historians are beginning to show that the scientific demonstration of evolution had to wait upon taxonomy having reached a certain stage of maturity, and soon after it reached that stage the evidence for evolution became irresistible.[4] Taxonomists before Darwin achieved considerable success, as measured by their identification of particular taxonomic groups and by their improvements in taxonomic methods—those groups and methods that continued to be judged valid by later generations. Whether their achievements had the character of theory-free observation is an issue just beginning to be examined.

My previous studies in the history of systematics have been in one way or another episodic, as is this one. Using limited segments of the animal kingdom—barnacles or insects or radiates (coelenterates and echinoderms)—has convinced me of the value of pursuing selected episodes in technical detail.[5] My excursions into the systematic thought of a few leading evolutionists suggest that the links between Darwinism and taxonomy were by no means as straightforward as one might expect.[6]

A particularly curious feature of biology in the post-Darwinian period was the low esteem that beset the very fields of study that had given birth to evolutionary theory and offered the richest ground for developing it. Early in this century a number of leading biologists were expressing regret at the scorn and neglect suffered by systematics.[7] From the 1940s through the 1960s, Ernst Mayr carried on a fierce campaign, with notable success, to raise the status of systematic biology.[8] He wrote:

> One might have expected that the acceptance of evolution would result in a great flowering of taxonomy and enhancement of its prestige during the last third of the nineteenth century. This was not the case—in part for almost purely administrative reasons. The most exciting consequences of the findings of systematics were studied in university departments, while the very necessary but less exciting descriptive taxonomy, based on collections, was assigned to the museums.[9]

Foolish the historian who would ignore such a suggestion! Because nearly all taxonomic research takes place in museums, we should expect that the history of systematics cannot be properly understood until the effect of this location is taken into account. Thus a promising direction for orienting an investigative slice should be not across one point in time, nor along one group of animals, but within the precincts of one museum.

Harvard's Museum of Comparative Zoology in Cambridge, Massachusetts, seemed an ideal site for exploration into the post-Darwinian decades, as it is small enough to be manageable, yet large enough to be of international importance. Coincidentally, it had its beginning just at the moment Darwin's revolutionary book was published. In this century it has been the home of a number of influential evolutionary biologists, most notably Mayr himself, who served as director from 1961 to 1970, yet the museum was founded by one of evolution's most implacable foes, Louis Agassiz.* After his death in 1873, his son Alexander took charge of the museum and remained influential in its affairs until his own death in 1910. The two Agassizs thus promise a degree of continuity for the fifty-year period after the *Origin of Species*. Finally, the M.C.Z. is especially attractive for a historian because intelligent care has been devoted to the preservation

*His forename is pronounced "lewee," not "lewis." His surname has the first syllable emphasized and rhymes with "bag." The last syllable is pronounced "see," the "z" silent. The name thus sounds quite different from the surname "Agassi," which has the second syllable emphasized. The possessive form is of course "Agassiz's."

and order of its archives, an enlightened policy which has not been in force in many of its sister institutions.

Like a novelist using one family to explore the foibles and nobility of human nature, I have selected some episodes from the history of this museum to explore the richness and limitations of systematic zoology. A full institutional history has not been my intention, neither have I attempted a balanced record of the scientific activities of the Agassizs and their associates; some of the issues I have identified should challenge others to contribute to the history of systematics and its institutions.

The story can begin nowhere else than within Louis Agassiz's vivid dream in the 1850s of the new discipline he called "comparative zoology." The complex undertaking that reflected that vision collapsed surprisingly quickly, but his museum continued, thanks largely to the sudden new wealth of Alexander Agassiz, loyally committed to carrying on his father's plans. I have long been aware that the younger Agassiz accepted evolution, but I was surprised to discover that his views on the nature of classification and the value of museum collections were not only radically opposed to his father's beliefs but just as radically out of step with those of his own contemporaries. The impression of continuity and community enterprise created by the M.C.Z.'s *Annual Reports* left me unprepared for the isolation and neglect I found afflicting the collections during Alexander Agassiz's directorship. Rather than leaving the museum in the gloomy state things had reached by 1910, when Alexander Agassiz died, we are permitted a peek forward to the rescuing knight, Thomas Barbour, who became director in 1927, forming a coda to the era of the Agassizs. I do not touch upon the rescue of systematics itself, however, which is a later and entirely different story.[10]

In 1884 Alexander Agassiz, after years of pouring his efforts and fortune into the M.C.Z., complained, "I have allowed myself from sentimental reasons to carry out plans which are not my own and [in] which I had but little interest, practically sacrificing any views or intuitions of my own . . . nobody should undertake another man's work if he has any he can do himself."[11] Certainly the hopes and dreams which began the M.C.Z. were not Alexander Agassiz's but his father's. It was Louis Agassiz, driven by the fervent belief that natural classification approaches the Mind of the Creator, who had insisted on the need for a museum in Cambridge.

The Museum of Comparative Zoology, along with every such museum of natural history, embodies the dream of Linnaeus. With a marvelous mixture of humility and arrogance, he believed that God had appointed him to be a second Adam, responsible for naming (and knowing) every kind of created thing. Linnaeus inspired his students with the vision of a great authoritative catalogue of the whole diversity of the world, all arranged in one hierarchical classification. The task soon proved larger than he had ever imagined, but his ambition continued to be cherished by subsequent generations of naturalists, and it is very much alive today.

Acknowledgments

A grant from the National Science Foundation under the aegis of the American Academy of Arts and Sciences supported the research for this book, and I am grateful to them for their confidence and patience. I thank also Victoria University for supplementary funding and the University of Toronto for research leave. Officers and staff of the Museum of Comparative Zoology at Harvard, particularly Eva Jonas and others in the Library, were unfailingly helpful. The M.C.Z.'s Mollusk Department, beginning with the late William J. Clench's encouraging hospitality to a high school student thirty years ago, has assisted me in countless ways over the years; I thank Marion Britz, Richard Johnson, Kenneth J. Boss, and most especially Ruth D. Turner for repeated welcomes and numerous favors.

I have been very fortunate in the helpers I have employed at different times, chief among them Ann Blum and Sharon Kingsland, whose good ideas I have absorbed and whose keen interest renewed my own. Charlotte M. Porter, Margaret Monis, and James Ireland have also made me gifts of their insights while working for me.

I am grateful to Alan T. R. Powell, Trevor H. Levere, and Jed Z. Buchwald for generous gifts of concrete help, and also to many other supporters, including Arthur Cain, Stanley Weitzman, Richard P. Vari, Roy Pearson, Ron DeSousa, Ralph Dexter, Roger Hansell, Gordon McOuat, Tim de Jager, and the late Marie Prince Jones. A number of people at the University of Chicago Press, chief among them Susan Abrams, were kind as well as helpful; I am especially indebted to David Hull, whose constructive criticisms were as speedy as my revisions were slow.

Throughout this project I have stood on the shoulders of Ernst Mayr and Edward Lurie. They have both been very patient and generous to me in various ways and have made astute suggestions on the manuscript, some of which, alas, I have neglected. Above all, however, I am obliged to these men for their writings, from which I have derived, again and again, information, insight, and inspiration.

I acknowledge with thanks the cooperation of several institutions, including the American Philosophical Society, the University of Rochester, the American Museum of Natural History, the Boston Museum of Science, the Academy of Natural Sciences of Philadelphia, the Archives of the Smithsonian Institution, and the Massachusetts Historical Society. I am particularly grateful to Cornelius Conway Felton, Jr., Charles P. Lyman, A. Hyatt Mayor, Anna Prince Jones, and Mrs. Samuel Hallowell for permission to quote from the unpublished writings of their ancestors. Without such a liberal attitude as theirs, historical scholarship would be crippled.

In my quotations from unpublished letters and diaries, I have amended spelling and punctuation, and expanded abbreviations, sparingly but silently, where helpful for clarity.

1

"In the Prime of His Admirable Manhood"

When an invitation to deliver the Lowell Lectures in Boston reached Louis Agassiz in 1845, he was nearing thirty-eight years of age, living in his native Switzerland, directing a small museum, and teaching natural history in the town of Neuchâtel.* This was a man whose profound and pious love of nature was coupled with a passionate vision that biology must adhere to new standards of scientific rigor, exemplified by Döllinger's embryological research and Cuvier's comparative anatomy. Yet more than ideas impelled him; a hungry ambition to accomplish vast projects drove Agassiz, as early as his undergraduate years, to collect about himself an array of coworkers, from colleagues inspired by his plans to youths taken into his home. A printer in Neuchâtel worked only on Agassiz's productions; he retained a full-time artist, Jacques Burkhardt; and his students pursued research he set out for them. Besides a monumental survey of fossil fishes and observations on glaciers that swelled into the dizzying picture of an entire continent once buried under ice, he and his helpers were studying the embryology of fish, geographical distribution, and the classification of echinoderms.† His colleagues looked on in wonder.

*Lurie's *Louis Agassiz: A Life in Science* is a model of historical scholarship, based on a wide range of sources and written with balance and insight. My statements about Agassiz are based upon Lurie's book unless otherwise credited.

†Most people use the zero plural "fish" (and "starfish" and "crayfish") most of the time, switching to "fishes" only when we want to emphasize the plurality (as in "loaves and fishes"). Systematic zoologists, conscious of the multiplicity of kinds of fish, deliberately choose "fishes" when referring to more than one species. I have steered a middle course, keeping the zero plural for an undifferentiated crowd ("barrel of fish"), though several species may be present. Of course, the generalized type uses the singular form ("the rights of man," "the behavior of fish"), but zoologists nowadays are trying to avoid the sin of typological thinking—doubtless a good move.

His friend's fears that he was undertaking more than any man could handle proved distressingly well-founded. By the mid-1840s he was running into trouble. Some of his collaborators accused him of appropriating their research without proper acknowledgment. His wife, the artistic and delicate Cécile, moved to Carlsruhe to her brother's home, taking with her their two daughters, Pauline and Ida (their son, ten-year-old Alexander, lived at school). At the same time, the publishing projects were pushing Agassiz toward bankruptcy. His American hosts knew nothing of these worries, only that eminent scientists spoke highly of Agassiz's achievements, but he and his friends had engineered this fortunate opportunity.

Agassiz throve on the flattering reception which greeted him wherever he traveled in the United States. Large public audiences were sympathetic to his message that the structure of living things was explainable only as the handiwork of God. In his Lowell Lectures, titled "On the Plan of Creation in the Animal Kingdom," he pointed to recent work in comparative embryology, paleontology, and anatomy to show that coherent relations of similarity permeate nature, relations no material necessity could explain. To Agassiz, patterns of similarity were sure evidence of a planning Mind. That conviction allowed him to invest the findings of comparative anatomy with a spiritual message that was received gratefully by specialists and the general public alike.

So impressed were the businessmen and intellectual leaders of Boston with Agassiz's personal charm and scientific attainments that they installed him in a new professorship in a new branch of Harvard University. He saw his position in the Lawrence Scientific School as the opportunity to point American biology toward future glory, and he was tireless in promoting his vision. Whether explaining points of natural history to young people or dining with the social and literary elite, Agassiz preached ceaselessly on the great things that could be achieved if only money were made available for buying specimens and books, employing assistants, creating curatorships, sending students on collecting trips, printing illustrated volumes; plan followed plan at a dizzying rate. He hustled hard, and, thanks to the lively economic climate as well as his political skills, he was given a good deal of what he asked for. Fortune was kind to him personally as well; after the death of the wife he had left behind, he won the devotion of a woman of exceptional character and intelligence, Elizabeth Cary. His three children found in her a loving stepmother, and Agassiz became a family member of the New England aristocracy.

Whatever we may feel about Louis Agassiz—admiration for a charismatic lecturer who inspired two generations of Americans to value natural history, or disdain for an egotist who appropriated the work of others and refused to give up an obsolete worldview—we cannot deny that he made a difference. The events of his life, familiar to readers of Edward Lurie's fine biography *Louis Agassiz: A Life in Science,* affected the lives of contemporaries and successors. The Agassiz phenomenon, for better or worse, is not the story of a man but of the enterprise in which he was engaged, an enterprise that flew under the flag of Science.

In his heyday, about 1854 to 1864, what Agassiz achieved was a synthesis of a remarkable and interesting kind. He did three closely connected things: he articulated an ideal for a newly coherent field of study, he founded an institution as the locus and material instrument of that field, and he trained a generation of young practitioners to carry on his vision. His ideas, published in 1857 in his "Essay on Classification," his Museum of Comparative Zoology, founded in 1859, and the students who worked on his collections in the 1850s and early 1860s formed a tightly integrated network of ideas and practice. The failure of this enterprise, when the band of eager students scattered and as evolution began to supplant his worldview, was evident to many of his contemporaries, but the outward structure of Agassiz's world remained in place. After his death, tactful memorials, plus the continuation of his museum, blurred recollections of Agassiz's rocket-like flight and crash.

In the years of the M.C.Z.'s conception, birth, and infancy, Agassiz and his students experienced the intense energy of shared belief (fig. 1). During that exciting period, the kind of group effort going on in Cambridge was the same phenomenon which, if successful, earns a special place in the history of science. Is there a geneticist who has not heard of the "fly room" at Columbia University in the 1920s, where the undergraduate Sturtevant shared cramped space with graduate students Bridges and Muller, surrounded by milk bottles of *Drosophila,* and where the foundational texts of a new science were written. William Bateson had already coined the word "genetics," and many others were conducting breeding experiments or examining chromosomes, but it was in T. H. Morgan's laboratory that the discipline of genetics was created. We often use the word "discipline" in loose and varying senses, to mean either the content of a field of study or the social network of its practitioners, but it is really only those nodes that nicely combine ideas and social structures that deserve to be called a discipline.[1] Had Agassiz managed to maintain his synthesis, we would honor him as the founder of a discipline.

Sometime during the 1850s the germ of Agassiz's brilliant synthesis began to grow. He was collecting, buying, and begging specimens, intending to produce a handsome series of descriptions of North American animals which would exemplify the superiority of thorough and thoughtful work over the superficial descriptions naturalists too often let pass. He was making use of student volunteers as well as paid assistants to sort and study clams, turtles, fishes, and other specimens sent to him by the barrelful from American lakes and rivers by his many admirers. At the same time he was giving much thought to the intellectual goal of natural history and beginning to envision a special new kind of museum. At first the theme of his museum was little more than the greedy feeling that a few representative specimens of each species were not enough, that he must have masses of material. His thoughts on what he would later call "comparative zoology" at first only repeated the beliefs standard since early in the century—that a natural classification is one that distinguishes organisms' "types" or plans of structure, which are revealed not only by anatomy but by patterns of embryological development.[2]

Students who joined Agassiz in the 1850s were not put through a structured set of lessons but became his apprentices. The goal of their work was always to define a natural grouping of species rather than merely describing individual species one after another, and the backbone of his method was comparison. Joseph LeConte, who arrived with his cousin Lewis Jones in 1850, recalled that Agassiz "pulled out a drawer containing from five hundred to a thousand separated valves of Unios [shells of freshwater clams] of from fifty to a hundred different species, all mixed together."[3] Rather than ask the students to "identify" them by comparing the shells to published descriptions, Agassiz challenged them to make their own judgments about how many natural units these shells might really belong to. LeConte was ever afterward grateful for Agassiz's contagious enthusiasm, magnetic personality, and insistence that his students learn to think for themselves rather than taking either the printed word or Agassiz himself as authoritative. (The shells of the family Unionidae were an excellent choice if Agassiz's purpose was to make students distrust all but their own eyes, for they are plastic to environmental pressures, and published descriptions made poor allowance for their variability.)

However long their hours at their assigned tasks, Agassiz's students could never get the impression that zoology consisted only of the indoor study of dried specimens. He took Jones and Leconte with him to Florida to study the growth of coral reefs. He arranged to get his assistant Henry James Clark the latest high-quality microscope with which to study the developing eggs of turtles.

In summertime Agassiz would be found at a cottage by the seashore in Nahant, on the tip of a peninsula just north of Boston. There his father-in-law built him a simple laboratory, close to tide pools full of the sea urchins, sea anemones, and starfishes he was investigating. The table for his micro-

Figure 1. On the steps of Louis Agassiz's Quincy Street home, Cambridge, about 1855. The famous professor, in a stovepipe hat, can be recognized with certainty. The figure down the step from him may be the geologist Jules Marcou. The group at his other side, clockwise from his elbow, may be Henry James Clark (above), Jacques Burkhardt, Joseph LeConte or Jeffries Wyman or Theodore Lyman, and George Adam Schmitt [J. H. Blake to T. Barbour, 4 November 1936, Harvard University Archives]. A retouched version of this early photograph appeared in Samuel Eliot Morison, *Development of Harvard University,* facing p. 381. (By permission of the Museum of Comparative Zoology Archives, Harvard University)

scope "stood on a flat rock sunk in the earth detached from the floor" to avoid vibration.[4] His wife also learned to love the "Radiata." She wrote,

Nothing can be prettier than the smaller kinds of jellyfishes. Their structure is so delicate, yet so clearly defined, their color so soft, yet often so brilliant, their texture so transparent, that you seek in vain among terrestrial forms for terms of comparison, and are tempted to say that nature has done her finest work in the sea rather than on land. Sometimes hundreds of these smaller medusae might be seen floating together in the deep glass bowls, or jars, or larger vessels with which Agassiz's laboratory at Nahant was furnished. When the supply was exhausted, new specimens were easily to be obtained by a row in a dory a mile or two from shore, either in the hot, still noon, when the jelly-fish rise toward the surface, or at night, over a brilliantly phosphorescent sea.[5]

The dory belonged to her teenaged stepson Alexander, who had inherited from his German mother an artistic sensibility and hand.[6] He would continue in later life to collect and to make delicate drawings of marine animals. Also in the Nahant laboratory, besides Agassiz's artist, Jacob Burkhardt, were students and assistants. Alexander's friend and classmate Theodore Lyman wrote in 1856 that "every day I spend about six hours in Agassiz' laboratory, where are sometimes the most fearful smells that ever attacked the human nose. . . . Today came his bosom friend Dr. Holbrook, the herpetologist. It was funny to see Prof drag him in and show him all his plates, with immense glee."[7]

Agassiz and his helpers were at work during the mid-fifties on a promised series, to be lavishly illustrated, supported by subscription, called *Contributions to the Natural History of the United States of America.* The title suggests appeal to a broad range of readers, while in fact Agassiz's plan was to publish research done by himself or his associates that would meet the highest standards of professional zoology. Those who expected "Natural History" to mean narrative about the out-of-doors, or catalogues of local fauna to help amateurs identify their collections, might be perplexed and disappointed by the abstruse descriptions of anatomy and embryology that would constitute Agassiz's contributions, but he knew that such descriptions were what the scientific elite abroad would most respect. For this reason his new series would need an introduction which would make the subscribers appreciate and sympathize with the kind of science Agassiz wanted his students to do. Between January 1854 and July 1856 he labored to explain the philosophy of contemporary zoology as he perceived it. The resulting "Essay on Classification" filled half of volume 1 of the *Contributions,* published in 1857, and was reprinted as a separate book in 1858 and again in 1859.

The "Essay on Classification" of 1857

Agassiz argues that classification is interesting and important because when done correctly it represents our search for the natural types or plans which link together the various forms of life scattered across the planet. Naturalists had learned to distinguish kinds of similarities, calling a feature homologous when the resemblance seemed to express a deep-seated essential "affinity," calling it analogous when the resemblance was based on function rather than affinity. The affinities formed the basis for taxonomic groups:

> Not only is the wing of the bird identical in its structure with the arm of man, or the fore leg of a quadruped, it agrees quite as closely with the fin of the whale, or the pectoral fin of the fish. . . . The same agreement exists between the different systems and their parts in Articulata, in Mollusks, and in Radiata, only that their structure is built up upon respectively different plans, though in these three types the homologies have not yet been traced to the same extent as among Vertebrata. There is therefore still a wide field open for investigations in this most attractive branch of Zoölogy.[8]

As in his Lowell Lectures, Agassiz wanted to impress his readers with how extraordinary a thing it is that there are homologies linking embryonic forms of one species to embryos and adults of other species, including extinct fossil forms. He sought for pattern in geographic distribution. Thus embryology, paleontology, and biogeography added new dimensions of complexity to the basic network of taxonomic affinities. Forms separate in time and space as representative, or parallel, or "prophetic," were thus linked together.

Agassiz's view of the central theme and fruitful direction for future research was perfectly consonant with most of his professional peers. The young English zoologist T. H. Huxley, for example, said in 1858, "The biological science of the last half-century is honourably distinguished from that of preceding epochs, by the constantly increasing prominence of the idea, that a community of plan is discernible amidst the manifold diversities of organic structure."[9]

The existence of deep bonds of essential similarity, all the more striking when discovered lurking unexpected within larval stages, rudimentary organs, or subtle skeletal features, meant, said Huxley, that the scientist should not be content to arrange living things according to some arbitrary system of categories, however convenient, but must search for the most natural classification. Another contemporary of Agassiz, his Harvard colleague the comparative anatomist Jeffries Wyman, taught and conducted his research in the exact same framework, although his modest personal style was the opposite of Agassiz's.[10]

Agassiz also proclaimed in the "Essay" an explanation of what zoological affinity really means. The essence of his explanation was likewise the standard one. In 1845 the English geologist and zoologist Hugh Strickland had said in a lecture:

> On comparing together the innumerable species of organized beings, we find their structure to present every possible degree of variation, from an almost perfect identity to the utmost amount of difference. . . . These agreements and differences are not however devoid of laws and principles; they admit of being classed under certain general heads, and we thus discover the traces of Divine workmanship not merely in the structure of an individual organism, but in the mutual relations of those organisms, the due combinations of which constitute the Natural Systems of Botany and Zoology.
>
> . . . When we say that Affinity consists in an essential agreement of structure resulting from a fixity of purpose in the mind of Creative Wisdom, it must not be supposed that all affinities are equally strong, direct, and palpable.
>
> . . . *Affinities* are expressions of the real and elementary and esoteric Plan of Creation which the Author of Nature has been pleased to follow.*

Writing a decade after Strickland, Agassiz declared that to many of his colleagues "the name of God appears out of place in a scientific work."[11] The German and French-speaking scientific culture in which Agassiz had been trained denied the appeal to the Creator which Strickland and others in the English-speaking world were allowed. That exclusion has now become so firm a rule in scientific writing that modern biologists who read Agassiz may have trouble accepting that he was "really" a scientist. Worse yet, Agassiz's commitment to the direct divine creation of species reminds us of the self-styled "scientific creationists" of our own day. But Agassiz should not be lumped with biblical literalists. From his first studies of geology and paleontology, he rejected any attempt to prejudge the reading of the book of nature by a literal interpretation of the Bible as being contrary to the principles of scientific inquiry as well as dangerous to faith. A student noted, in April 1860, "Splendid lecture by Prof this morning on the absurdity of believing that Adam and Eve were the first created and the only ones. It was a masterly lecture and was listened to with great attention."[12] Agassiz was deeply pious, but he was no fundamentalist. He joined his

*Strickland, "On the structural relations of organized beings," pp. 354–55, 358, 364. Agassiz does not cite Strickland, nor need he have, since these views were common assumptions.

American wife as a member of a Unitarian congregation, a denomination that denies the divinity of Christ.

In the "Essay" he insists on the propriety of discussing God, and explains that he is especially led to do so in reaction to "the discussions now carried on respecting the origin of organized beings."[13] He is not referring to Darwinism, since at this time the ideas of Alfred Russel Wallace and Charles Darwin were known only to their intimate confidantes, but to a book by Baden Powell, professor of mathematics at Oxford, which had just appeared. The idea of an organic rather than a miraculous birth of species was in people's minds well before Reverend Powell's *Essays on the Spirit of the Inductive Philosophy* appeared in 1855. When the anonymous *Vestiges of the Natural History of Creation* appeared in 1844, the reviewer in the *American Journal of Science* declared, "Although we cannot subscribe to all of the author's views, we would strongly recommend the work to our readers." The notorious *Vestiges* reached its tenth edition by 1853. There also appeared a flurry of materialist writings associated with the 1848 unrest and revolutions throughout Europe. It is fair to say that Agassiz built his museum as a fortress against evolution.

Agassiz believed himself to be exploring the facts of nature without preconception, following wherever they might lead. It seemed to him that evidence of thought, planning, and intelligent design were manifest in all the correlations and affinities between species he and his colleagues were discovering, and he concluded that a thinking Creator must be the cause. He declared, " . . . as long as it cannot be shown that matter or physical forces do actually reason, I shall consider any manifestation of thought as evidence of the existence of a thinking being as the author of such thought."[14] Of course, it could be argued that Agassiz was reasoning in a circle, for the patterns and correlations that were to him evident manifestations of thought could easily be interpreted, by an investigator with no prior belief in a creating deity, as the consequence of regular and natural, though still unknown, processes.

Plans for a Museum

Soon after his appointment at Harvard, Agassiz began to talk about his need for a museum. It was a familiar and accepted fact that a teacher of natural history needed a cabinet of specimens for demonstration. The comparative anatomist Jeffries Wyman quietly built up a very creditable collection in Boylston Hall from the time of his appointment in 1847. But moderation was no part of Agassiz's nature. By 1854 he was spinning visions of a museum in Cambridge to rival the leading institutions of Europe. What a rash idea! The imperial fleets of Britain, Germany, France, even Holland, had carried specimens at state expense from around the globe and thus built up nationally funded collections over many years. If Agassiz's ambition would not let him be satisfied with a modest cabinet, surely a reasonable person would have advised him to contribute to the growth of the museum that had been accumulating at the Boston Society of Natural History since 1830. There the energies of a number of dedicated amateurs combined very effectively with the work of professional and semiprofessional researchers, including Wyman, Thaddeus William Harris (who had lectured on natural history at Harvard and produced a much-admired entomological report commissioned by the Commonwealth of Massachusetts), Humphreys Storer, William Barton Rogers, and Josiah Dwight Whitney. The BSNH published a journal, held regular formal meetings, attracted bequests, owned a respectable cabinet of specimens, and in 1847 had purchased and moved into larger quarters.[15] One could easily imagine its collection growing, through the efforts of Agassiz and his students, to excel the best in the nation, that of the Academy of Natural Sciences in Philadelphia.

Agassiz did become an active member of the Boston Society, and he encouraged his students to speak at its meetings and to help arrange its collections. But he could be satisfied with nothing less than a museum entirely his own, one that would command the attention of Europeans as well as his new countrymen. He convinced philanthropists, college administrators, state legislators, and hundreds of ordinary citizens to contribute cash in response to his ideas, appealing to their piety, their patriotism, their hunger for culture, and their hope for practical benefit from science. He may not have persuaded all these people that they really needed a new zoological museum, but he did convince them that the great Professor Agassiz needed one, and at the peak of his rhetorical powers, that was enough. Hundreds of people responded to his magnetic personality and his description of his dream with gifts of money, specimens, and labor.[16]

After the publication of the "Essay," his vision of what would make his museum special became sharper. In his fundraising speeches, references to the Creator were prominent, but only in connection with the intelligibility of systematic relations. Natural history museums were commonly expected to stimulate their visitors to pious appreciation of God's powers by exhibiting the delicate beauty of a butterfly's wings, the amazing height of a giraffe, perhaps even the wonderful construction of a fish's skeleton, as well as through the very mass of material showing the quantity and variety of life on earth. Agassiz's plan was utterly different. Indeed, he even began to insist that many of his exhibits would seem unattractive to an ignorant public. Agassiz wanted his new museum to do something other natural history collections did not. His museum, by its arrangement, would illustrate patterns of organic similarity, as shown by morphology, embryology, paleontology, and geographic distribution. Because any "intelligent and intelligible connection between the facts of nature" is "direct proof of the existence of a thinking God,"[17] visitors who understood the *arrangement* of his collection would be led to admire the Creator.

For Agassiz, however, the primary job of his museum was not to serve the public. It was to provide material for scientific research of the most professional kind, and as any reader of the "Essay" knew, to Agassiz this meant exploring several dimensions of comparisons. The numbers and kinds of specimens to be acquired, and their arrangement, must be very different for the investigation of embryology and growth, geographical distribution, comparative morphology, even the natural variability within each species. Close study of individual adult anatomy was the modern way to evaluate affinity, and this would require whole specimens preserved in alcohol to supplement the usual dried skins, skeletons, or shells. This museum must have specimens of immature individuals, including embryos, again in alcohol, since correlations might be sought between the early form of one species and the adult form of another. Fossils would be arranged in association with related extant forms, instead of forming a separate paleontological collection as in most museums. Time and again Agassiz declared to his friends and supporters that the museum he was planning would be unlike any in existence, that it would make possible improvements of knowledge those others could not. All this meant that his museum would cost much more than an ordinary natural history collection, because space, glass jars, alcohol (for which he negotiated tax exemption), and fire proofing were expensive.

With its rows of glass jars, Agassiz's museum would resemble museums of comparative anatomy like John Hunter's in London, where respiratory organs of various animals formed one series, digestive organs another, and so on. Wyman's collection in Boylston Hall was called the Museum of Comparative Anatomy. Agassiz emphasized what he saw as the novelty of his plan when he discussed with his benefactor Francis Calley Gray the question of what his museum should be named. Most accurate would be "Museum of Comparative Zoölogy, Embryology and Paleontology," but this was unwieldy, so he proposed "the title of *Comparative Zoölogy*, which I think is that likely to prevail for our science, as it grows."[18] The field of study he had in mind was enormous, but it had a focus in the method of comparison. No, more than the method, its focus was the conviction permeating the "Essay on Classification" that significant similarities link all living things into an intelligible network.

From 1855 through 1859, Agassiz and his students worked in a rough wooden building, "somewhat better than a barn and not quite so good as a house,"[19] a simple two-storey wooden box (see fig. 4 below) crammed with a jumble of specimens: carefully boxed treasures purchased from European specialists, along with things Agassiz had collected, or bought in a fish market, or begged acquaintances to send him. In November 1859* his students, assistants, and friends helped carry all this stuff into a new four-storey square brick building (see figs. 19 and 20, below).

Agassiz's plan for the new museum, from the first, involved his students. He conceived their positions to be something like postdoctoral fellowships:

> I have thought that a number of Curatorships . . . might perhaps be founded by some of our wealthy citizens, which would furnish a small income to students who have already taken their degree, and who, wishing to prosecute further their studies under my direction, might thus come by the means of remaining in Cambridge by assisting in the arrangement and preservation of the collection.[20]

Although such positions were never endowed, he did sometimes use general museum funds, or even his own money, to hire students as assistants.

The Solitary Fish and the Unique Lobster

Several of the young men who came to study natural history with Agassiz in the 1850s and early 60s experienced an initiation rite they remembered vividly all their lives. The recollection of one, Nathaniel Southgate Shaler, is typical:

> When I first met Louis Agassiz [1859], he was in the prime of his admirable manhood [fig. 2]. . . . His face was the most genial and engaging that I had ever seen and his manner captivated me altogether. . . . Agassiz's welcome went to my heart,—I was at once his captive. [He assigned me] a small pine table with a rusty tin pan upon it. . . . Agassiz brought me a small fish, placing it before me with the rather stern requirement that I should study it, but should on no account talk to any one concerning it, nor read anything relating to fishes, until I had his permission so to do.

*Shaler later writing his memoirs recalled the move as early 1860, but contemporary letters and diaries establish the time beyond doubt.

Figure 2. Louis Agassiz in 1859. This photograph was taken in London, which he visited after the ground-breaking ceremony for his new museum, collecting books, specimens, and ideas. (By permission of the Museum of Comparative Zoology Archives, Harvard University)

Agassiz, though working nearby, left his pupil alone for a week.

> At first, this neglect was distressing; but I saw that it was a game, for he was, as I discerned rather than saw, covertly watching me. . . . At length on the seventh day, came the question "Well?" and my disgorge of learning to him as he sat on the edge of my table puffing his cigar. At the end of the hour's telling, he swung off and away, saying, "That is not right."[21]

To another student the professor gave an explicit hint that this was a riddle with a solution, saying, "'You haven't even seen one of the most conspicuous features of the animal, which is as plainly before your eyes as the fish itself; look again, look again!' and he left me to my misery."[22]

Agassiz's unusual style of teaching became famous, and his "case study method" has passed into the annals of education.* Writing about this experience in later years, students recalled it as a lesson in giving close observation to detail and in trusting one's own eyes rather than turning to authorities. Doubtless the exercise did have these general effects, but it contained as well, I think, a more particular and novel claim about biological classification.

The lesson Agassiz wanted to impress upon his pupils through their baptism-by-fish is fully set forth in Agassiz's "Essay on Classification." There at the outset he poses to his readers the following thought-experiment:

> Suppose that the innumerable articulated animals [crustacea, spiders, insects], which are counted by tens of thousands, nay, perhaps by hundreds of thousands

*Scudder's story was reprinted and circulated after Agassiz's death by the committee raising money in his memory, the Teachers' and Pupil's Fund; it also appeared in Jules Marcou's biography *Life, Letters, and Works of Agassiz* (2:94), and in Lane Cooper's *Louis Agassiz as a Teacher.* A. S. Packard said, "When I went to study with him . . . he gave me a specimen of a dried moth, and I kept at it for a fortnight" (Campbell, "Biological teaching," p. 127). Samuel Garman may have had a similar experience: "Soon after his return from the Hassler Expedition, in 1872, Professor L. Agassiz placed before me, his pupil at the time, a specimen of one of the Batoidei [skates and rays], with the remark, 'See what you can find out about it'" (Garman, "On the lateral canal system of the Selacia and Holocephala"). Both Morse and Verrill kept a diary, but neither recorded such an initiation. Verrill, however, later recalled having begun with an assignment to articulate a skeleton, and recalled Hyatt having been assigned to sort mixed lots (Cooper, *Louis Agassiz as a Teacher,* pp. 45–47). Morse reported having been kept for a week working on the common edible clam (*Mya arenaria*) (Cooper, p. 61n.). Solitary exercises using a number of specimens rather than one are reported by Joel Asaph Allen (*Autobiographical Notes and Bibliography,* pp. 8–9) and by Joseph LeConte (*Autobiography,* pp. 128–29). I am grateful to Edward Lurie for calling my attention to the Robert Scholes article, "Is there a fish in this text?" which is a brilliant commentary on Ezra Pound's use of the fish story. Scholes anticipated me in seeing the exercise as a riddle, and in pointing out that rather than being an inspiring model of teaching it is a disturbing indoctrination into a kind of Platonism.

[the count is now around a million species], had never made their appearance upon the surface of our globe, with one single exception: that, for instance, our Lobster (Homarus americanus) were the only representative of that extraordinarily diversified type,—how should we introduce that species of animals in our systems?[23]

Would the zoologist need categories, like class, order, and family, if the world contained lobsters but nothing else like them?

The answer, which Agassiz hoped to impress upon his students by means of their initial lonely exercise, was that even a single individual embodies the layered structure of the hierarchy of taxonomic categories. He told the readers of his "Essay" that "the individuals of one species . . . exhibit characters which, to be expressed satisfactorily . . . would require the establishment, not only of a distinct species, but also of a distinct genus, a distinct family, a distinct class, a distinct branch."[24]

A young man already experienced as a bird-watcher or butterfly collector would think he ought to study the specimen for details of the sort naturalists use to distinguish one species from another. One student, Samuel Scudder (who likely had already read Agassiz's "Essay"), had the wit, after a night of racking his brains, to ask, "Do you perhaps mean that the fish has symmetrical sides with paired organs?" Bilateral symmetry was characteristic of the whole vertebrate type. "Thoroughly pleased," Agassiz beamed, "Of course, of course!" and then discoursed "upon the importance of this point." Scudder knew well that the lesson was not one of brute observation but of interpretation. "Agassiz' training in the method of observing facts and their orderly arrangement, was ever accompanied by the urgent exhortation not to be content with them. 'Facts are stupid things,' he would say, 'until brought into connection with some general law.' "[25]

What Scudder fails to explain to us in his reminiscence is just what general law the fish was meant to illustrate, but this Agassiz himself spelled out in the "Essay." Certain features make the solitary lobster a member of the genus *Homarus,* others make it a member of the species *Homarus americanus*. Even an isolated individual lobster has the plan of structure of the entire *embranchement* Articulata (approximately equal to our phylum Arthropoda), and the characteristics by which we define the class Crustacea, and the special features of the order Decapoda, and the form of its particular family. Using a fish, like the drum *Haemulon,* the one Agassiz gave Scudder, changed the names and definitions of the groups, without altering the idea. The fish has the plan of structure of the entire *embranchement* Vertebrata and the characteristics by which we define the class Pisces, the special features of its order, and the form of its family. The point was that the framework of the taxonomist was not invented by man but built into every living thing.

The lobster thought-experiment in the "Essay" was an imaginary version of the students' fish assignment, but whether the literary device grew out of a teaching technique, or whether inventing his thought-experiment was what impelled him to impose the fish exercise, I am not sure. The two may have evolved together. It is easy, however, to guess why he chose the organisms he did. Barrels of pickled fishes crowded his workspace; young amateur naturalists usually thought them uninteresting, but they were one of Agassiz's specialties. Why, then, not ask readers of the "Essay" to imagine a fish? Perhaps because to cancel in our minds all other vertebrates would eliminate ourselves, the sapient but bony observers. All right, then, but of all invertebrates, why a lobster? Perhaps because even those readers who have never cared about bugs and starfishes will be happy to have a whole crustacean put before them, as Agassiz had doubtless observed many an evening in Boston while dining with the exclusive Saturday Club.

Though the solitary fish has become legendary, in fact it was always followed by a second assignment, one which constituted the key portion of these students' experience—to put in order a large number of specimens. (This assignment, rather than the challenge of a single specimen, was the initiation rite recorded by some of those who came to Harvard in the years before Agassiz composed the "Essay.")[26] This would seem to be a more obvious way to introduce a student to the principles of classification. It is little more than the Enlightenment version of the stories of Adam and Noah. In the mid–eighteenth century the Comte de Buffon had laid out beautifully the full version of this ideal of total collection in the introduction to his monumental *Histoire Naturelle,* where we are given an imaginary and unsorted museum being contemplated by an imaginary thinker:

> But when specimens of everything that inhabits the earth have been collected; when, after much difficulty, examples of all things that are found scattered so profusely on the earth have been brought together in one location; and when for the first time this storehouse filled with things diverse, new, and strange is viewed . . . [27]

He believes that when a collection thus perfectly complete is made, all gaps will be filled, and system makers like Linnaeus will be embarrassed by the evident artificiality of their genera, orders, and classes.

Buffon's fantasy lurks always in the semiconsciousness of every conscientious taxonomist. They ask as they work, "When other specimens arrive to supplement the limited sample I have before me, will the new group I am proposing, and the characters I use to define it, remain useful?" The taxonomic ideal is not merely to classify what has already been assembled but to find groupings that will stand the test of time—time during which museums will be progressing toward completeness. Will experience prove this species "good," or will it merge with another as exploration proceeds; will this family still seem natural when every form resembling it becomes known?

Yet what Buffon and perhaps all taxonomists imagine, a perfectly complete collection, has never been assembled and never will be. By asking us to accept the image as theoretically possible, however impracticable, Buffon is begging the question; he is assuming that some finite set of "examples" could indeed represent everything that inhabits the earth. This assumes the discreteness of currently observed natural kinds, and omits as irrelevant the previous inhabitants of the earth. The alert skeptic should insist that anything can be represented only by itself, not by an example, so that a complete collection of the present inhabitants could be nothing less than the entire biomass of the earth.

But the solitary lobster turns Buffon's thought-experiment upside down. Instead of a vast number of forms, Agassiz posits a single form. How ironic that while Buffon had claimed that no taxonomic system could succeed at encompassing all of nature's diversity, Agassiz claims to need the full series of taxonomic ranks to describe just one! A skeptic should object that classifying a single organism is like assigning a street number to an isolated farmhouse or composing a symphony for a penny whistle. Classification deals with comparisons, thus requiring a minimum of three units so that one may ask, "Is *A* more like *B* than it is like *C*?"

Thomas Henry Huxley was another eloquent advocate of the pedagogic value of focusing on an individual specimen, and it might not have been a coincidence that he, too, chose the example, in an 1861 lecture, of a lobster.[28] Comparing its segments one to another allowed him to introduce the idea of a plan of structure with different modifications. But in contrast to Agassiz, Huxley introduced the idea of grouping into kinds, from genera up to class, only after comparing the rock lobster, crayfish, prawn, shrimp, crab, king crab, wood louse, water flea, and barnacle.[29] In the real world, instead of the fantasy world of thought-experiments, the concept "Crustacea" would not have been developed had those kindred forms not existed. And in the real world of his museum, Agassiz admitted as much in practice. After three days with a single fish, Scudder was given another *Haemulon,* and another, "until the entire family lay before me, and a whole legion of jars covered the table and surrounding shelves."[30] Only with the Buffonian fantasy put into play could he seek to define the characters of the genera by inspection and comparison, as taxonomists actually do.

Louis Agassiz neglected to warn either the readers of his "Essay" or his students that by this seemingly innocent thought-experiment he was about to lead them far outside the boundaries of zoology, trespassing into philosophical regions whose ontological bogs and epistemological swamps had swallowed up better men than he. Debates had echoed across the walkways and seminar rooms of ancient Athens, with followers of Plato explaining that we recognize this thing as a "horse" because of its sharing in a transcendent horseness more real than any mortal example (an idealist view called, oddly for us, "realism"), while Aristotle ridiculed analyses whose logical binary divisions did not mesh with well-acknowledged kinds like "birds" and "fish." Those medieval philosophers who insisted that "horse" was nothing but a name and that reality belonged only to material individuals like this or that horse called themselves "nominalists." In the seventeenth century, at the height of the scientific revolution, what remained of Greek idealism was firmly expelled by John Locke who presented the nominalist view compellingly. The species and idea "horse" is just a shorthand way for us to talk about the things all horses have in common. We deceive ourselves when we begin to think that the essence of horseness really exists. Agassiz does not mention, however, and apparently makes no use of, the many serious thinkers who had already discussed the issues he was raising.

Ernst Mayr, one of the most influential systematists and historians of systematics of our day, has claimed that "the replacement of downward classification (logical division) by grouping (upward classification) in the post-Linnaean period was a major philosophical advance."[31] Against that yardstick Agassiz's lobster would appear to be an entirely retrograde move. He insists that more inclusive groups, from genus up to *embranchement,* are every bit as real as the species. And each species at its creation had "full existence" even before its members began to weave for themselves and their descendants material interconnections by mating and generating offspring.[32] (He adds that the reality we associate with material existence of course belongs to individuals only). With such views Agassiz cannot avoid being classified with the idealists. But I do not find such philosophical labels much use in understanding exactly what Agassiz was trying to say, nor why it was important to him. If he had been knowledgeable or competent in philosophy, he would certainly have disavowed any deep resemblance of his views to pre-Christian ones, because each layer in the hierarchy of groups was the free creation of the mind of God. Cataloguing Agassiz as if he were a philosopher is not a helpful exercise. Also, Mayr's yardstick is misleading, because practicing naturalists, from Cesalpino in the seventeenth century, Linnaeus and Jussieu in the eighteenth, to Cuvier and certainly Agassiz himself, never did work by purely downward or upward classification but always by a mixture of the two. When Linnaeus sought artificial characters for neat definitions of his classes and orders, and when Cuvier sought fundamental physiological causes governing his *embranchements* and classes, they began by inspecting the range of actually known and already ordered living forms.[33] Even Agassiz, who believed he was reading the thoughts of God, did not think he could guess them using reason or intuition, but believed himself and fellow naturalists to be discovering them empirically.

Agassiz's Novel Analysis of Categories

We should distinguish, as Agassiz asks us to do, between the meaning of the categories and the particular groups one considers natural. It may be helpful to recall the distinction, insisted upon by Mayr, between "category," meaning level of taxonomic division, and "taxon," meaning a particular set of organisms.[34] The distinction is a simple one. If we were speaking of geography, examples of *categories* would include "nation," "province," and "city," and examples of *taxa* would be Canada, Ontario, and Toronto. In zoology, "order" and "species" are categories, while Decapoda and lobster are taxa. A field guide or a taxonomic monograph is about taxa, but Agassiz's "Essay" was about categories. Anyone who classifies uses categories, but a geographer seeking to describe *natural* rather than man-made features would choose categories like "island" or "mountain" rather than "state" or "nation." Agassiz was seeking to understand natural categories. (Notice, though, that geography is not so easily pressed into natural categories as I have implied. Is Australia a big island or a small continent? Is a seasonal brook a river? Practical mapmakers lose no sleep over such questions, and neither did most zoologists.)

Agassiz believed fervently in the naturalness of the main taxa of Cuvier, the Vertebrata, Mollusca, Articulata, and Radiata, so that he stubbornly resisted Rudolf Leuckart's proposal to replace those four with six primary groups (Vertebrata, Mollusca, Arthropoda, Vermes, Echinodermata, and Coelenterata). It might even be true, as I once imagined, that Agassiz's defense of these taxa motivated his search for the meaning of categories.[35] Nevertheless, Agassiz explicitly states that the correctness of his new definitions of categories does not depend upon the correctness of any particular taxa he uses as examples. He points out that Cuvier's original Radiata had already been altered by the removal of microscopic animals and intestinal worms, and he declares that whether the remaining radiates were a natural group must be decided by inspection.[36] The reviewer in the *American Journal of Science,* James Dwight Dana, noted with approval that Agassiz had made this distinction.[37]

We can construct a definition—more correctly, a diagnosis—of any particular taxon, once insight and experience have led us to recognize it. For example, Echinodermata, embracing animals as unlike as sea cucumbers and brittle stars, are defined by their unique internal vessels that provide pressure to their tiny suction-cup "feet"; Hemiptera, ranging from bedbugs to cicadas, are characterized by particulars of their sucking mouthparts. But how do we decide whether Hemiptera is a family or order? Why should we call Echinodermata a class rather than an *embranchement?* Is there a way to define categories like "family," "order," or "class," which will hold true across the whole animal kingdom? It has long been notorious among naturalists that there is not. Linnaeus had defined classes of plants by stamens, and orders within those classes by pistils, but he had frankly admitted that the groups so defined were artificial. For the natural groups, it was obvious that the glue holding hundreds of beetles together as one family was utterly unlike whatever it was that made various snakes members of one family. When a naturalist faced the question of whether a taxon which had grown in size with the progress of collecting, such as a genus which had come to contain hundreds of species, ought to be raised to family level, he would appeal for his decision to convenience or aesthetics, if he gave any justification at all. Agassiz set out to change this state of affairs.

While composing the "Essay," Agassiz convinced himself that he had uncovered the true meaning of natural categories, a secret which had eluded all previous zoologists, botanists, and philosophers. Scarcely concealing his elation, he wrote to the curator of the Smithsonian collections, Spencer F. Baird,

> I shall not wait till it is *published* to send you my Chapter on Classification. The results are so practical that even my students of one years standing with these rules are able to trace for themselves in lots [sets] of unlabelled specimens of any class I put into their hands, the natural limits of genera and families and they actually do it better than our old practiced Zoologists. So you see it will tell in the progress of science.[38]

What a nice example of how central students were to his enterprise! In this case they were playing the role of the intelligent ignoramus of Buffon's thought-experiment, handed a magic tool by Agassiz.

What were these powerful new "results" which would "tell in the progress of science"? He announces in the "Essay" that he has found the essence of categories:

Branches or *types* are characterized by the plan of their structure;
Classes, by the manner in which that plan is executed, as far as ways and means are concerned;
Orders, by the degrees of complication of that structure;
Families, by their form, as far as determined by structure;
Genera, by the details of the execution in special parts; and
Species, by the relations of individuals to one another and to the world in which they live, as well as by the proportions of their parts, their ornamentation, etc.[39]

They do look plausible, if a little vague. Agassiz tries for forty pages in the "Essay" to bring his reader to see them as he sees them, but all he can do is allude to exemplary taxa of beetles and turtles and bats, and especially to the *embranchement* Radiata with its distinctive radial symmetry. In elementary lectures, he would rapidly sketch in chalk a simplified sea urchin, jellyfish, and sea anemone, arguing that the classes Echinoderm, Acaleph, and Polyp should not be called "modifications" of the radiate plan but different "ways and means of executing" it (fig. 3). What he could not do was expand upon these definitions.

To recognize how useless and empty these definitions of categories are does not take the hindsight of a later generation—just some experience in natural history. The nature of the living world is such that there are no such criteria governing near and distant relations of similarity. The more distantly organisms are related, the fewer features they have in common, but there are no firm constraints limiting the sort of features an evolving lineage may retain, or reacquire, or develop independently. Even members of the same species, through accidents of birth, may differ in structural characters like numbers of fingers or limbs or vertebrae. Races of domestic dogs or pigeons differ in features that would distinguish whole genera or even families of wild carnivores or songbirds. Laying eggs versus bearing young alive would seem to be a feature of high importance, and it coincides with the great division of mammals versus birds, yet there exist members of generally egg-laying groups—certain snakes, sharks, and fishes—which manage the trick of live birth. Throughout the history of zoology, every attempt at rational or orderly classification had foundered on the bursting uncontainability of life. Strickland, combating those who thought they saw numerical regularity in taxonomy, said that "the natural system is an accumulation of facts which are to be arrived at only by a slow inductive

Figure 3. Louis Agassiz lecturing on one of his favorite subjects, the homology of the "Radiates": starfish, sea urchins, sea anemones, and jellyfish. (By permission of the Museum of Comparative Zoology Archives, Harvard University)

process, similar to that by which a country is geographically surveyed."[40] What then had induced Agassiz to make one more try at this philosopher's stone, wisely abandoned by others?

He was sensitive to the changing times, and he recognized that views he cherished were under threat. As we saw, evolution was very much on his mind when he composed the "Essay." Though he didn't know what Darwin was up to, he was familiar with the same community of taxonomists Darwin drew from. Both men shared the same belief, that those taxonomists who felt there was a profound naturalness of the "natural system" were on the right track. In Darwin's words,

> Such expressions as that famous one of Linnaeus, and which we often meet with in a more or less concealed form, that the characters do not make the genus, but that the genus gives the characters, seem to imply that something more is included in our classifications, than mere resemblance. Some authors look at it merely as a scheme for arranging together those living objects which are most alike, and for separating those which are most unlike; or as an artificial means for enunciating, as briefly as possible, general propositions. . . . But many naturalists think that something more is meant by the Natural System.[41]

Darwin's views on taxonomy can be extracted and twisted into unnecessary confusion by those who like to think in terms of philosophical boxes and want to label him "idealist" or "nominalist." They think they know where he belongs when he writes in the *Origin of Species:* "I look at the term species, as one arbitrarily given for the sake of convenience to a set of individuals closely resembling each other. . . . The term variety, again, in comparison with mere individual differences, is also applied arbitrarily, for mere convenience sake."[42] Aha! Obviously a nominalist. Yet this is the same Darwin who tells us in another chapter:

> From the first dawn of life, all organic beings are found to resemble each other in descending degrees, so that they can be classed in groups under groups. This classification is evidently not arbitrary like the grouping of the stars in constellations.[43]
>
> . . . all past and present organic beings constitute one grand natural system, with group subordinate to group.[44]

It was critical to Darwin's theory that Strickland, Huxley, and Agassiz were right to believe that they were discovering, not inventing, natural groups. Once his branching tree of modification with divergence is understood, it follows that classification, which carves nested groups out of nature's real continuum at appropriate places, can be both natural and artificial at the same time.*

Darwin recognized that many of his contemporaries were content, as Strickland was, with a supernatural cause for the naturalness of taxa. Gently but firmly, Darwin scorned that explanation as vacuous:

> But many naturalists think that something more is meant by the Natural System; they believe that it reveals the plan of the Creator; but unless it be specified whether order in time or space, or what else is meant by the plan of the Creator, it seems to me that nothing is thus added to our knowledge.[45]

Of the two views Darwin mentions, probably the majority of naturalists in the 1850s would have professed that good classification is not purely man-made and that, insofar as a taxonomic arrangement is truly natural, it does express the Creator's plan. The key to Darwin's success in winning many naturalists over to a belief in evolution, including many who withheld assent from the efficacy of natural selection, was precisely that he did replace the vague idea of a divine plan with a detailed explanation of why living things are so strongly classifiable.

Agassiz did not have to read this still-unpublished criticism to recognize the weakness Darwin identified. He bent his efforts to spell out in detail for the first time what is meant by "plan of the Creator," in order to strengthen the creationist view. If he could demonstrate that the fabric of the living world is woven on warp thread of pure thought, that is, relationships neither required by material necessity nor imposed by human classifiers, then divine causation would be more directly implied. Natural theology argued that adaptation of form to function pointed to the all-powerful Engineer, but Agassiz would provide a loftier argument. He declares his ambition frankly in the "Essay":

> I confess that this question as to the nature and foundation of our scientific classifications appears to me to have the deepest importance, an importance far greater indeed than is usually attached to it. If it can be proved that man has not invented, but only traced this systematic arrangement in nature, that these relations and proportions which exist throughout the animal and vegetable world have an intellec-

*This point has been made very clearly by M. J. S. Hodge, who points to Darwin's "chapter in the Descent of Man on 'the Races of Man'. We may not be able to decide in some particular cases whether a settlement is a town or only a village, he says. In some cases, the decision to deploy the designation *town* rather than *village* is made arbitrarily. But this uncertainty and arbitrariness does not entail disbelief in the real existence of the settlements themselves. As Ghiselin [*Triumph of the Darwinian Method*] has brought out, Darwin's position can be characterised in the terminology of today by saying that he was a realist regarding species and varieties as taxa, but a nominalist concerning the species category itself" ("Darwin, species, and the theory of natural selection," p. 247).

tual, and ideal connection in the mind of the Creator, that this plan of creation . . . was the free conception of the Almighty Intellect, matured in his thought, before it was manifested in tangible external forms . . . we have done once and for ever with the desolate theory which . . . leaves us with no God but the monotonous, unvarying action of physical forces.[46]

It was the icy gloom of atheism Agassiz feared. Never mind that evolutionists from Lamarck to Baden Powell had insisted that the scientific discovery of secondary causes does not necessarily destroy, indeed could exalt, the power and glory of the First Cause. For Agassiz, a distant Creator content to let His machinery grind on without providential care could give no comfort.

But he knew that simply reasserting the naturalness of taxa would not be enough. In a moment of rare candor he admitted that

a system may be natural, that is, may agree in every respect with the facts in nature, and yet not be considered by its author as the manifestation of the thought of a creator, but merely as the expression of a fact existing in nature—no matter how—which the human mind may trace and reproduce in a systematic form of its own invention.*

Typical of Agassiz to slip this point by in a footnote. It should have devastated him. It is, of course, profoundly true that one can create classifications intended only to represent observed resemblances, disavowing any inference as to the causes of those resemblances. Many taxonomists in our own day feel their job is complete when they have made groups that summarize their data.† Agassiz had in mind those taxonomists who believed themselves good Baconian empiricists (though Francis Bacon would have shuddered at their professed lack of interest in finding causes), the ones who looked at natural classification, Darwin said, "merely as a scheme for arranging together those living objects which are most alike, and for separating those which are most unlike." Whether it is desirable, or indeed, possible thus to separate fact from theory is highly doubtful, but in any case Agassiz marched on in his text as though he had never written this significant footnote. He blithely asserts that if you agree with his claim that classifications are not artificial, you must see that they are "translations into human language of the thoughts of the Creator."[47]

*L. Agassiz, "*Essay on Classification,*" pp. 8–9, n. 7. Mark Ridley calls attention to this footnote in a stimulating discussion in his *Evolution and Classification: The Reformation of Cladism,* pp. 107–8.

†And some are citing Agassiz as their precursor! Olivier Rieppel, in his article "Louis Agassiz (1807–1873) and the reality of natural groups," argues effectively that he is a poor hero for modern empiricists, since his system depended upon the thinking divinity.

That weighty footnote tells us at the very least that Agassiz knew he must do more. He must show not only that taxa are natural but that they bear a clear stamp of intellect, that the pattern of similarities cannot be explained by physical forces or necessity. He attempts this in a number of different ways throughout the "Essay," of which the first and most original is the solitary lobster, bearing the message that the hierarchy of categories exists in a single individual. His definitions of the categories, had they worked, would have established that what we experience as categories of our thought, when we do zoology correctly, are realities which already existed; categories of Someone Else's thought, frozen in organic form. That was why he wanted to locate the naturalness of the horse species, the family of grasses, the class Crustacea, not in the *taxa* horse, grass, and crustacean but in the *categories* species, family, and class.

To epitomize Agassiz, as is so often done, by the phrase "a species is a thought in the mind of God" seriously misrepresents him, insofar as it suggests that the species was a privileged category. What he was saying rather was that God's thought consisted of layers of analysis, themes, and variations corresponding to each category from species up to *embranchement.* This pushed idealism to such an extreme that it cracked under its own weight.

Of all of Louis Agassiz's ideas, his category definitions fared the worst. He must have become aware of their deficiencies, perhaps from the efforts of his students trying to apply them, even before the "Essay" went to press. He pleads in one of his many interesting footnotes:

> It is almost superfluous for me to mention here that the terms plan, ways and means, or manner in which a plan is carried out, complication of structure, form, details of structure, ultimate structure, relations of individuals, frequently used in the following pages, are taken in a somewhat different sense from their usual meaning, as is always necessary when new views are introduced in a science. . . . I trust the value of the following discussion will be appreciated by its intrinsic merit, tested with a willingness to understand what has been my aim, and not altogether by the relative degree of precision and clearness with which I may have expressed myself, as it is almost impossible in a first attempt of this kind to seize at once upon the form best adapted to carry conviction.[48]

His wishes and hopes were no substitute for the substance his definitions lacked. One of his American peers most sympathetic to his undertaking, James Dwight Dana of Yale, tactfully admits in his long review of the "Essay" that Agassiz has not achieved what he claimed:

> Even if the principles may require a fuller expansion and more precise definition to meet all the difficulties in this most difficult department of science, Prof. Agassiz has the honor of pointing out the right way.[49]

> . . . Many, while admiring the clear-sighted vision . . . will find difficulties in applying the scheme. We feel them ourselves, and shall need to give the system a more thorough study, before we can fully appreciate all the bearings of the principles.[50]
> . . . The *application* of the views brought forward by Prof. Agassiz, we suspect will give the greatest occasion for diversity of judgment. . . . there will probably be a division among naturalists as to the signification of the principles laid down.[51]

There was not so much a division of opinion among naturalists, however, as a stunning silence, suggesting unanimous inability to make concrete sense of Agassiz's category definitions.

When *The Origin of Species* first appeared, Agassiz's initial response was simply to restate his position more dogmatically:

> Individuals alone have a material existence, [while] species, genera, families, orders, classes, and branches of the animal kingdom exist only as categories of thought in the Supreme Intelligence, but as such have as truly an independent existence and are as unvarying as thought itself after it has once been expressed.[52]

It was a statement of belief, not an argument capable of convincing anyone not already convinced.

Agassiz's admission in the "Essay" that "it is almost impossible in a first attempt of this kind to seize at once upon the form best adapted to carry conviction" implied that he would strive to improve his formulation, but no revised version of his definitions appeared. Quite the contrary: he soon fell notably silent about them. His last piece of writing, the article "Evolution and the Permanence of Type," which appeared posthumously in 1874, while repeating his assertion that there are no transitions between types does not allude to the natural category argument. He concluded it by regretting that "the young and ardent spirits of our day" are giving "themselves to speculation rather than to close and accurate investigation," regrettable because the "more I look at the great complex of the animal world, the more sure do I feel that we have not yet reached its hidden meaning."[53] He could deny that evolutionists had found the answer, but he could no longer maintain that he had found one. His former students, in their obituary notices and memoirs of their late master, tactfully refrained from explaining in full the message of their baptism-by-fish.

Students Committed to Comparative Zoology

The young men who came to study with Agassiz found his general worldview perfectly congenial to theirs. Most of them brought with them a traditional religious faith, and many shared his abhorrence of materialism.[54] Leaving aside his novel claim about categories, the religious and transcendental aspects of his teaching complemented what his students learned from their other professors and from their reading. Agassiz's colleagues Jeffries Wyman and Asa Gray, though both were leaning toward evolution even before Darwin's book appeared, interpreted the order and beauty of nature as evidence of divine design. Wyman's publications revolve around the search for abstract patterns of homology—evidence to him of "forces of polarity" explaining form.[55]

Though Agassiz allowed divine action a more direct explanatory role than suited the taste of many of his peers, and though his attempt to define taxonomic categories was peculiar, for the most part his teaching placed his students well within the mainstream of what his peers saw as the compelling research questions of the day. His students used and admired the textbooks of Richard Owen, the eminent English anatomist, with their emphasis on ideal archetypes. Patterns of resemblance in embryology and morphology, one of the issues of systematics emphasized by Agassiz, were attracting bright and ambitious young men like T. H. Huxley and Ernst Haeckel, and they were the focus of Wyman's research.

In 1863 and 1864 a series of articles in the *American Journal of Science* on the "principle of cephalization" by James Dwight Dana greatly excited several of Agassiz's students.[56] Dana claimed that the concentration of body parts and functions toward the anterior end of an animal, with reduction of the posterior, is a mark of superior rank in every group, with our own species representing the "acme of cephalization." Parallel forces of degradation or concentration of life force give rise to "representation"—similarities—echoing across unrelated groups. One of Agassiz's students elaborated this theory in 1865 into a transcendental scheme far more speculative than anything in Agassiz's "Essay."[57] Wyman's ideas of polarity, equally transcendental, were also adopted by several of the Harvard students.[58] Agassiz's students did absorb much of his worldview, but he was only one source of their knowledge, and much of the idealism he upheld so aggressively was common coin of zoological thought of the day.

The New Museum

Probably the most novel and important difference between the M.C.Z. and any existing collection was that Agassiz's new museum was a training ground for a new generation of professional zoologists. It was not enough for Agassiz to have a vast quantity of specimens, he must also have students, for they had a central role in his museum, an intimate, symbiotic function in his plan. Granted free access to large series of specimens, his students would learn how to conduct original scientific research. At the same time, only the devoted labor of a corps of young workers would make possible such an ambitious plan for the collections as he outlined. Agassiz's activities in Switzerland, though ending in financial disaster, had taught him that his own charisma and leadership could transform colleagues and younger helpers into a powerful unit, a coordinated community of researchers. By providing not only inspiration and direction but room and board, Agassiz had built in Neuchâtel what one participant called a "scientific factory," churning out original research.* Thus Agassiz knew from experience that he could get productive workers at bargain rates. Unquestionably, an eager young man given the responsibility for part of a great museum would learn more of the principles of zoology than he would from any lecture course or textbook.

The pattern for such a symbiosis had been laid down among the jars and boxes of specimens Agassiz accumulated in the early 1850s as the nucleus of his future museum. In the series of temporary storerooms provided by Harvard, Agassiz did more than accumulate specimens—he put to work among those specimens students and former students. Henry James Clark was unpacking and arranging fishes for him in 1853.[59] Henry Augustus Ward arrived from Rochester in the spring of 1854, eager to study, and was assigned to unpack specimens: he counted 353 kegs and barrels in one shipment.[60] Frederic Ward Putnam, eighteen years of age, was given curatorial responsibilities over Agassiz's enormous fish collection in 1857. These and others found that studying with Agassiz meant getting your hands dirty.

Of course, what transformed Agassiz's growing collection into an institution with any prospect of outliving him was money. Edward Lurie provides details of Agassiz's fund-raising efforts. Harvard helped a bit, but a professor's need for materials to teach classes of a few dozen advanced science students could not justify an institution as large as Agassiz's ambition demanded. Wealthy businessmen, charmed by Agassiz's ideals, sincerity, and industry, organized appeals to their fellow philanthropists. Most important, Agassiz attracted the support of the legislature of the Commonwealth of Massachusetts. In his appeal for government support he made use of the ideas of his "Essay" in simplified form: "When we thus trace the relations which exist between organized beings, and reach higher and higher generalizations, it is not our thoughts that we put in nature,

*Lurie, *Louis Agassiz,* p. 109, cites Carl Vogt, one of Agassiz's Neuchâtel workers, who wrote: "Es war, wenn ich mich so ausdrücken soll, eine wissenschaftliche Fabrik mit Gütergemeinschaft" (*Aus meinem Leben: Erinnerungen und Rückblicke,* p. 194); "It was, if I may so express myself, a scientific factory with common ownership."

which we read out of it. It is in fact God's thoughts as manifested in tangible realities which we attempt to decipher."[61]

Construction of the new four-storey brick building began in the summer of 1859. Its design was utterly plain, its most remarkable feature being a temporary back wall, its blankness testifying to Agassiz's confidence that additions would soon be built (figs. 19 and 20, below).

Now his well-practiced techniques of integrating his teaching with his museum went into high gear. The wooden building that had contained his collections—and where a few of his student-assistants lived—was moved and set down adjacent to the new brick museum. Emptied of specimens, the old "Agassiz Museum" building was made over into living quarters for more museum assistants and renamed "Zoological Hall."* When he invited young naturalists to Cambridge, as he often did, Agassiz could offer them free lodging as well as free tuition in return for their services as helpers at the museum (see figs. 4–6). A great collector is one who knows how to catch men as well as fish.

Excited to see the museum of which he had so fondly dreamed now filling up with specimens, Agassiz infected his students with his delight. One wrote home,

> Yesterday we received at the Museum eighteen boxes of fossils from Europe forming part of an extensive collection that Prof bought. . . .
>
> Besides receiving the above, we have . . . three barrels from Prof's son collected

*This building had begun its existence as "Engineer Hall," constructed early in 1850 for the Lawrence Scientific School, housing also Agassiz's collections on the upper floor. In 1855 Harvard let Agassiz take over the whole building (Lurie, *The Founding of the Museum of Comparative Zoology*). F. W. Putnam recalled that when he arrived in Cambridge, in February 1856, the building was known as "the Agassiz Museum" and James Mills, a student-assistant, was living in a ground-floor room, which Putnam moved into when Mills left a year or so later. "The Laboratory, as we always called the room opposite the entrance, was where the Professor had his table with his microscope at the further window. At this window [Henry James] Clark also worked. At the next window Burkhar[d]t the Professor's artist, had a table where he was constantly drawing specimens under the Professor's direction. Near the center of the room was a large stove and at one side of this was a long table where the Professor lectured to us. We students sat on one side of the table with our pipes. The Professor generally had his cigar which he occasionally puffed while he lectured, and he often asked one of us: 'By chance have you a match?' He generally sat on the corner of the table within easy reach of the blackboard which ran the length of the wall back of the table. The board was always well covered with drawings before the Professor finished the lecture. During the lecture we would ask questions of the Professor, and he would often call on us to explain some specimen we had been studying. These lectures and talks were our delight. They were given at almost any time. The Professor would say to me, 'We will have a lecture,' and then I would call the other students and we would gather about the table" (three typescript pp. accompanying photograph of the Agassiz museum of 1857 or 1858) (Harvard University Archives).

Figure 4. Zoological Hall, drawn by James Henry Blake in 1868, a year after he joined the M.C.Z. as student and assistant to Louis Agassiz. In another location it had housed Agassiz's growing collection, and then, when the first stage of the M.C.Z. was completed in 1859, it served as dormitory and clubhouse to his assistants and students. (By permission of the Museum of Comparative Zoology Archives, Harvard University)

in Acapulco. . . . you can't imagine what a pile of specimens there were. All of us were engaged yesterday in sorting them over. . . .

Prof was jolly enough over them. As we would take them out of the barrels, it was fun to hear his ejaculations.

"Ah, what a beautiful Pleuronectes [flounder]!"

"Look at that gentlemen, isn't it splendid?"

And he would hold it up for us to see more fully.[62]

Agassiz cultivated each student's devotion with loving concern. By his interest in the details of their living arrangements, including advice on how to use their spare time and money, he placed himself in the role of parent to these young people, some of whom were indeed still in their teens, none past their early twenties. The feelings of one student were typical of the prevailing euphoria:

Prof. Agassiz is 52 years old and a better man never existed. Good natured and clever. . . . There is no better man in this world than Prof. Agassiz. We all feel toward him like a son to a father. . . . [He is an] advisor and friend, one who is pure, good, noble in every attribute. . . . I could never begin to tell you what a truly good

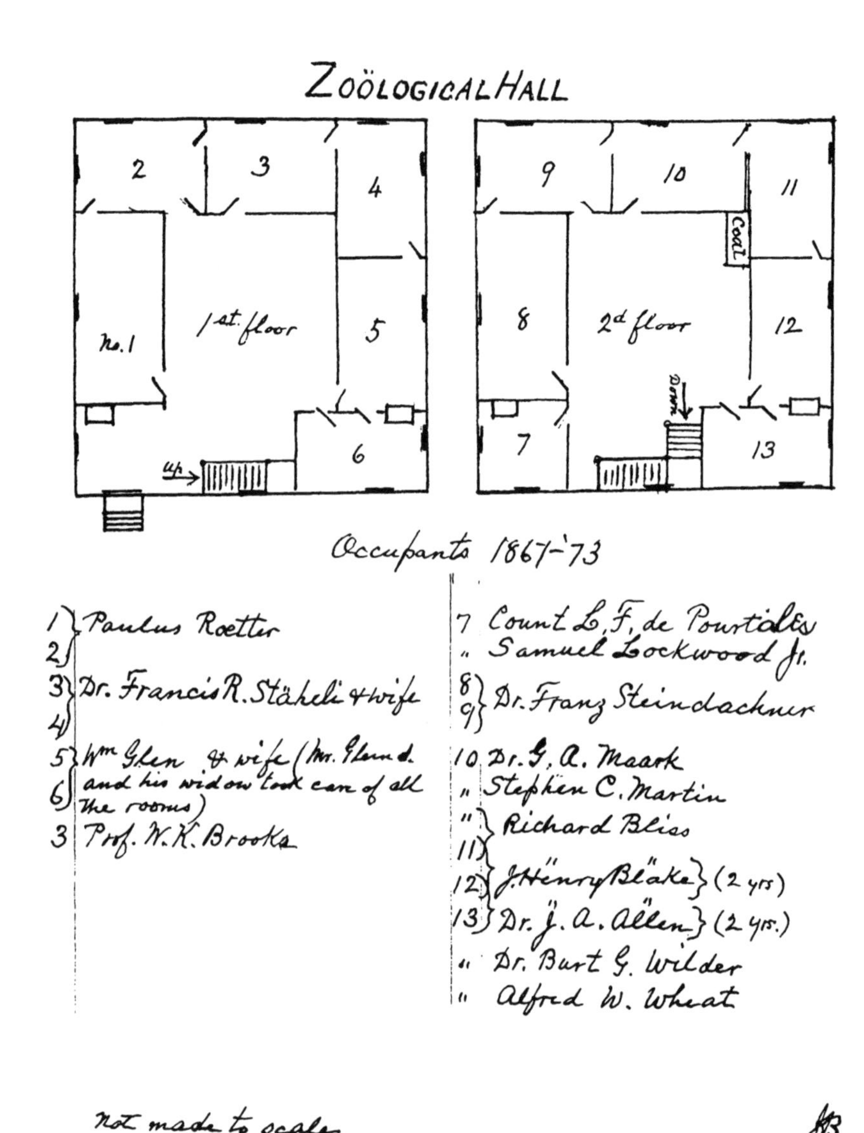

Figure 5. Residents of Zoological Hall around 1870, as recollected by James Henry Blake. Dr. Staehli left his position as anatomical assistant in 1870, and Brooks did not arrive until the fall of 1873. George Augustus Maack worked on fossil vertebrates in 1870 and 1871. (By permission of the Museum of Comparative Zoology Archives, Harvard University)

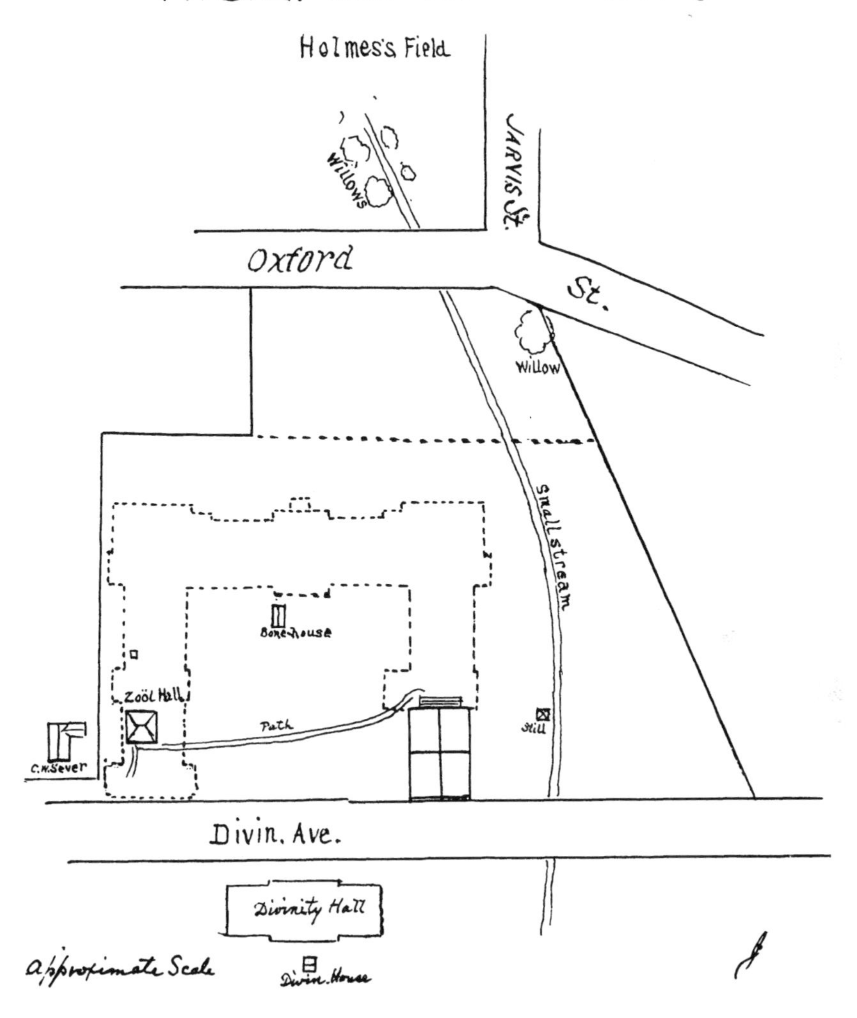

Figure 6. Blake's recollection of the location of Zoological Hall during the M.C.Z.'s early years. In 1876, when the Peabody Museum of Archaeology and Ethnology was constructed, the wooden building was moved across Oxford Street to serve other purposes. When Blake, aged eighty-nine, gave Thomas Barbour his sketch (fig. 4), he accompanied it with this map of the first stage of Agassiz's museum, showing in dashed line the structure which was completed in 1913. Blake remembered outbuildings for distilling alcohol and preparing skeletons, as well as the fact that the path from Zoological Hall ran to the museum's back door. (By permission of the Museum of Comparative Zoology Archives, Harvard University)

man Prof. Agassiz is. He shows it in a thousand ways.—He is one of the best men that ever lived.[63]

Another student's diary records: "Prof. Agassiz talked with me on private business and walked with me to my boarding place. He is almost *too* kind and generous."[64] Most important, he made each student feel that his diligence in scientific work gave him joy:

I have been engaged for the past two weeks in Genera, have taken, for example, the Genus *Ceritheum* and detected seven or eight genera in it. It is a great work and Prof is highly pleased. . . . by the way he talks I should judge he intended that I should write something extensive on it.[65]

Agassiz was careful to let his students know, in private conversations and in formal lectures, what a historic undertaking they were lucky enough to be sharing. One wrote home to his mother, in January 1860,

Prof says he intends that each student shall have a key for every lock in the building, cases and all, so that we can go in any time.[66]

Prof wished to make a few remarks to us so we all took chairs and sat down around him. He commenced his remarks by saying that we were together to pursue the study of Natural History in its true sense, earnestly and scientifically. That we had an opportunity never offered to anyone before. . . .

He said that in Europe no one can get access to specimens unless he collected them himself. . . .

. . . He wished to impress upon us the fact that we had a duty of profound interest to perform. That this museum was unlike any other museum ever founded. That it should be carried on in such a liberal manner as to set an example to others. . . . And in order to make it great, he depended on us to work with a will and an interest.[67]

What Agassiz told this group in 1860 was essentially true: they were indeed uniquely privileged. Not that his museum would finally become so remarkable—in a few years it would settle down to take its place as one more decent collection in the international network of museums—but that they were getting a very special educational experience. The conditions Agassiz had created in Cambridge succeeded extraordinarily well in educating young men to become professional biologists.

The Roll Call of Agassiz's Students

Something very special happened during the early years of the museum. These future naturalists shared the excitement of pursuing the study of natural history not only "in its true sense, earnestly and scientifically," but *together*. Agassiz had other students in later years, including the men and women who attended the experimental summer school on Penikese Island off Cape Cod in 1873. Of all the people who called themselves his pupils, the best known are these nineteen:

Alexander Agassiz, director, M.C.Z.
Joel Asaph Allen, curator, American Museum of Natural History
Edward Asahel Birge, professor, University of Wisconsin
William Keith Brooks, professor, Johns Hopkins University
George Brown Goode, director, United States National Museum
Charles Frederick Hartt, professor, Cornell University
Alpheus Hyatt, custodian, Boston Society of Natural History
David Starr Jordan, president, Stanford University
Joseph LeConte, professor, University of California
Charles Sedgwick Minot, professor, Harvard Medical School
Edward Sylvester Morse, professor, Imperial University of Tokyo
Alpheus Spring Packard, Jr., professor, Brown University
Frederic Ward Putnam, director, Peabody Museum of Archaeology and Ethnology
Samuel Hubbard Scudder, distinguished entomologist
Nathaniel Southgate Shaler, dean, Harvard Graduate School
William Stimpson, director, Chicago Academy of Sciences
Addison Emory Verrill, professor, Yale University
Charles Otis Whitman, founder, Marine Biological Laboratory at Woods Hole
Burt Green Wilder, professor, Cornell University

There were, of course, numerous other people who had some contact with Agassiz.[68] Among those who definitely were his students and did become scientists, though achieving less reknown, were Albert Bickmore, Henry James Clark, Walter Faxon, Walter Fewkes, Samuel Garman, Theodore Lyman, William H. Niles, Orestes St. John, and Henry A. Ward. We should note that some on my list had rather limited association with their famous teacher, though their memoirs and obituaries imply a mystic virtue to the connection. Birge, Brooks, Minot, Jordan, and Whitman were students for only a few months in 1873, and Goode was in the M.C.Z. only for part of a year, when Agassiz was ill and delegated all his lecturing to Shaler. Subtracting those six, we are left with thirteen men. Eleven of those thirteen cluster together, overlapping with one another, during the four-year period 1859–62 (see fig. 7). These students attended at the birth or infancy of the museum.

(One might quibble with some of my assessments: does not Bickmore belong on the first list, as founder of the American Museum of Natural History, and was Stimpson really such an accomplished zoologist? This switch would only strengthen my argument, since Bickmore was a classmate of Hyatt, Scudder, Packard et al., while Stimpson was earlier. My estimate of Faxon, Fewkes, and Garman may seem mean, since their publications after lifetimes of service in museums were substantial, but it is my impression that their contemporaries would not have classed them in the first rank. I have excluded William James, because his contact seems to have been limited to his adventures as a volunteer assistant, not a student, on Agassiz's 1865–66 collecting trip to Brazil.)

In looking back to that extraordinary cohort of about a dozen young naturalists, it is hard to avoid thinking of them in terms of their later careers. Morse's biographer writes,

> To call the roll of Agassiz's student-assistants for the year 1859–60 is to recite the names foremost in American natural history forty years later, authors of standard textbooks, curators of splendid collections, directors of great museums in every large city of the Eastern seaboard. If youths of genius gravitated naturally to the feet of Agassiz, equal weight must be given to Agassiz's wise selection. Certainly it was

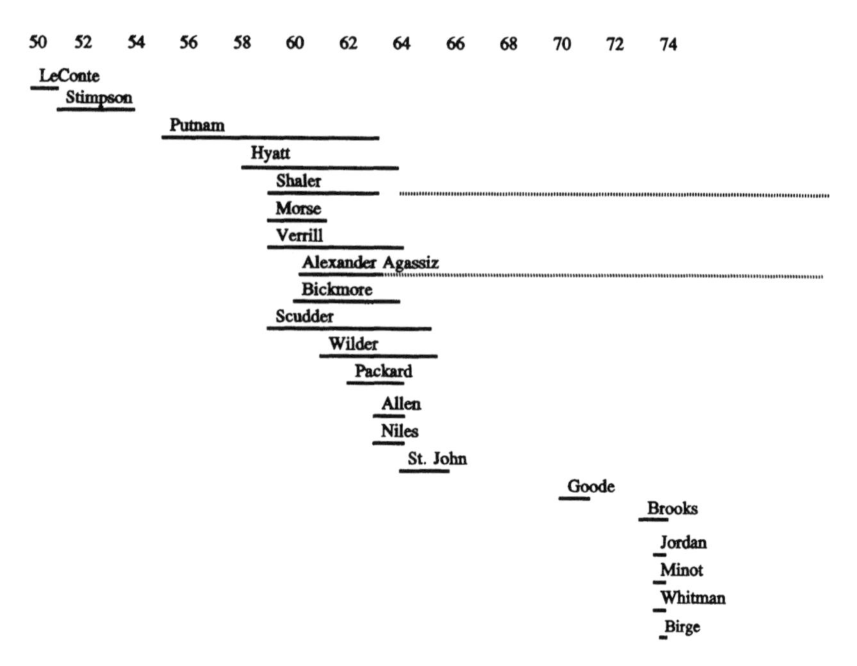

Figure 7. Notable students of Louis Agassiz. The period from the founding of the M.C.Z. in 1859 to the "Salem secession" of 1864 was particularly productive.

not chance which produced the constellation made up of Verrill, Hyatt, Ordway, Shaler, Morse, and Putnam.*

I agree it was not chance that produced them, but neither is there any evidence that Agassiz picked and chose among candidates for study. Almost all who came to him had some abilities or talent worth developing, and several were bright; none displayed genius.

I doubt that Agassiz's legendary solitary fish was really the key to his pedagogic success. Probably more important was the second stage in a student's training under Agassiz, which would last months or years instead of days or weeks: confronting several hundreds or thousands of specimens, perhaps an entire section of the museum's collection, to sort and arrange and study. So not only were the students soon in a position to name new species, but they could also recognize new genera and groups of genera, revising and correcting the classificatory schemes to be found in the literature. I think what was central to the scientific productivity of Morse and Shaler, Verrill and Scudder, Packard, Hyatt, and Putnam was their feeling of being placed at the frontiers of scientific research, and that with the materials and ideas Agassiz was giving them they could make original discoveries of fundamental importance. It was not only in the privacy of a letter to Baird that Agassiz declared that his students could rival their elders; he also encouraged them to believe it themselves.

It was partly by nourishing his students' expectations that Agassiz could mobilize their energies to his aid. Their sense of excitement was fostered as much by their interactions with fellow students as by each student's relations with his professor.

During the very days that Agassiz's devoted band carried his accumulated collections from Zoological Hall into the new building, the *Origin of Species* was coming off the press.† The irony of timing seems acute to us, but only because we know that Darwin's views triumphed. They did triumph, but the process took over a decade. To us it may seem clear that Agassiz's ideas were rendered obsolete by Darwin's great book. We might, then, imagine his students experiencing a rapid and bitter period of disillusionment. (In spite of Agassiz's own unbending opposition, every one of his

*Wayman, *Morse,* p. 69. Ordway is an anomaly in Wayman's list, apparently included because of Morse's positive references to him. A promising student of Agassiz's, he seems not to have returned to science after his service in the Civil War and published nothing.

†There was a ground-breaking ceremony on 14 June 1859; specimens were transferred from Zoological Hall to the new building 15–21 November 1859, and there was an elaborate inauguration day on 13 November 1860. The *Origin of Species* was published on 24 November 1859.

students, including his own son, did eventually accept evolution.) In fact, though, most of his students, with Alexander Agassiz perhaps the only exception, found the religious and transcendental aspects of his teaching more attractive than the directionless cruel workings of natural selection. Their shift away from their teacher was a gradual one.

Agassiz's students were perfectly aware that this new version of the old transmutation idea had earned the respect, and perhaps the allegiance, of Gray and Wyman. They knew of Agassiz's debate with William Barton Rogers at the American Academy of Arts and Sciences, and several of them witnessed his debates with Gray at the Boston Society of Natural History.[69] Like Huxley's clash with Bishop Wilberforce, the consequence of these public confrontations was not the conclusion that evolution was certainly true but that it deserved more serious consideration than mere denial, mere repetition of old evidence without fresh thought, which was all Agassiz was able to deliver.

Agassiz's efforts to keep his students on his side against Darwin were initially successful. His tactic was to emphasize that scientific men should care for facts, not speculations, and that the stability of species is a well-authenticated fact. He recommended that they read the *Origin,* advising them to notice how far Darwin had "deserted the inductive method."[70] The students did read the *Origin* for themselves in 1860 and 1861, with no notable effect.* The program of scientific research into which they had been initiated in the museum remained coherent and attractive. It informed their work not only while they studied with Agassiz but even after they left Cambridge.

A lively and productive atmosphere of scientific research was sustained in the museum's first few years. The students held regular meetings of the Agassiz Zoological Club, which Agassiz encouraged them to do with no faculty member present, to read and to criticize one another's work. They went across the Charles River for meetings of the Boston Society of Natural History, impressing themselves with their ability to hold their own among their elders. Those who came too late to feel the excitement of carrying specimens into the new building could join in unpacking consignments from Zanzibar and China, Switzerland and Australia. Two lads arriving together in February 1862 were initiated in the same fashion as their predecessors:

> We were then given other subjects to study, Mr. Niles taking up crinoids, while I was given a large miscellaneous collection of fossil gastropods to assort and arrange ac-

*Verrill recorded discussion of Darwin's book with Hyatt and two other students on 5 February 1860 (Private Journal, Harvard University Archives).

cording to their relationships. We were both happy, having been assured by our great teacher that we were making good progress in "learning to observe."[71]

All students, whether paid as museum assistants or not, were working on questions of zoological relationship, using museum specimens to explore the meaning of natural classification according to Agassiz's ideas. Albert Ordway, when he went off to fight in the Civil War, left behind a pile of notes on the blue crab *Callinectes,* the product of his assignment to arrange the Crustacea in the museum's first year; these papers were edited for the Boston Society of Natural History's *Journal* by William Stimpson, a former student and occasional museum helper. The young Samuel Scudder, already devoted to entomology, was assigned by Agassiz to analyze a family of fish, the Sciaenoids (which included his baptismal drum).[72] Scudder also arranged specimens of Orthoptera (crickets and grasshoppers) for public display, and also created drawings for exhibition that showed how various genera of the family Pieridae (white butterflies) had distinctive patterns of veins in their wings.[73] In his published reports, Scudder attempted to apply Agassiz's definition of the categories, writing of the "different modes by which the plan of articulation is carried out" in the *embranchement* Articulata, agreeing with Agassiz that these constituted three classes and that there were three natural orders of insects depending on their "degrees of complication" of structure.[74] A. S. Packard prepared a museum exhibit to demonstrate the details of structure of the bumblebee family.[75] He carried out research on the embryology of these insects based upon specimens preserved in alcohol paying special attention to the bees as representing "the highest types of articulates."[76] Besides publishing natural history notes and taxonomic synopses, Packard also wrote on insect morphology using Agassiz's concept of "synthetic types."*

Addison E. Verrill, who had arrived planning to be an ornithologist and was put in charge of the museum's birds and mammals for several years, was also given the job of putting in order the corals and other polyps. That work led him to propose new subdivisions of the Polypi, which he described to his fellow students at the Agassiz Zoological Club; he then set about developing his analysis of polyps for more formal publication, discussing his progress with Agassiz at several stages. His consultations served to warn him that his teacher could be selfish and petty. On one occasion, Agassiz instructed him to describe the new species in the collection but leave the new genera for Agassiz to describe. Verrill wrote in his diary, "I think the last part much the best, and would rather have the privilege of

*Packard, "On synthetic types in insects," but later Packard preferred Dana's term "comprehensive type" for the same phenomenon (*Guide to the Study of Insects,* p. 48).

defining the genera myself and then I don't care who describes the species, but I suppose he thinks the same himself."[77] Whatever such incidents did to his view of Agassiz's character, Verrill seems to have remained satisfied with the concepts Agassiz had taught him. Verrill discussed with a fellow student how the permanence of species, which they could see daily in their museum work, showed Darwinism to be absurd; he made use of Agassiz's definitions in deciding whether the subdivisions he perceived were suborders or families; he was satisfied he had found embryonic and synthetic types, and on New Year's Eve of 1861 he explained to the Agassiz Zoological Club the "parallelism" between two families of corals: "nearly every genus of the first has its analogue in the second."[78]

To convinced evolutionists, the dogmatism in Agassiz's "Essay on Classification" is evident, contrasting unfavorably with the open-minded tone of Darwin's book. In lecturing his students, however, Agassiz did say that his view was only one of those possible, and that his was not the final word. Personally convinced that his mature perceptions were sound, Agassiz nevertheless communicated very clearly both in his lectures and in the "Essay" that there was much that remained to be learned; that there were still many scientific questions to which neither he nor anyone else yet knew the answer but which some dedicated young man could address and perhaps solve for the benefit of science and to his own credit and fame.

His students responded to these intellectual challenges. They learned from him, for example, that the fundamental idea underlying the type Mollusca was not as well defined as the plans of the Radiata, Articulata, and Vertebrata; they learned that the homologies which must exist if the group was a natural one—between squid and clam, snail and brachiopod, tunicate and bryozoan—had not yet been satisfactorily worked out. His own most recent contribution to science, which they heard him expound again and again, was the tracing of homologies within the Radiata. Someone must do the same for the mollusks. So Nathaniel Southgate Shaler, aged twenty, presented a paper to the Boston Society of Natural History in December of 1861 calling attention to the problem, noting Agassiz's improved definition of the general idea of the Mollusca, and offering an analysis of brachiopod morphology as a step toward finding the hidden homologies.[79] Shaler's scheme was immediately contradicted by his classmate Edward Sylvester Morse, who had a different proposal about brachiopod morphology but whose interest in the problem had the same source:

> After becoming acquainted with the perfect unity of plan in the Radiata and the connecting series of homologies, running through the whole branch, (as demonstrated by Prof. Agassiz in his private lectures) my interest was excited, to discover, if possible, a like symmetry of development in the Mollusca. . . . I could not but

believe that in the Mollusca some plan lay hidden, which when unfolded, would as definitely convey their type, and unite them all.[80]

Morse carried this interest with him when he left Cambridge in January of 1862. Before long he learned that his former classmate Alpheus Hyatt was likewise working on an improved definition of the *embranchement*.[81] The museum's mollusk collection had been in the charge of Hyatt, Morse, and Shaler.

In the mid-1860s all of Agassiz's students were still working within his framework of thought, and some were moving in even more conservative directions.* During the 1870s they one by one accepted evolution. Several also accepted natural selection as well, but with varying degrees of reservation.[82] Alpheus Hyatt is usually credited with being the first of this cohort to ally himself with evolution because of his 1866 article "On the parallelism between the different stages of life in the individual and those in the entire group of the molluscous order Tetrabranchiata."[83] Whatever may have been his private views at this date (he is reported to have accepted evolution during his first year with Agassiz),† his article makes no reference to transmutation, nor does it depend upon any evolutionary concept. Instead, Hyatt was doing exactly the kind of exercise Louis Agassiz had championed, that is, tracing parallels between embryological development and fossil series, with a bit of Dana's cephalization and Wyman's polarity thrown in for good measure. Whether he privately held evolutionary views and was not yet ready to commit himself in print, or was still thinking in transcendental rather than physical terms, cannot be proven from his published statements of the 1860s.

*Evidence for the evolutionary views of four of Agassiz's students was assembled by Ralph Dexter, "The impact of evolutionary theories on the Salem group." He there stated (p. 148) that "E. S. Morse immediately accepted Darwinism," mentioning his other article on Morse, "An early defense of Darwinism"; but that "early" defense was in 1888. In 1864 Morse published an article "Variability of species" wholly supporting Agassiz's views, citing "that wonderful profusion of God's handiwork everywhere" and calling evolution "irreverent; for it is restricting the Almighty to follow this chance and illegitimate way to create those forms in which we discover only the manifestations of a thinking God." Morse insisted that departures from type do not consist of "class, ordinal, family, or generic differences, only individual variation; and this fact is a distinct contradiction to the Darwinian theory." Certainly he did become an avid Darwinian, but not immediately. His biographer Wayman (pp. 220–25) puts his conversion in the period 1870–73. Dexter has since revised his view (personal communication) that Morse "immediately" accepted Darwinism, noting instead that Morse and Packard were trying to convince Agassiz of evolution on Penikese Island in the summer of 1873.

†A. G. Mayer, Hyatt's son-in-law, reported that "within a year of the time when he began his studies under Agassiz he became an evolutionist and an admirer of Lamarck" ("Alpheus Hyatt," p. 133).

I have yet to see evidence of any of Agassiz's students publicly endorsing evolution anytime in the 1860s. By 1873, however, Morse was doing so aggressively, and those who attended the Penikese summer school in 1873, like a twenty-two-year-old schoolteacher named David Starr Jordan, knew that most of the younger members of the teaching corps believed in evolution.[84] Joseph LeConte's first indication of his conversion was in a memorial address after Agassiz's death in December 1873.[85] By 1876, two-and-a-half years after Agassiz's death, Morse declared that he and his compatriots were up-to-date with modern biology. His address, which was attended by the English champion of evolution T. H. Huxley, contains the statement:

> Agassiz's earnest protest against evolution checked the too-hasty acceptance of this theory among American students. But even the weight of his powerful opposition could not long retard the gradual spread of Darwin's views: and now his own students, the last to yield, have, with hardly an exception, adopted the general view of derivation as opposed to that of special creation.[86]

Modern readers who are convinced of the soundness of Darwin's reasoning are tempted to believe its acceptance was quick, and thus focus on Morse's implication that Agassiz "could not long retard" the spread of the theory in America. Yet "not long" was over a decade. Further, if we are tempted to imagine the debate over the theory as a source of strife between Agassiz and his students, we should notice Morse's testimony that Agassiz's students were "the last to yield."

Whether one calls that slow or quick, it does mean that the entire "Essay" was rendered obsolete—not just Agassiz's audacious attempt to make categories real but his entire thesis that systematic relations show evidence of intelligent contrivance. A growing tendency to believe in evolution would perhaps have elbowed Agassiz's ideas aside even if Darwin and Wallace had never lived. Agassiz had taken what had once seemed the pious consensus of naturalists and had pushed the view to its logical limits. Stretched so far, that view yielded up a Creator with a highly suspicious resemblance to a taxonomist.

2

"I Have Been Disappointed in My Collaborators"

We are all members of so many social structures—family, employment, state—that it is easy for us to grasp those interactions which constitute human institutions, although their complexity would leave a visitor from another planet stymied. We live daily with the paradox that we are simultaneously constrained and empowered by social structures. I have described a scientific discipline as a combination of intellectual content and social structure. To describe the cognitive content we would turn to textbooks and journal literature, as I have turned to Agassiz's "Essay on Classification," though it is not the publications themselves but the thoughts in the minds of individuals who write and read such publications that constitute the cognitive dimension of the discipline. The social dimension would include among many other factors the educational structures that expose a new generation to the literature, the system of readers and editors that gets research published in a certain form, the laboratories, universities, and museums where research goes on, and the network of personal liaisons that speed communication and affect evaluations. If we move from abstract model to real life, it would be impossible to describe in complete detail the operation of any particular scientific discipline, wherein each role is played by a person of unique character and intellect, whose actions have causes but are not determined. Fortunately, there is no need for a full description. Humans can quickly understand essentials that would take years to explain to a visitor from space.

That visitor, though, might instantly notice a dimension humans seem prone to overlook (perhaps because our languages make relationships and processes into things), namely, the vital dimension of time. Time is especially essential to science. The theory at the center of a scientific discipline, as recent philosophers of science quite properly emphasize, is not simply a static body of accepted knowledge; it is a program for research. A scientific discipline consists of an interpretation of recent success and, most important, a vivid expectation of future developments. It must generate interesting questions and suggest feasible ways to answer them. As months tick over into years, the cognitive dimension of a discipline is a constantly changing vantage point from which recent achievements are evaluated and future returns estimated. Its social dimensions also exist as a moving plane in time. Most people who commit their energies to a scientific field of study do so in the expectation that the discipline will in time reward them, at the very least with the private intellectual satisfaction of discovery; usually the discipline is called on socially as well, to provide an audience—a network of readers able to appreciate, criticize, and encourage. A young person without means must also judge whether a particular discipline promises a living. Just as the intellectual viability of a discipline depends upon whether some percentage of its puzzles turn out to have solutions, so, too, its social viability depends upon whether some percentage of its apprentices can realize their hopes for a career. In both cases, the real issue is not an analysis of the discipline by some disinterested commentator but the subjective assessment of the members.

For the young men who joined Louis Agassiz in his new museum, one of his great attractions was that he seemed to understand all this so well. He made comparative zoology seem like Renaissance astronomy and physics, when centuries of observation and the recognition of laws by the likes of Kepler and Galileo readied the field for Newton. Zoology was on the verge of its most exciting period. Here was a man who had proven able to convert the charms of his favorite subject into monetary support, first getting a professorship for himself, then attracting the endowment for a new institution, the M.C.Z. Agassiz explicitly encouraged his students to see into the future with him, giving them the expectation that they could make their living as zoologists. If the M.C.Z. grew according to his dream, Agassiz would be able to offer them "a good appointment and a good salary," he told them.[1] At the very least, his certification of their training would get them positions elsewhere.

Most of Agassiz's students did go on to careers in zoology, and later looked back with gratitude to the training they had received in his museum. But those retrospections were too late to repair the break in the thread of expectation that happened four years after the museum's founding. By December of 1863 most of his students had resolved not to accept employment under Agassiz. The jobs they scattered to in New Haven, New York, Boston, and Salem they found without his help. Their defection seems to have had little if anything to do with the inherent weaknesses in the "Essay on Classification," the publication of the *Origin,* or the disruption of the Civil War.

The joyful spirit which graced the M.C.Z. in 1859 evaporated quickly under the glare of Agassiz's volatile personality. For many students, it was a blow to discover that their interests were after all not the great professor's highest priority. The first incident in their disillusionment, a seemingly trivial one, Morse's biographer called the "Butter-pat Insurrection."[2] The students living in Zoological Hall were affronted by the manners of William Glenn, a museum employee whom Agassiz had put in charge of Zoological Hall, and one day in July 1860, when Glenn spoke roughly to their cook, the students turned to Agassiz for help.* They were shocked to find that negotiation and compromise were not in the professor's vocabulary. They might have been more prepared for his censure if they had realized that the man they complained of was so skillful in his museum job as preparator of skeletons that Agassiz felt he would be hard to replace. One of the students, Albert Ordway, went to Agassiz expecting to explain the affair:

> To Ordway's grief and astonishment, Prof after having listened to Glenn would not hear one single word from us, simply saying that if it did not suit us we could leave. . . . This from Prof, who we have loved and adored as a Father! Have done everything to oblige and please him! and now when we run to him as a family would to its head for redress and advice, he tells us to hold our tongues—that he will not hear a word from us. Is this morally right? Is this consistent with what he has taught us in his lectures? No, it is not, and now Prof. has also fallen in our eyes from a God to a man—even to the lowest and most unreasonable of all.[3]

Agassiz for his part was furious at the students for their ingratitude in finding fault with the home he had provided them, and he let them know how easily he could do without them by taking the museum keys away from three of the students. The students were all "highly indignant in regard to the Row"; they decided "that if much more injustice was heaped on us, we would leave the Museum."[4] However, the whole affair was in the nature of a family argument, and in fact resulted in no one's leaving. The immediate cause of trouble was eliminated by meals no longer being prepared in the dormitory; the students ate in Cambridge boarding houses while continuing to live together in Zoological Hall. Their advisors, and their good sense, made them realize that Agassiz's museum could still offer them the best start available in North America for a career in biology. Their

*Triggered by comments about the service of butter at meals, the quarrel involved personal insults and struggle for power between Glenn and the students. The identification of "Katie," the cook, with Mrs. Glenn (in Wayman's *Morse,* p. 126) seems to me doubtful.

personal frustration at Agassiz's behavior they dealt with in classic student fashion:

> In the evening . . . we held a meeting to organize our Society of Zoologists. We appointed committees, had some curious parliamentary proceedings and after our chairman had run away and had been again brought back by an escort we finally adjourned, and then had rich fun in Putnam's bedroom exercising ourselves.[5]

The students formed a secret league called something like "the Society for the Protection of American Naturalists against the Oppression of Foreign Professors."*

But if both students and teacher had had their feelings hurt and learned that they were not after all in Camelot, the mutual advantages of the relationship acted to heal the injury. Within a few months the students were having a friendly supper with their professor, and when another emotional contest threatened—over the students' place in the formal opening ceremonies of the museum—it was quickly smoothed over.[6]

Another source of strain upon personal relations within the museum, as letters of the time and later recollections testify, was the Civil War. Agassiz wished his enterprise exempt from that tragic conflict, and even tried to tell his young men to bury themselves all the deeper in the scientific work of the museum and avoid reading newspapers—a futile gesture that only offended the students' sense of patriotism:

> Intensely exciting news this morning. All the previous news confirmed. Lincoln calls for 75,000 men. Saturday Prof after lecture told us what we should do. That we must work and show the people that amid such exciting times we should not go wild with the general excitement but work. That we must not read papers, etc. This morning Ordway was reading a paper and Prof said we must remember what he said on Saturday, and then went out of the room. He immediately came back and said if we did not think as he did, he should forbid every paper from coming into the building. This was perfectly outrageous on Prof's part and we expressed our minds freely on the subject. Hansen spoke to him about it afterward, and Prof confessed that he was hasty in his expressions.[7]

If Agassiz and his students seemed now better able to forgive each other their extremes, the fact remained that each student was pulled by conflicting demands of duty to country and to Science. In 1862 Agassiz reported

* "Do you remember the trouble I had the day of the inauguration of the Museum? Then & there I learned that Mr. Putnam was president of a secret society among my students, a society organised 'to protect american students against the oppression of european naturalists.'" Louis Agassiz to Theodore Lyman, 7 May 1867, M.C.Z. Archives; Lurie, *Louis Agassiz: A Life in Science,* p. 314.

that ten of his assistants had left the museum to take up arms;[8] one of these would lose his life.[9] Nevertheless, the Civil War represented only a temporary inconvenience for the operations of the museum. Of the promising students who had heeded the call to arms, almost all would return to the museum.

One direct effect of the Civil War more damaging to the museum than the temporary depletion of workers from its staff was the sudden change in the value of the American dollar. Committed to purchases of European specimens and books, Agassiz found himself desperately short of funds. Several of his students thought Agassiz hypocritically generous with promises while feigning poverty in the event.

Louis Agassiz's integration of teaching with curatorial work seemed worse than inefficient to his son Alexander, who believed in the businesslike discipline of balanced books, a stable, loyal workforce, and visible production (fig. 8). He wrote to his friend Theodore Lyman in April 1862:

> To turn from politics to a subject not less disgusting the Museum I will speak perfectly openly (*entre nous*). I do not think matters go as they should at the Museum. The students who ought to work and are *paid for working,* font les Messieurs and think it is below their dignity to do anything except to study for their own advantage. Instead of arranging their part of the collection they are preparing themselves for their degrees and of course neglect everything else. The consequence is we have no work done, when I speak to father about [it] I can get no satisfaction. He has a thousand excuses always ready to give in their favour, so that all my attempts at introducing any kind of system or order have most signally failed. The Museum has little enough money and it certainly ought not to be wasted in paying small salaries to a large number of young men and not to get a thing in return. I believe it would be much more advantageous to pay one or two men *well,* get good men, and have the work properly seen to. Instead of which Father has to spend the whole day in showing the assistants how to do their work.[10]

What Alexander saw as inefficiency, however, his father doubtless saw as an investment for the future. If he were to hire experienced naturalists, how could his museum be arranged according to his new ideas? If he were to hire obedient clerks, how would he train the next generation of zoologists to spread his principles far and wide? His students' discoveries would redound to his credit and build the reputation of his museum.

The incident that proved fatal to the integrity of the M.C.Z. was the disappointment of a former student, Henry James Clark, a man whose career had seemed enviable when the museum was new. Coming to Cambridge in 1850 at the age of twenty-four to study with Asa Gray, he had graduated from the Lawrence Scientific School in 1854 and was married in that year.

Figure 8. Alexander Agassiz (standing, thumb in watchpocket), with Theodore Lyman (left) and Jacques Burkhardt. The two younger men were very fond of the artist Burkhardt, who had worked for Louis Agassiz from the days of the "scientific factory" in Neuchâtel in the late 1830s through the Thayer Expedition to Brazil in 1865. (By permission of the Museum of Comparative Zoology Archives, Harvard University)

Agassiz hired him in 1855 to work on the *Contributions to the Natural History of the United States.*[11] Clark's diary records great friendliness between him and his employer, who told visitors that Clark was the "most accurate observor in the country."[12]

The first two volumes of the *Contributions* appeared in October 1857; the name "Louis Agassiz" was the only one on the title page. In the preface, though, were full acknowledgments to his helpers:

> Mr. James E. Mills, of Bangor, (Maine,) has worked out for me the special characters of the families of the Testudinata; and Dr. Weinland has helped me in revising the anatomical characters of the order, in accordance with the principles laid down in the First Part of the work; while Mr. H. James Clark has assisted me from the beginning of my investigation of the embryology of these animals, and drawn, with untiring patience and unsurpassed accuracy, most of the microscopic illustrations which adorn my work. I owe it to Mr. Clark to say, that he has identified himself so thoroughly with my studies since he took his degree in the Lawrence Scientific School, that it would be difficult for me to say when I ceased to guide him in his work. But this I know very well,—that he is now a most trustworthy observer, fully capable of tracing for himself the minutest microscopic investigation, and the accuracy of his illustrations challenges comparison.[13]

Indeed, the second volume, which described in painstaking detail the development of turtle eggs, consisted not just of Clark's work and ideas but of Clark's own words.

Early in 1859, Clark began to feel uneasy about his status.* He went to his employer to have a long talk, not complaining about the published *Contributions* but "in reference to my future stay in Cambridge and in regard to what was to be the connection between him and me. I gave him to understand that I wished to go on *freely* and *independently* in my investigations."[14] Agassiz seemed to meet all Clark's concerns, assuring him that his future as a scientist in his own right was secure, "and he even proposed that I should become the *author* of one of the volumes of the 'Contributions.' I am to have the new professorship of *Embryology* when it is founded, which will happen now soon."[15]

Over the next few months the legal structure of the proposed new museum was being organized. There was a ground-breaking ceremony on 14 June 1859, and a few days later Agassiz departed to visit relatives and colleagues in Europe and purchase books and specimens for the museum.

*My narrative on Clark's battle with Agassiz is based upon material assembled by David Lokari, whom I have to thank for urging upon me the interest and importance of this story. Further thanks are owed as well, for it was Mr. Lokari's youthful labor and enthusiasm that laid the foundation for the well-organized archives which now exist in the M.C.Z.

Clark attended the sunny ceremony, pleased with the result of a recent interview with Agassiz.[16] He had been awarded an official position in the new institution. Clark noted in his diary, "At last an important addition to my income. This afternoon Agassiz informed me that from this time I am to have ***$500*** a year as assistant Curator of the Museum of Comparative Zoology, to be paid quarterly."[17] Agassiz also told Clark that he had given the president of Harvard a letter stating "that as I am the 'only man in America' who understands his views . . . he wishes, in case he should be lost by shipwreck or otherwise on his intended journey, I might be appointed to take his place as Director of the Museum."[18] (How thoroughly Clark had made Agassiz's views his own would become obvious when Clark gave the Lowell Lectures in 1864.)[19]

There is every reason to believe that Agassiz was sincere in what he told Clark. He had, indeed, given President Hill such a letter, although it was the position of curator, the scientific head of the museum, that he specified for Clark, suggesting Theodore Lyman as director, the administrative and financial head.[20] Even the usually sour Alexander Agassiz liked Clark and noticed his loyalty to the museum. But Agassiz returned safely from his voyage, and Harvard failed to fund a professorship of embryology. Agassiz then arranged for the appointment of Clark to the Lawrence Scientific School as adjunct, that is, unpaid, professor. Clark now gave lectures on embryology to the students in the museum. To them he was "Professor Clark." Undoubtedly, Agassiz did expect to find a benefactor to transform this formal title into a salaried position.

It was perhaps as early as 1861 that the fabric of loyalty and career expectations that bound Clark to Agassiz began to unravel. It dawned on Clark that his extensive researches on jellyfish embryology, written for the third and fourth volumes of the *Contributions,* would appear under Agassiz's name, for Agassiz rebuffed his suggestion that portions of the work should be distinguished as Clark's own.[21] When the volumes actually appeared late in 1860, the preface showed him how far Agassiz had been willing to go. Clark was thanked for his "valuable assistance" and was acknowledged as having "investigated for himself some special points of . . . structure, which are noticed as his contributions in the proper place."[22] The credits sounded generous, yet Clark again in the text found page after page of his own research, and his own words, which any reader must take for those of Agassiz. Still, it was true that Agassiz had supervised Clark and paid for his time. Clark dared not complain, for his "family was dependent upon what I got for working upon those volumes. I could not *afford* to quarrel then."[23] (Clark and his wife had a total of eight children.)

His brother, traveling in Europe, made sure the leading zoologists learned how the *Contributions* had been pasted together, but Clark himself was silent.*

Harvard's president, Thomas Hill, and Jeffries Wyman, professor in the Lawrence Scientific School and member of the M.C.Z.'s governing board (called its "Faculty"), encouraged Clark to believe that his professorship would become salaried before long. So Clark had a reason not to confront Agassiz. Apparently Clark decided to proceed on his own to carry out and publish a substantial piece of research quite independently of his former teacher. Possibly he imagined he could thus bring Agassiz to recognize his maturity. Agassiz would later explain to Hill his own sense of betrayal by saying that beginning in late 1861 Clark became "fitful in his performance" of the work Agassiz was paying him to do, and that Agassiz only realized how Clark's time had been spent "when I accidentally learned from my Son, that Mr. Clark had completed an extensive monograph with numerous plates of the family of Lucernaria, respecting which he had never whispered a word to me."[24] Very likely Clark did take care that this research was done without the advice or consultation of Agassiz, in order that there could be no question about its authorship, but it is most unlikely that he was actually conducting it in secret. His diary records extensive periods of work on *Lucernaria* in the museum in 1861 and 1862, including several trips to the seashore to collect live material in company with Alexander Agassiz. Indeed, he reported publicly on the progress of these studies at the Boston Society of Natural History as early as March of 1862.[25] Busy as Agassiz was, though, he may well not have paid attention to Clark's activities. And now that Clark clearly had a career independent of Agassiz in mind, his zeal in doing Agassiz's assignments may indeed have cooled, which would explain why the two men could have quietly differed in their estimations of how much time Clark owed Agassiz.

When Agassiz realized what Clark had been up to, his reaction was not at all the respect Clark might have wished. Instead,

> I took occasion to ask him to show me his work, which he did. I made no comment but silently estimated for myself the amount of labor bestowed upon this investigation, which could only have been made in the daytime [because the microscope required strong natural light]. . . . It explained to me fully the reduction of the attention which ought to have been paid to my work during the preceding year.[26]

*David Weinland, who like Clark is acknowledged in the preface to the *Contributions*, apparently went to some trouble in Germany to claim credit for Clark and himself. A. Agassiz to Lyman, 21 July 1862; Lyman to A. Agassiz, 20 October 1862, M.C.Z. Archives.

Agassiz thought of the liberality with which he had supplied Clark with books and microscopes and decided that he could no longer place any trust in Clark.

It is my guess that it was this event between Clark and Agassiz that caused Agassiz to call together the whole museum community on 12 December 1862. Verrill recorded it in his diary in detail:

In the morning Prof. gave us a private lecture. Pres. Hill and Prof. Pierce being present and adding remarks. He alluded to the low financial condition of the museum and expressed his intention of resigning unless presently relieved. He then spoke of the students and stated that he could not be responsible for their hopes and aspirations. He spoke about some of us commencing publishing and gave a rather personal and, as it seemed to me and others, uncalled for account of cases of plagiarism toward him from some of his former students; and stated that there was a paper in preparation which he had not seen, but was sure was full of improprieties, and he would like to have a chance of taking it up and criticising it! Various other remarks of similar personal character were made, and he closed by stating that he should insist on seeing all papers prepared by students at the museum before publication. The drift of these remarks appearing to be, that he was jealous lest some of his students should rob him of some of his honor and glory by not giving him credit for everything that *he had ever done,* (published or not) and a little besides! I wonder if there may not be some chance, judging from my own experience, of the students having reasons for making the same complaint of him! Hill's remarks were in about the same strain and Pierce joined in preaching to us about plagiarism and seeming to take it for granted that we had all or part been guilty of it and bringing up before us the awful tribunal of European usage and judgment!!![27]

Agassiz had not been able to suppress recollections from Neuchâtel in the 1840s, when he had felt himself the victim of plagiarism at the hands of his collaborators Edouard Desor, Carl Vogt, and Carl Schimper.[28] Like Clark, they had agreed to contribute portions to his grand publication schemes, but as the studies came off the press they had begun to demand recognition as authors rather than acknowledgments as mere assistants. Both Vogt and Desor had managed to affix their own names to studies Agassiz considered his. Carl Vogt had even researched and published an entire study without Agassiz's knowledge, just as Clark was now doing with *Lucernaria.* What had begun as collaboration in a spirit of scientific enthusiasm had ended in bitter quarrels. One of these dogged him across the Atlantic, in the person of Desor, who brought an embarrassing public accusation against Agassiz in Boston in 1857. A private tribunal had fully vindicated Agassiz, but one scientist who knew both men and the situation well judged them "about equally unright."[29]

Though the manuscript evidence which would allow a modern jury to settle every disputed point raised by his assistants—from Schimper, Desor, and Vogt in Neuchâtel to Weinland and Clark in Cambridge—does not exist, it is perfectly clear that Agassiz's sense of intellectual responsibility ran counter to the assumptions of his coworkers. For Agassiz now to harangue his museum students shows how little he understood how he himself contributed to his own troubles.

Carl Vogt remembered Agassiz in the Neuchâtel period as enthusiastic about grand plans and exceptionally energetic and resourceful in collecting the specimens, books, and collaborators appropriate to his ambitious research projects. He would invariably be too busy to persist with the dull work of description that was necessary to the completion of his schemes, so his assistants found themselves carrying the load. Vogt's recollections of Agassiz's talents and failings agree with what we can reconstruct of his American career. Even his solidly loyal supporter Shaler later recalled that

> it was his custom to set one of us to work on a group of animals concerning which he had some knowledge. . . . For a while I felt that I was following on the trail which he had broken, and then . . . I began to teach him a bit that he did not know. He was as eager to receive as to give, and what I supplied went into his memory as his own discoveries, which in a way they were, for the direction of the work came from his mind.[30]

In private conversation and in his lectures, Agassiz would cast before his students and assistants ideas for research, and when an idea was taken up he retained a proprietary interest in it, feeling that the student, and the scientific world, should give him full credit for his fertile thoughts. Agassiz was certainly capable of getting pleasure from the success of his students, but if one seemed disloyal, apparently he would reconstruct their early relationship in his mind, and as time went by, remember his own contributions so vividly and the junior's abilities so faintly that he would imagine, in all sincerity, that the work had been essentially his own.

Of course, this was not how Agassiz viewed his disagreements with his collaborators. He told the gathering in December of 1862 that it was he who had been wronged. Although for Clark the only issue between himself and Agassiz was the authorship of portions of the *Contribution*, Agassiz was harboring feelings of injustice of his own, engendered by Clark's independent work on *Lucernaria*. Agassiz hinted privately to President Hill that Clark had stolen his ideas:

> I was so intimate with him that I never did anything relating in any way to scientific matters without mentioning it to him and sometimes discussing at length the various investigations which might be most advantageous to the progress of science. Such were on my part our relations when I accidentally learned from my son, that

Mr. Clark had completed an extensive monograph with numerous plates of the family of Lucernaria, respecting which he had never whispered a word to me, which surprised me the more, since not long before I had been myself interested in those animals and had spoken to him about them, especially mentioning particular views to which I had arrived concerning them and which I have stated in my last volume of "Contributions."[31]

The substance of Agassiz's innuendo was rather thin, for his conclusions in print were that these hydroid-like creatures were really jellyfish attached by a stalk. Clark took this further, arguing that *Lucernaria* could stand as a generalized type in relation both to jellyfish and hydroids.[32]

The evolutionist Ernst Haeckel, perhaps annoyed by the encomiums heaped upon Agassiz after his death, would later pronounce a damning assessment of Agassiz's career.

Louis Agassiz principally owed his exceptional and wholly predominant situation among American naturalists, not to the scientific value of his own work, but to the marvelous talent he had for appropriating to himself the work of others, to the rare mercantile aptitude which he knew how to deploy to make large amounts of capital conspire to make his ideas become real, and finally to the prodigious spirit of organization which allowed him to create collections, museums and the most grandiose institutes. Louis Agassiz was the most ingenious and most active swindler who ever worked in the field of natural history.*

After we subtract the vituperation, we may concede that Haeckel's assessment has some foundation. When allowance is made for the contributions of his collaborators, the volume of original published research that may be credited to Agassiz is very much smaller than that of most scientists of comparable reputation.

The students gathered to hear Agassiz's petulant tirade against former and potential "plagiarists" were, of course, greatly offended by his tone, especially since it could not have occurred to them that his remarks may have been aimed not so much at them as at Professor Clark, ten years their

*Haeckel, *Ziele und Wege der heutigen Entwicklungsgeschichte:* "Louis Agassiz war der genialste und thätigste Industrieritter auf dem Gesammtgebiete der Naturwissenschafte. Das er sich dabei nicht selten zu einer bedenklichen Höhe des Schwindels verstieg, war nur natürlich" (p. 80). "Es ist hier nicht der Ort, näher auf den grossartigen Humbug des neu-americkanischen "Gründers" einzugehen" (p. 81). To leave no mistake about his seriousness, Haeckel emphasized the first sentence by extra spaces between the letters. This appeared also in the *Revue scientifique* the next year, where "industrieritter" is translated "chevalier d'industrie." This term did not signify a captain of industry but a con man. Immediately upon reading this, Alexander Agassiz wrote to Haeckel calling him a calumnious slanderer whom he wishes he'd never met, and declaring all intercourse between them at an end (A. Agassiz to Haeckel, 11 December 1875, M.C.Z. Archives).

senior. Indeed, when Clark later did break with Agassiz, Verrill could not guess the cause. Several students assumed that Agassiz was alluding to Verrill's polyp study, to which Agassiz had contributed a number of his own observations as well as the many drawings he had paid Jacques Burkhardt to make. Verrill had presented his results at a meeting of the Boston Society for Natural History less than a month before the plagiarism lecture, and immediately after it he submitted his manuscript to Agassiz for his approval and was surprised that Agassiz seemed displeased that Verrill had taken the speech personally. Instead, he hinted to Verrill that he had had in mind the work of some other person or persons.[33] As for Clark, who had been listening to Agassiz's warning with a clear conscience because his *Lucernaria* work, nearing completion, was entirely his own, he gave his manuscript not to the director and curator of the M.C.Z. but to the editors of the *Journal* of the Boston Society of Natural History.

Agassiz was not intending that his own name should be listed on his students' publications, either as author or joint author, even when he had given them substantial quantities of material as well as ideas. He told them that all he wanted was to be able to check the papers over to see that any contributors of specimens, whether individuals or the museum, were properly acknowledged. When Verrill's polyp paper finally appeared, in the handsome new *Memoirs* of the Boston Society, it fairly dripped with credits.[34] Even some species Verrill was describing for the first time were given the names Agassiz had chosen, with "Agassiz" listed as authority.

In reality, Agassiz wanted something more significant than appreciation from his students; he wanted to publish their manuscripts. As early as February 1860, Agassiz's student Morse had heard that "the Museum will soon commence publishing articles and papers in Natural History and it is Prof's intention to have all the students prepare papers for it."[35] At that time the museum had been so far from having any money to spare for publications of its own that Theodore Lyman was asking the Smithsonian to publish his monograph on brittle stars.[36] In December of 1862 Lyman and Alexander Agassiz were still discussing the merits of the Smithsonian versus the Boston Society as vehicles of their museum results. Lyman wrote,

> I know Pa [Louis Agassiz] would cry for a periodical of our own, first to give reputation to the Museum, second to get exchanges of books. To which I reply, the Museum can get enough reputation in other ways; and it is much cheaper to *buy* exchanges of books. Moreover a periodical implies (a) money (b) contributors:—the first we have not: the second, hum! Peutêtre!—Moreover, again, there are three times too many periodicals already. . . . So no "Annals" of *our* "Museum" says Yours, Ted[37]

But Agassiz's young advisors were fighting a rising tide, and when his February appeal to the Massachusetts legislature was rewarded in April 1863 with $10,000, they came round to his view.

To succeed in this enterprise, though, Agassiz would need more control over his students. He had suggested to Verrill after seeing the polyp manuscript that it be published by the museum, but Verrill had gone ahead with his arrangements with the Boston Society. In fact, Verrill had thought Agassiz's suggestion "absurd," perhaps because at that point any publication of the M.C.Z. existed only in Agassiz's dreams, or perhaps because Verrill thought an institution's official "Catalogue" would demand an analysis more comprehensive than his. In any case, there were several projects underway besides Verrill's which Agassiz presumably had his eye upon. Putnam had a manuscript review of a family of fish, the etheostomids (darters) that had been nearly ready for several years; Scudder had done some assigned work on another group of fish that Agassiz had pronounced publishable; Scudder and Packard had done much research on various insect groups in connection with their arrangement of the museum's exhibits. Not surprisingly, Agassiz's warnings about plagiarism only helped drive these potential authors away.

Agassiz's lecture on plagiarism signaled a far more serious issue than the Butter-pat Rebellion. His plans depended upon cords whose strands were complementary interests. The students' original research was not so much a commodity as a social structure. Fresh, good quality research based upon the museum's collections could attract reputation both to the students and to the institution, but reputation is a thing not easy to measure and its benefits are often delayed. The students had to take on trust Agassiz's assurances that they could rise in the world by following his advice. Such trust must have been strained by their encounters with his volatile nature. But the plagiarism lecture touched more directly the implied contract they had with him: that if they did good work for him, he would see that their careers would benefit. Now he was denying responsibility "for their hopes and aspirations."

Meanwhile, the distrust between Agassiz and Clark seethed unexpressed through Christmastime of 1862 and into the depths of winter. Finally, Clark could stand the tension no longer. He meant to confront Agassiz only about some overdue back pay but found he could not suppress his sense of deeper injustice. He wrote in his diary,

> 20 March 1863—At Museum. Had a high worded quarrel with Agassiz in which I denounced him for keeping my family in distress and want for the last six months. I think it is time I had put a check on his overbearing way of speaking to me. No

honorable man ought to allow his assumption of superiority, especially when it is done in such an insolent manner as he exhibited today. No wonder he is losing friends in the community.[38]

Most likely it was Louis Agassiz's version of the quarrel we find recorded by Alexander Agassiz:

I am sorry to have to tell you of a very disagreeable occurrence in the Scientific family of the Museum. The other day for what reason I cannot find out, but simply in the course of conversation Clarke attacked Father in the most abusive manner, telling him that he did all he could to keep him under and that he never had made the slightest acknowledgement of what Clarke had done for him, that he was under infinite obligations to him and that he would not endure any longer to be treated with such neglect. He was as white as a sheet while saying this and Father, supposing he was unwell or out of his head, merely smiled and turned away.[39]

How Agassiz's smile must have infuriated Clark! Though correct in thinking that the great man was losing friends, Clark was to learn that he had far underestimated Agassiz's power in his own realm.

After a few more days of mutual sulking, Clark received what Alexander Agassiz, and certainly its author too, felt was a "very moderate letter":

Cambridge, March 24, 1863

Sir,

I think it is best not to allow our interview of last week to pass wholly without notice from me.

From the time you came to me as a student nine years ago, I have fostered your scientific interests in every way. I have exerted myself to obtain for you a desirable position as a scientific man, and have done everything in my power to secure for you not only the opportunity for original investigations, but also for lecturing and obtaining the means of support, often at the cost of no small personal sacrifice.

It is not the first time that I have given to young men all the advantages which my scientific position enable me to offer them and have found after some years of mistaken confidence that they considered me under obligations to them for results which were drawn from my own observations.

I did not expect that your name would be added to the list and while I now close irrevocably a personal intercourse which I could no longer base upon confidence, I would add that I do it without the slightest feeling of irritation and that it is my earnest hope that you may win in an independent career the best honors that could ever have been wished for you by your old teacher.

LS AGASSIZ

P.S. Under the circumstances I expect you immediately to return to me the books, lenses, etc. you may have that belong to me and also the keys of the Museum and the books of the Museum library.[40]

Clark turned, in this crisis, to Jeffries Wyman and Asa Gray for advice.[41]

One strike against Clark was that, although Agassiz had given him to believe that he was an assistant curator of the museum, the appointment had never been made official; what he had thought was a salary from the museum came from funds in Agassiz's personal control (partly subscriptions to the *Contributions*). So, technically, he had no formal status within the museum.[42] The secretary of the museum's Faculty, Oliver Wendell Homes, confirmed this disturbing fact, at the same time offering Clark his personal sympathy and encouragement.[43]

Clark did immediately return the items demanded, but he expected President Hill to uphold his right as a duly appointed professor, adjunct or not, to use the lecture room of the museum, which was, after all, the site of natural history teaching within the Lawrence Scientific School. Ordinary standards of fairness and justice told Clark that he should not be refused access to whatever books and specimens in the museum belonged to the institution and not to Agassiz personally. Instead, Harvard's governing Corporation—the President and Fellows—did their best to stay out of the feud altogether.* They informed Clark that Agassiz held the final power over use of the museum facilities.

Since he must now seek his scientific career elsewhere, Clark resolved at this point to win back what he thought Agassiz had appropriated from him. With Wyman's advice, Clark carefully constructed from his journal and manuscripts a report of exactly which portions of the *Contributions* were his, and he demanded that Agassiz let his claims be arbitrated by a private tribunal (as Desor's had been).[44] When Agassiz refused, Clark printed a three-page broadside entitled "A Claim for Scientific Property." Dated 6 July 1863, this soberly worded document was sent to zoologists and scientific societies throughout America and Europe.

Agassiz decided not to dignify Clark's charges with a rebuttal, but he did demand that his colleagues acknowledge the offense to his honor by excluding Clark from meetings of the Harvard faculty. At first they refused Agassiz's request, but this decision was overruled by the Corporation, so Clark no longer had a foothold in Cambridge. (He did find teaching jobs, at the Agricultural College of Pennsylvania, the University of Kentucky, and the Massachusetts Agricultural College.)†

The students were still not directly affected by Clark's "claim" though they must have taken his case as a warning, further weakening their sense of

*E. R. Hoar wrote to President Hill, "I fancy we had better let it work itself clear with as little interference as possible" (Harvard College Papers 30:183, Harvard University Archives).

†Lurie, *Louis Agassiz;* Tuckerman, "Henry James Clark." He died in 1873, six months before Agassiz, at the age of forty-seven.

contract with Agassiz. But Agassiz himself, of course, was very directly affected, having been assaulted and embarrassed, however ineffective Clark's actions seemed. He could convince himself that he was above "the slightest feeling of irritation," but now that he had a brick building, an endowment, governing boards to report to, and an established entity depending upon his management for its future, he had something beyond himself to protect. Though the connection was nowhere stated, it seems clear that Clark's "Claim for Scientific Property" precipitated the new set of "Rules and Regulations" Agassiz drafted for assistants in the M.C.Z. The museum Faculty formally adopted them on 5 November 1863.

The new regulations spelled out a policy of tight control over research, reading in part:

> No one connected with the Museum is authorized to work for himself in the Museum during the working hours fixed for Museum work. Whatever is done by any one connected with the Museum, during that time, is to be considered as the property of the Museum, but due credit is to be given him by the Curator in his Annual Reports. . . . No one is authorized to publish, or present to learned societies, anything concerning his work at the Museum, without the previous consent of the Curator. All such contributions are to be submitted to the Curator for examination.[45]

These rules did not address the apportioning of credit or authorship in collaboration of a senior scientist and his research assistant, for Agassiz felt sure that he had been more than generous in his acknowledgments to Clark for work on the *Contributions*. He imagined that he had handled the affair successfully by ignoring Clark's pretensions. Where he thought he needed more control, however, was shown by Clark's work on *Lucernaria* and by his students' publications of their own research: people he had thought were under his direction could proceed all too independently.

Agassiz had a special interest in the disposition of any manuscripts produced within the walls of the museum, because of the series he hoped the museum would publish, the *Illustrated Catalogue* and the *Bulletin*. He lost Verrill's polyp revision to the Boston Society of Natural History as well as Scudder's work on Orthoptera and Packard's on bumblebees—both outgrowths of their arranging the museum's collections for exhibitions.[46] He publicly complained, by way of explaining his delay in using the grant from the legislature: "I have been disappointed in my collaborators. Investigations made under my direction, in the Museum, and which I looked upon as material for the catalogue, have been, without my knowledge, published elsewhere."[47]

The reaction of the students to the rules of November 1863 was immediate and vehement. Viewed objectively, these regulations were no more restrictive than those governing similar institutions. In the context of

Agassiz's previous promises of liberalism, his lectures against plagiarism, and Clark's "Claim for Scientific Property," however, they hit a tender nerve. Verrill wrote to Morse, "Has anyone told you about the French, Louis Napoleonic-tyrannic-Papalistic set of regulations." Putnam complained to Agassiz that the rules were not "in accordance with the broad and liberal spirit that we had been taught to consider as the one which would govern the Museum." Scudder's assessment was that "the Museum is guarded against any infringements of propriety on the part of the students, [but] the students are not guarded against any infringement of propriety on the part of their superiors. . . . neither do I think the laws of intellectual property are either very carefully or justly laid down."[48] At first Agassiz tried to mollify the indignant students, but then he began to feel insulted by their rebellious attitudes. Verrill describes one discussion he attempted to have with Agassiz in December 1863: "He flew into a great passion and accused us of all manner of ill usage and wickedness."[49] Unwilling to accept the new rules, Packard, Putnam, and Verrill left Cambridge. Hyatt, Morse, and Scudder had left earlier, and they shared their fellow students' views on the situation. A few months later Bickmore had a different fight with Agassiz and was summarily dismissed.* The scientific grapevine hummed with sympathy for the rebels.

To Agassiz's exasperation, his students seemed to think they had a right to do as they pleased with their research. He wrote to Putnam, who had been paid to care for the fish collections:

> On looking over the papers you left in the Museum, I miss the Catalogue of the Etheostomoids and the Synonymy of the Balistes etc. Allow me to remind you that that work was done for the Museum while you received a compensation from it. The *authorship* of the work is yours, as secure as anything can be in literary and scientific matters, but it is nonetheless the *property* of the Museum. If you have not yet understood the difference between authorship and property this circumstance will make it clear. . . . I beg therefore that you return the papers belonging to the Museum which you have carried away.†

*Dexter, "Salem secession." Since Bickmore's departure was against his will (he thereby lost not only any hope of a career in the M.C.Z. but left the Lawrence Scientific School with no degree after four years) and at Agassiz's insistence, he should perhaps not be counted among the rebels, though he did share their sentiments. His own manuscript autobiography (in the library of the American Museum of Natural History) makes no allusion to the circumstances of his leaving Cambridge. Alexander Agassiz wrote Lyman on 25 July and 15 August 1864 about the firing of Bickmore (M.C.Z. Archives).

†Agassiz to Putnam, 24 March 1864, Letterbooks 2:397, M.C.Z. Archives. Several years later Putnam agreed to Agassiz's request that he finish the etheostomid monograph for the M.C.Z.'s *Illustrated Catalogue,* but he never found time to do so (Dexter, "Historical aspects of F. W. Putnam's systematic studies on fishes," p. 133).

Disgusted by the conceit and ingratitude of his assistants, Agassiz confidently assumed he could easily recruit and train replacements for them all.[50]

Central in the Clark affair and in the mutiny of the student-assistants was the issue of scientific work as a form of property. Negotiating with a prospective assistant made wary by rumors, Philip Uhler, Agassiz wrote,

You ask next whether you shall be permitted to describe new species or elaborate monographs from your own collection, placed in the Museum, to be published in the Transactions of learned Societies?—Of course you shall; the contrary would be defeating the very object of the Museum. . . . I have claimed the privilege of looking over such descriptions before they went to press and also of *publishing the same,* as far as the means of the Museum would permit, *in the illustrated Catalogue of the Collection now under preparation.* With the young men I have educated I have had the further object in view of directing their first attempts at authorship by such a revision. But all this has been taken as a presumptuous interference with the rights of independent investigators. . . .(My emphases)

Struggling to translate his anger into rationality, Agassiz further explained in a postscript that his difficulties had arisen

from the claim made by some of those formerly employed in the Museum not only to the *authorship* of everything they have done in the way of investigation, which nobody would dispute, but for *absolute property* in it, to the extent of claiming the right of publishing wherever and whenever they chose whatever they had done in the Museum as well as outside of it, including as theirs the suggestions from me on which some of these investigations were actually founded. The fact is that I have always given so freely to all my pupils what could be of any use to them from my own scientific capital that they have accustomed themselves to draw from me what they wanted without ever thinking that they even owed an acknowledgment for it. To all that there must be a limit and when I undertook to draw the line I met with the most determined manifestation of ill will and egotism, to which I have put an end by dispensing with the services of those who had positively misbehaved in that respect or showed no inclination to take the right view of the case. I thus write freely that you need feel no apprehension and no necessity of guarding yourself against possible difficulties. You can take for granted that there is no probability of any arising with me.[51]

Employees of research institutions today accept as a reasonable compromise of interests what Agassiz was claiming. Publication in the museum's own periodical rather than in the journal of a scientific society could still advance their own reputation, if their authorship was clear. He would have been wise to have somehow converted the *Contributions* into a journal, giving at least joint authorship where substantial pieces of work were done by others, salaried or not. But joint authorship was not yet a familiar practice. At least a bit of this solution did occur to him, for his son's research on starfish was printed in 1864 with a title page identifying it as the start of the fifth volume of Louis Agassiz's *Contributions* (subscribers had been promised ten) with Alexander Agassiz clearly credited as sole author. The problem was, as so often in human affairs, one of timing. Agassiz's young associates had gone to other editors at a time when the M.C.Z.'s periodicals had not yet begun publication.

Agassiz's description of scientific research as property that can be laid claim to or stolen was appropriate, reminding us that scientific knowledge does exist in the real world of human affairs, not in some other world of abstract truths. It did matter to the potential careers of Clark, Verrill, and the others in what form their works were published, and it did matter to the director of a fledgling museum what work it produced, especially in its own journals.

More curious than the concept of science as property is Agassiz's frequent allusion to science as capital. This is a distinct and more interesting designation. The language we saw in his letter to Uhler recurred in a letter to Dana: "Do you really believe me so poor in intellectual capital that I must steal the work of my students?"[52] The metaphor was at the heart of his public explanation of the source of his trouble with his students. Agassiz claimed that his method of teaching propelled his students so rapidly to the limits of knowledge in their particular field that they would grow conceited. He wrote in his *Annual Report:*

> And now begins for me the most difficult part of my task . . . when they are to be made to understand to what extent they have been working with borrowed means, which honesty requires they should pay back. The difficulty is the greater on account of the fact that the intellectual capital to be restituted belongs to him whose duty it is now to claim it back himself, and at the same time make the student understand that it is for his own good that he must settle his intellectual accounts.[53]

Now, "capital" is not a synonym for "property"; it denotes a form of property such as money, land, or equipment, capable of being employed in the generation of wealth. Louis Agassiz dined with textile mill owners and with bankers whose money built railroads and mines; he was familiar with the language of investors. Had he meant merely "property" he would have said so; any doubt of this is resolved by a statement of his in another *Annual Report.* Describing the M.C.Z.'s unsorted collections of specimens, including many duplicates which might be bartered for specimens in other museums if only he were given workers to identify and pack them, Agassiz wrote that the immense accumulation of material now stored up in our building may be considered not only as a great scientific fortune fully realized and our own, but even as a source of ever increasing scientific wealth, if we succeed in preserving the whole and making it available for exchanges. At present, it is like an immense capital lying unused, and we lack the means to put it out at interest, to distribute our riches and make their value felt.[54]

Regardless of when it first occurred to Agassiz to *call* his scientific knowledge and ideas his intellectual capital, he had always *employed* them as actual, and not merely metaphorical, capital. His scientific attainments, coupled with his ambitions and plans, could be used to generate income: hard money, not figurative wealth. There is a manuscript of his, written at the age of fourteen, listing the studies he wished to pursue and the books he would need, concluding, "For all this I ought to have twelve [gold] *Louis*."[55] As an advanced student in Paris, still unemployed but with one book on fishes already published and plans for a grander one announced, he learned that the new college in Neuchâtel would like to appoint him professor but had as yet no funds. Agassiz effectively pressured the wealthy citizens of the town by suggesting how highly his scientific abilities were valued in Paris. With no private fortune of his own to invest, Agassiz nevertheless quickly built up in Neuchâtel his "scientific factory," including a printing house and a corps of artists, clerks, and naturalists to keep the presses busy. Though this enterprise had collapsed in debt, Agassiz knew that he still possessed a powerful source of income in his own scientific knowledge and reputation. His ambitious plans to compare the geology of North America with the evidences of an Ice Age he had been tracing in Europe appealed to the King of Prussia, who agreed to pay for a two-year journey, while Agassiz's colleagues convinced wealthy Bostonians that they should pay to hear this eminent Swiss scientist give a series of lectures. Once established at Harvard, Agassiz used the grandeur of his plans for future scientific activity to attract funding from private and state benefactors. To the continual exasperation of the Museum's business managers, Agassiz's policy was always to convert any cash received into scientific assets, such as books, specimens, or the wages of assistants. He counted on building an institution whose collections and activities would be so obviously worth continuing that new sources of support must always be forthcoming.

Agassiz was laying his knowledge and resources at the feet of his students, not as an altruistic gift but as capital put out for investment. He expected their most significant research to be published in his projected catalogue, so as to add to the stature of the museum. He had spared no expense to get the best microscopes for Clark, not out of a selfless quest for knowledge of the innermost secrets of life but partly to generate manuscripts so impressive that some benefactor would feel bound to bring them to publication. When his students put their own career ambitions ahead of his hopes, Agassiz felt robbed. Understandably so, for where there is money to be had for the cultivation of science, scientific ideas that are figuratively property are literally a form of capital.

We should not blame all Agassiz's troubles on his difficult personality. He lived in an age before the large-scale funding of science by government agencies or charitable foundations, and when the pattern of advanced education for science in university laboratories was only just emerging in Germany. In effect, Agassiz at Harvard was trying to invent the graduate student, a generation before the German Ph.D. degree was finally imported to the United States. He saw the need to attract and keep talented students through a program of financial aid, and he wanted them to work in the best modern laboratory and library. Most particularly, he aimed to teach them the process of original research and discovery rather than merely to impart a body of learning to them. The pattern of advanced education with which Agassiz was experimenting is now standard in science, but he encountered difficulties as a pioneer.

It is common practice now, within a range varying according to the personality of the scientist, for the director of a research laboratory—the professor who provides key ideas or supervision and obtains grant money providing the necessary equipment—to be named as joint author with the student or assistant who carries out the daily drudgery of scientific research and who writes up the results. Innumerable kinds of contribution, and a range of feelings of fairness or injustice, are now papered over by the current practices of joint authorship. Agassiz's attempts came to grief in part because there was no common model already established to mediate the expectations of either teacher or pupil.

Agassiz told himself that the immature behavior of half-trained pupils could not harm him. He would take care not to be so indulgent and trusting again and make a fresh start. Yet, in fact, the special new museum of his vision had, indeed, failed in an important way. Lurie has rightly said that "some vital spark had been extinguished at the museum with the departure of that first devoted band. . . . Progress and maturity brought stability and a new order to the Divinity Avenue building, but it was an atmosphere that lacked the excitement of pioneering days."[56] More than an inspired mood had been lost, though. Never again did Louis Agassiz have significant numbers of students as assistants in his museum, nor was the experience of that first group, learning the principles of taxonomy by immersion in its practice, ever duplicated.

All the men who declined assistantships in the museum had been students for several years and were feeling ready to move on in their careers. Afterward, in the 1864–65 academic year, six of the pre-mutiny band were still working with Agassiz: Shaler and Hyatt, who had been present at the founding; Niles and Allen, who had come in the fall of 1862; and Hartt and St. John, of the next year's class. No new students joined these six, however, either that year or for several years thereafter. In the fall of 1863 there had been, in spite of the Civil War, twelve Lawrence Scientific School students under Agassiz's direction, seven of them living in Zoological Hall; the mutiny cut the enrollment in half. In 1866, though the war was over, the dormitory was empty. In the ten years that remained to Agassiz between the mutiny and his death in 1873, he was either seriously ill or away from Cambridge a good deal of the time, and he rarely lectured. He never managed to rebuild a community of enthusiastic students in his museum.

Lurie suggests that the replacement of student labor with full-time permanent staff represented "progress and maturity" of the museum, and in one sense he must be right.[57] It would indeed have been a daring innovation had Agassiz persisted in his practice of giving new students not only a single pickled fish but his entire collection of some family to assort and analyze. Close as Agassiz had come to inventing the graduate school, he could not sustain it.

Since one source of the mutiny was Agassiz's personality, it is easy to imagine that a man of more tact and insight could have avoided discord. Before being satisfied with such a judgment, though, we should understand that the reason the situation demanded tact and insight was the imperfect development of appropriate social structures in a period when professional science as we know it was only beginning to take shape.

3

"Our Work Must Be Done with Much More Precision"

In April 1865 Louis Agassiz's classroom was the saloon of a vessel of the Pacific Mail Steamship Company, the *Colorado,* bound for San Francisco via Rio de Janiero. Agassiz and his party would leave the ship in Brazil, where their goal was the upper reaches of the world's greatest river, the Amazon. The politics of science, pressures of fund-raising, and illness had worn Agassiz down to the point where he realized he desperately needed a holiday. Now he was anticipating with intense and childlike delight his first visit to the vast Amazon River, something he had wanted to see ever since the late 1820s when, as a student in Munich, the task of describing a collection of Brazilian fishes had launched his career. This vacation tour had by now assumed the character of a major scientific expedition, to the surprise of no one familiar with Agassiz's ambitious character.

Every afternoon on shipboard Professor Agassiz would give a "*long*" lecture[1]; "a couple of leaves from the dining-table with a black oil-cloth stretched across them serve as a blackboard."[2] The talks were meant for the twelve men who were the workforce of his expedition, but his audience included as well almost the entire company; the captain and some of the officers, a bishop, and the passengers, including a few ladies, "all of whom," noted his wife, "seem to think the lecture a pleasant break in the monotony of a sea voyage."[3]

One afternoon his topic was "the objects of scientific explorations in modern times." His voice had a remarkably warm and earnest quality, with an accent Americans found charming. His purpose was to build commitment in his corps of helpers by convincing them of the scientific importance of their work. What he said was this:

When less was known of animals and plants the discovery of new species was the great object. This has been carried too far, and is now almost the lowest kind of scientific work. The discovery of a new species as such does not change a feature in the science of natural history, any more than the discovery of a new asteroid changes the character of the problems to be investigated by astronomers. It is merely adding to the enumeration of objects. We should look rather for the fundamental relations among animals; the number of species we may find is of importance only so far as they explain the distribution and limitation of different genera and families, their relation to each other and to the physical conditions under which they live. Out of such investigations there looms up a deeper question for scientific men, the solution of which is to be the most important result of their work in the coming generation. The origin of life is the great question of the day.[4]

He told them frankly in a later lecture:

I am often asked what is my chief aim in this expedition to South America? No doubt in a general way it is to collect materials for future study. But the conviction which draws me irresistibly, is that the combination of animals on this continent, where the faunae are so characteristic and so distinct from all others, will give me the means of showing that the transmution theory is wholly without foundation in facts.[5]

Aware that his dogmatic pronouncements against Darwinism had only damaged his own reputation, and seeing Darwin's theory steadily winning adherents, Agassiz was turning for fresh ammunition to his first love, ichthyology. Thus while it is true that this visit to Brazil had a recreational purpose, and was financed for the sake of the growth of his museum, we may take Agassiz at his word that his fondest hope for its theoretical achievement was that its collections would prove the evolutionists wrong.

Half of his corps of assistants were employees of his museum in Cambridge (Jacques Burkhardt, artist; George Sceva, preparator; J. G. Anthony; F. C. Hartt; O. St. John; J. A. Allen). The other six were volunteers, young men in search of adventure, whose parents were perhaps glad to send them out of reach of the War between the States (A. V. R. Thayer, son of the man who financed the trip; Newton Dexter; Edward Copeland; Thomas Ward; Walter Hunnewell; and William James). James would later teach physiology in the museum and then become famous as a psychologist and philosopher, but just now he was a rather undirected and unpromising twenty-three-year-old.

This excursion to Brazil, called the Thayer Expedition because financed by the Boston banker Nathaniel Thayer, was in many ways extraordinarily successful. Because Agassiz sent his helpers on separate side trips, overland

and by canoe, the territory covered was much greater than the route he and Mrs. Agassiz traveled.* Back to Cambridge were sent hundreds of barrels and kegs of specimens of all sorts, adding to the wealth of Agassiz's museum. His only disappointment was having to leave Brazil at the end of fourteen-and-a-half months, when the money ran out.

Agassiz's great advantage, which he knew how to use to full effect, was his extraordinary charisma, which would win over people from all walks of life, making them feel that here was a man so simple and sincere in his passionate love of nature that to support him was the most worthwhile endeavor imaginable. Consequently, Agassiz's resources kept multiplying as he traveled. The specimens were shipped back to Boston free of charge by various freight companies. The emperor of Brazil not only ordered that collections be made for Agassiz, he also did some collecting himself. Most important, an officer of the Brazilian corp of engineers, Major Coutinho, who had been engaged for several years in a survey of the Amazon, was detailed to accompany the Agassizs.

The daily labor of the expedition was assembling and preserving animals, mostly fish. Like most traveling naturalists, Agassiz made use of the knowledge and skill of the local inhabitants:

> The Indians here are very skilful in fishing, and instead of going to collect, Mr. Agassiz, immediately on arriving at any station, sends off several fishermen of the place, remaining himself on board to superintend the drawing and putting up of the specimens as they arrive.[6] . . . Mr. Agassiz has a corps of little boys engaged in catching the tiniest fishes, so insignificant in size that the regular fishermen, who can never be made to understand that a fish which is not good to eat can serve any useful purpose, always throw them away.[7]

Many of these same lovely little Amazonian fishes are now the staples of the aquarium trade.

It was much to Agassiz's advantage that he could set up a floating laboratory:

> Our arrangements were very convenient; and as the commander of the steamer allowed me to encumber the deck with all sorts of scientific apparatus, I had a number of large glass dishes and wooden tubs in which I kept such specimens as I wished to investigate with special care and to have drawn from life[8] . . . that is, from the fish swimming in a large glass tank before my artist [Jacques Burkhardt; fig. 9].[9] This is

*A map of the expedition, made in consultation with the Agassizs, is included in the French edition. M. M. Dick's "Stations of the Thayer Expedition" has been superceded by Horácio Higuchi "An updated list of ichthyological collecting stations of the Thayer Expedition to Brazil (1865–1866)" (MS, available in the M.C.Z.'s fish department and at other collections containing Thayer material).

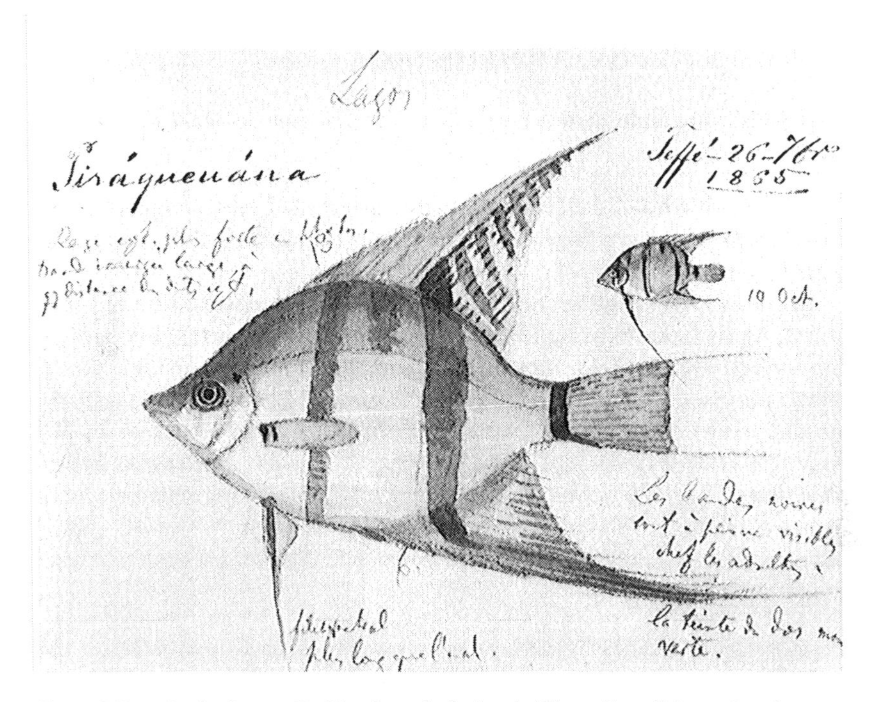

Figure 9. Drawing by Jacques Burkhardt made during the Thayer Expedition to Brazil, annotated by Louis Agassiz. The fish pictured here is *Pterophyllum scalare,* the "angelfish" of later aquarium hobbyists, who have created scores of fancy varieties, as if to taunt Agassiz's ghost with the malleability of species. Only four days before this drawing was made, Agassiz was boasting to a colleague that many of the fish he was collecting were "painted from life, that is, from the fish swimming in a large glass tank before my artist. I am often pained to see how carelessly colored plates of these animals have been published" (L. Agassiz and E. C. Agassiz, *A Journey in Brazil,* p. 220). (However, a few of Burkhardt's other fish sketches from the Amazon in the archives of the M.C.Z. were certainly based on dead specimens.) By making notes supplementing Burkhardt's eye, Agassiz aspired to superior exactness. (Horácio Higuchi informs me that a colored illustration of this species had already been published [Castelnau, *Animaux nouveaux* 2:11, fig. 3]). In contrast to later practice, Agassiz seems to have made no attempt to keep track of the particular specimens to which his and Burkhardt's observations pertained. Here, as with many other projects, Agassiz's ambition far exceeded his grasp, for no scientific use was made of this or the hundreds of other drawings done in Brazil.

I am greatly indebted to Horácio Higuchi for assistance in interpreting the annotations to this drawing. Proceeding clockwise from noon: *Lagos* "Lakes" (to note that it was not found in the river); *Teffé—26—7br—1865* (Town of Teffé [now Tefé] 26 September 1865); *10 Oct.* (presumably the date of collection of the smaller specimen); *Les bandes noires sont à peine visible chez les adultes—la teinte du dos moins verte* "Black bands less visible in adults, color of back less green" 6. (count of spines in dorsal fin) *filet ventral plus long que l'anal* "the filament of the pelvic fin is longer than the filament of the anal fin" (although the artist has not had space to show it). *Rangé[e] ext[erne] plus forte en haut & bas. Bande immergée large, à q[uel]-q[ue] distance des dents ext[erieurs].* "Outer row [of teeth] stronger [than inner row, both] on the upper and lower jaw. A wide, embedded row, separated by a gap from the external teeth." *13* (count of spines in dorsal fin) *Piráqueuáua* (vernacular name used by local natives); this as well as the locality and date are not in Agassiz's hand, and were likely supplied by Major João Martins da Silva Coutinho. (By permission of the Museum of Comparative Zoology Archives, Harvard University)

no easy task, for the mosquitoes buzz about him and sometimes make work almost intolerable.[10]

Often their riverboat, the *Icamiaba,* was transformed into a cooperage, as barrels and kegs were assembled, filled with fish and alcohol, and hammered shut.

Besides supervising the collecting, recording, and preservation of specimens, Agassiz kept busy building and cultivating his important network of supporters. He made arrangements for his assistants' excursions, all of which proceeded without serious misadventure (although Joel Asaph Allen did return home early because of illness). He wrote letters to let his American friends know that he was flourishing, to tell his colleagues in Europe that he was still to be reckoned with among practicing scientists, and to assure all those who had contributed to the enterprise that he was enjoying remarkable success. William James watched the famous professor and wrote home,

I have profited a great deal by hearing Agassiz talk, not so much by what he says, for never did a man utter a greater amount of humbug, but by learning the way of feeling of such a vast practical engine as he is. . . . I delight to be with him. I only saw his defects at first, but now his wonderful qualities throw them quite in the background. . . . I never saw a man work so hard.[11]

Agassiz was writing letters in tones of high excitement, as he counted up the new species of fish he was finding. From all of South America only about a hundred different kinds of fish had been named, but his tally exceeded that in his first week of collecting. On September 8 he wrote to the director of the Amazonian steamship line,

I have passed from surprise to surprise, and . . . have scarcely had time to take care of the collections we have made, without being able to study them properly. . . . [During one week] we have collected one hundred and eighty species of fishes, two thirds of which, at least, are new . . . [and our total now is] already more than three hundred species. . . . You see that before having ascended the Amazon for one third of its course, the number of fishes is more than triple that of all the species known thus far, and I begin to perceive that we shall not do more than skim over the surface of the centre of this great basin.[12]

In early December he reported that he had a total of 1,300 species (counting the ones collected by his assistants in other rivers); by the end of the month the figure had grown to 1,442 species. (Such an exact number shows that he was keeping count, even though he would often round off the figure in his letters.) By the time Agassiz was finished, he was claiming "nearly two thousand."[13] This collection would create a revolution in ichthyology, he promised.

Certainly the Brazil fish had an immediate effect, not on science but on Agassiz's son and friends back in Cambridge. The kegs and barrels that began arriving from Brazil were a real strain to the little Museum of Comparative Zoology, already understaffed and overcrowded. His former student and loyal volunteer assistant Theodore Lyman recorded in his diary,

> Began today to go out to the Museum to help unpack the Brazil collections of Agassiz. Already there are 160 kegs & barrels! Alex and I went at them and got through six packages, distributing them approximately as species, in jars & fresh kegs, and putting in new whiskey. It makes one's hands rather flavorsome, though the condition of the things was generally good. We knock out the head, unpack the keg; then distribute by species in bottles or new kegs, and fill with fresh whiskey. Where we have many young fishes (put up in bottles usually) it takes a long while to sort them.[14]

In spite of his own lecture aboard the *Colorado*, teaching that the scientist must seek general relations rather than merely enumerating species like so many asteroids, Agassiz could not resist being excited by the sheer abundance of new forms. At that time, there were only a few hundred species of fish known in the United States and Canada, and around a hundred from South America. That his expedition might double or triple that number would surprise no one, but 1,800, or 2,000, or 2,200 Brazilian species was incredible.

To convince his peers in ichthyology, Agassiz would have to publish sober descriptions in the proper manner. This Agassiz failed to do, thus confirming the suspicions of those who thought his reports inflated.*

Still, the absolute number of species should not have been important, as Agassiz himself had insisted aboard the *Colorado*, in comparison to the great question of the day, the issue of evolution which he had said he hoped

*The South American fish fauna is, in fact, much more complex and numerous than that found in North America or Europe and does immediately impress an ichthyologist with its richness of species. There is good evidence, however, that Agassiz's field count of distinct species collected was exaggerated. Alexander Agassiz followed his father's labels as the barrels arrived in the Museum and published his tally in the *Annual Report* for 1866 (pp. 22–23). Individual localities are credited with 80, 100, 200, or even more species, whereas modern collectors spending months in one spot, using techniques like poison and explosives to make an exhaustive inventory, rarely find more than 75 species. I am indebted to ichthyologists William Fink, Richard Vari, and Stanley Weitzman for much assistance and advice.

to address with evidence from the Amazon. How did his findings relate to Darwinism? He wrote to a colleague in March 1867,

> My recent studies have made me more adverse than ever to the new scientific doctrines which are flourishing now in England. This sensational zeal reminds me of what I experienced as a young man in Germany, when the physio-philosophy of Oken had invaded every centre of scientific activity; and yet, what is there left of it? I trust to outlive this mania also.[15]

Yet a lecture or article explaining the significance of all those barrelsful of fish never appeared. A few months after returning to Cambridge, Agassiz did give three public lectures on his Brazilian research, but he spoke mostly about the evidences of glaciation. The next year he contributed several technical footnotes to the book his wife was putting together about this Brazil trip, but he added nothing on the evolutionary implications of his collections beyond the notes she had made of his shipboard lectures anticipating what he hoped to find.

Louis Agassiz was acting true to form, energetic when conceiving great projects, then shrinking from the drudgery of completing them. He did at first throw himself into the labor of taxonomic analysis, but soon fell back into the familiar pattern of politicking for science, mounting new schemes and expeditions and suffering severe illness. Yet in spite of his own silence, we can reconstruct with confidence how Agassiz saw in the Brazilian fishes evidence for the fixity of species.

Agassiz had told his companions that the first step in the investigation of the origin of species was to determine their exact geographic boundaries, to trace the pattern of their distribution.

> The origin of life is the great question of the day. . . . It must be our aim to throw some light on this subject by our present journey. . . . The first step in this investigation must be to ascertain the geographical distribution of the present animals and plants. . . . Fifty years ago the exact locality from which any animal came seemed an unimportant fact in its scientific history, for the bearing of this question on that of origin was not then perceived. To say that any specimen came from South America was quite enough. . . . In the museum at Paris, for instance, there are many specimens entered as coming from New York or Para; but all that is absolutely known about them is that they were shipped from those sea-ports. Nobody knows exactly where they were collected. . . . All this kind of investigation is far too loose for our present object. Our work must be done with much more precision; it must tell something positive of the geographical distribution of animals in Brazil. Therefore, my young friends who come with me on this expedition, let us be careful that every specimen has a label, recording locality and date, so secured that it shall reach Cambridge safely. . . . We must try not to mix the fishes of different rivers, even though they flow into each other, but to keep our collections perfectly distinct. You

will easily see the vast importance of thus ascertaining the limitation of species, and the bearing of the result on the great question of origin.[16]

Whether or not each of Agassiz's listeners could really see the bearing of distribution on the question of evolution, Darwin could. He considered geographic distribution to be so important that he had devoted two chapters in the *Origin of Species* to it.

Since it was Darwin's theory Agassiz intended to refute, we must remind ourselves of the role of geography in Darwin's argument. He began by setting out three general phenomena, which he called "laws" or "great facts."

First Great Fact: That neither similarity nor dissimilarity of inhabitants can be accounted for by physical conditions. Areas with the same conditions, such as deserts, mountains, or marshes, have different flora and fauna in the Old and New World.

Second Great Fact: That "barriers of any kind, or obstacles to free migration, are related in a close and important manner to the differences between the productions of various regions."[17] As the major continents, separated by water, have distinctive floras and faunas, and different marine organisms inhabit the two sides of the Isthmus of Panama, on a smaller scale we find the boundaries between species coinciding with rivers or mountain chains.

Third Great Fact: That there is an affinity linking together different species of the same great area. Even though within one continent there are many distinct species in different regions, they are found to be members of the same genera or families. Thus "the naturalist in travelling, for instance, from north to south never fails to be struck by the manner in which successive groups of beings, specifically distinct, yet clearly related, replace each other. He hears from closely allied, yet distinct kinds of birds, notes nearly similar, and sees their nests similarly constructed, but not quite alike, with eggs coloured in nearly the same manner."[18]

Darwin summed up his argument by saying,

We see in these facts some deep organic bond, prevailing throughout space and time, over the same areas of land and water, and independent of their physical conditions. The naturalist must feel little curiosity, who is not led to inquire what this bond is.[19]

In the modern reading of the phrase "deep organic bond," Darwin's statement plays the debater's trick of begging the question, for whether the bond perceived by taxonomists consisted of a chain of living stuff stretching backward in time was the question at hand. The adjective "organic," however, did not mean simply "living" as we sometimes use it today but referred to having organs, an organized structure. Agassiz would not have quarreled with the claim that species are organically connected, in the sense that it was their inmost structures that were similar. He himself had been calling attention to these very same phenomena over the past several decades. The hidden bond linking together the distinct salmon species of the Rhone and the Rhine, the sturgeon species of Lake Superior and the other Great Lakes, and the topminnow species of the Tennessee and Mississippi Rivers had indeed aroused his curiosity, but what excited Agassiz was exactly that the bond was so clearly an immaterial one.[20] These are fishes that do not interbreed or otherwise merge together; their only demonstrable connection is an abstract relationship of similarity, as he had explained in detail in his "Essay on Classification" of 1857.

Agassiz had argued that the same conditions that make islands such nice cases for studies of animal geography were what made freshwater fishes of interest. They are circumscribed in a narrower area than animals on land, he had taught, species of the headwaters differing from those of the middle or lower course of a river,

> thus forming at various heights above the level of the sea, isolated groups of freshwater animals in the midst of those which inhabit the dry land. These groups are very similar in their circumscription to the islands and coral reefs of the ocean; like them they are either large or small, isolated and far apart, or close together in various modes of association. In every respect they form upon the continents as it were a counterpart of the archipelagos.[21]

Agassiz had argued in his "Essay on Classification," repeating a claim he had been making for at least a decade, that the existence of distinct species in locations with apparently identical physical conditions contradicted the idea of evolution.[22] Never fully articulated, his idea evidently assumed that purely material causes were supposed to explain organic form. Under the same physical environment, the same living things would be generated. His reasoning seems to have been that closely neighboring portions of the same river cannot have significant physical differences, and so from material causes cannot be expected to contain distinct organisms. In addition, any kind of pattern or orderliness, such as the limitation of a particular genus to a particular region, suggested to Agassiz an ordering cause. Making sense of nature challenges our minds, so he felt that some other Mind must have set us the puzzle.

The only thing Agassiz found in Brazil that surprised him was the *number* of species. The *general sorts* of species and the overall picture of their relations to one another he was familiar with before he left. On the boat to Rio he had laid before his assistants two points:

1. South America has its own characteristic fauna. The ecological and morphological place filled in North America by sturgeons, perches, cyprinoids, and salmon will be occupied instead by goniodonts, chromids, cyprinodonts, and characins.[23]

2. His experience in Europe and North America led him to expect to find distinct (though congeneric) species in different rivers. "Suppose we first examine the Rio San Francisco. The basin of this river is entirely isolated. Are its inhabitants, like its waters, completely distinct from those of other basins? Are its species peculiar to itself, and not repeated in any other river of the continent? Extraordinary as this result would seem, I nevertheless expect to find it so."[24]

In this regard the Amazon far exceeded Agassiz's expectations. At every spot, he would find new kinds of fish, very similar and yet distinct from those of the previous collecting stations. Thus Agassiz was not only finding the very same "great facts" Darwin discussed in the *Origin,* he was seeing an even more clear and fine-grained instance of them, as if instead of the dozens of species on the Galapagos one had to do with an archipelago of islands stretching across hundreds of miles and populated by thousands of species of finches. He was enormously excited, not by the novelty of the phenomenon, for its general outline he had known perfectly well already, but by its intensity.

Because before he went to Brazil Agassiz already had not only a good idea of the patterns he would find there but an explanation of them which satisfied him, he never seriously considered Darwin's alternative explanation. The fundamental problem with Agassiz's reasoning had been exposed a century before by David Hume in his classic critique of the argument from design: it is based on a simple analogy and thus essentially assumes what it purports to prove. The general superiority of Darwin's explanation is that it employs only natural rather than miraculous causes. More concretely, in the case of geographic distribution, Darwin's theory explains *particular* patterns, whereas Agassiz could infer a designing intelligence from *any* pattern. Agassiz's version says that the Creator, having been pleased to make fishes that resemble one another so closely that we classify them as members of the same genera, or different genera but the same families, also chose to place them in the same river basin. God could have chosen to place them randomly, in which case we could not infer His existence, or He could just as well have populated all the eastward flowing rivers of the world with catfishes and all the westward flowing rivers with characins; that would be an orderly arrangement from which we could infer an ordering power. For Agassiz, the making of a creature's anatomical peculiarities and the selection of its home range are two quite independent acts of God. But Darwin unites the two under one and the same cause. Two species of fish resemble one another because they have inherited structures from a common ancestor, and they are living in the same area because that is where that ancestor lived.

But Agassiz's grand plan for putting the fishes of Brazil to use remained only a promise. Once again, he was wonderful at laying out ambitious projects but lacked the time and temperament to complete them. When Agassiz realized that he would never get around to serious work on the Thayer collection, he decided to delegate the labor. He enticed the Austrian ichthyologist Franz Steindachner to come to Cambridge with glowing tales of his rich collection of species new to science.* Steindachner came in 1870 and spent about twelve months working on the Thayer fish. In 1872 the Thayer fish sat undisturbed in their alcohol while Agassiz and Steindachner sailed around South America to the Galapagos and then to California on the *Hassler,* collecting yet more specimens.† Steindachner's publications, most of which appeared after Agassiz's death, supplied only ordinary taxonomic description of new species, not a critical analysis of their geographic distribution or variation. Agassiz's informal comments on the Brazilian fishes were published in the narrative of the Thayer Expedition written by his wife.

Allen's Law

At least one of the assistants Agassiz took with him to Brazil apparently gave some thought to the issues he raised. Joel Asaph Allen had been a student in the museum from 1862, one of those who lay low during the mutiny. Although illness, which handicapped him off and on for many years, had cut short his Brazilian travels, he had heard Agassiz's shipboard lectures and carried out collections according to Agassiz's stipulations. Not that this was his sole model for geographical study of species: he was al-

*Kähsbauer, "Intendant Dr. Franz Steindachner." Steindachner arrived in May 1870 and left in early December 1873, before the death of Louis Agassiz. These two-and-a-half years were not all devoted to the Thayer fishes, however. He was on the *Hassler* expedition (at Agassiz's request and expense) 4 December 1871–31 August 1872, and worked on the *Hassler* fishes for eight months thereafter. He took off the last six or seven months for a tour of the United States.

†Thousands of specimens went to Vienna when Steindachner returned home early in 1873. This series of duplicates from the Thayer Expedition was, he said, part of the original agreement by which the director of the Vienna museum had allowed him these years of leave (Steindachner, "Die Süsswasserfische," p. 499).

ready interested in birds and mammals and was greatly impressed by S. F. Baird's precise methods of geographical reporting.

After his return from Brazil, Allen sought improvement for his "chronic indigestion and intestinal troubles," first by returning to the family farm, then by a natural history excursion (financed by collecting for various specialists) through New York, Illinois, Indiana, Iowa, and Michigan. Louis Agassiz welcomed him back to the M.C.Z. in October 1867, placing him in charge of the birds and mammals.[25] More than any other employee, Allen appreciated and built upon Louis Agassiz's greedy collecting policy.

During this period, the literature on North American mammals and birds was in a state of increasing confusion. Did the wolves on the continent belong to one species or to several, and were they distinct from those of Europe? Likewise, what about the various bears, the martins, and the skunks? In the scientific literature, names seemed to be multiplying faster than knowledge. Allen put his finger on the problem: "*Specimens* have too often been described instead of *species*."[26] What was needed was information about the range of variability within each species. Probably no other museum could offer material as good as the M.C.Z.'s for the study of individual variation, Allen claimed (though he was not yet personally familiar with any other museum), because of its wealth of specimens "brought together by the Director in great part for this especial purpose."[27] Allen saw that he could make a contribution to knowledge and to the wealth of the museum at the same time:

> In order not only to increase the stock of duplicates for exchanges, but to bring together large series of specimens of a considerable number of species from a single locality, for the purpose of affording means of the investigation of the amount and character of individual variation, it was decided early in the year [1868] to collect extensively Massachusetts birds, and in following out the plan, some thirteen hundred specimens have been added during the past season.[28]

Though the investigation of variation was a project of obvious relevance to the issues raised by Darwin (who had devoted the second chapter of the *Origin* to "Variation under Nature"), it was also a project clearly outlined by Louis Agassiz before 1859. We need, he had written in the "Essay on Classification," to

> learn with more precision, how far the species described from isolated specimens are founded in nature, or how far they may be only a particular stage of growth of other species; then we shall know what is yet too little noticed, how extensive the range of variations is among animals observed in their wild state, or rather, how much individuality there is in each and all living beings. So marked, indeed, is this

individuality in many families . . . that correct descriptions of species can hardly be drawn from isolated specimens, as is constantly attempted to be done. . . . the individuals of some species seem all different and might be described as different species if seen isolated or obtained from different regions.[29]

Though Agassiz was not skillful at following this injunction himself, he did encourage Allen to adopt the problem. The following year Allen reported,

In accordance with instructions received from the Director, it is proposed to prepare a large series of skeletons of those species for which there is ample material, for the purpose of determining within what limits individuals of the same species may present variations in their osseus structure. This will give a basis for determining the value of specific identifications when made from single bones, to which unsatisfactory data paleontologists are frequently restricted.[30]

Thus not only in his teaching but in the operation of his museum Agassiz had stressed the phenomenon of variability within species, taking pride in being more aware of it than many taxonomists. This phenomenon took on a very different meaning in the context of evolution. Instead of being merely a source of error for careless taxonomists, variation became the steam driving the engine of wider change. In either case, the appropriate tool for its investigation was a museum, if its director was wise enough to value large series of "duplicates."

Allen carefully measured hundreds of specimens of birds and mammals from one locality to determine variability of characters. He found a range of fluctuation even for those characters usually used to differentiate similar species. He found extreme forms within one species so different that they would have been called two species if all the intermediates had not been known. He noted the different coloration of fox cubs in the same litter, and likewise of skunk kits. He traveled to Florida and collected for the M.C.Z. for three months in the winter of 1868–69. He traveled across the Midwest, noting again and again that forms distinct enough to seem like "good species" were connected in the intervening territory by a continuous intergradation, and noted that the intergradation was too gradual to be attributable to a zone of hybridism between two species.[31]

Allen knew that Spencer F. Baird had remarked that geographic races seemed usually to follow a certain pattern, with the smaller form more northerly; following Baird's example, Allen presented his own new data by constructing charts with all his measurements.[32] For hundreds of specimens, Allen recorded color differences, overall size, and the relative lengths of parts such as tail, ear, and foot, and found that these too seemed to vary north to south. Allen recognized his geographic studies to be "very suggestive," because if the intermediate forms of one of these varying species were to disappear, the remaining extreme forms would be considered, by usual taxonomic standards, to be two species. Thus, to admit the possibility of extinction seemed to entail admitting the multiplication of species. Allen referred approvingly to Andrew Murray's *Geographical Distribution of Animals* of 1866; Murray had become converted to evolution (though not to natural selection) by considering how the environment could produce local differences of race.

I have seen no evidence of Agassiz's reactions to the work of this former student and museum assistant, but, as with the geography of fishes, how "suggestive" Allen's findings were depended on what one assumed was likely. If the full range of local varieties, *A–F,* were part of the constitution of that species, carefully planned at its creation, and a catastrophe later intervened to exterminate races *B, C,* and *D,* the naturalist who identified *A* and *F* as distinct species would, in Agassiz's eyes, merely be mistaken. Both Darwin and modern biologists would require some further changes to occur within *A* and *F,* including some obstacle to free interbreeding, and it is exactly those kinds of changes Agassiz was assuming were impossible. If, however, one thought it likely that the species at its creation had been limited to form *C,* then whatever had caused its divergence could presumably cause further changes in *A* and *F.*

When Allen presented his "laws of geographic variability" to the Boston Society of Natural History in April 1872, Hyatt asked him what he thought the causes might be, and Shaler asked if darker colors in more humid regions could not be the result of natural selection, as an adaptation to a darker background. Allen replied that the bleaching effect of sunlight was well known and that the differences were not so well marked in young animals or just after a bird's molt. Besides, natural selection depended upon variations that mattered for survival, whereas these differences seemed too slight to be useful. The very fact of variation could be taken as evidence that these characters did not matter for survival, since, until the collector intervened, these individuals had been surviving. Allen said that characters that were obviously critical for survival, such as the woodpecker's bill or the wing of a swift, did not vary. He inclined to the belief that, by some direct mechanism, the heat of the environment, perhaps by affecting the circulation, could produce these regional differences.

Although Allen did not make a secret of what he saw as the implications of his systematic work, as the BSNH discussion shows, he did remarkably little to either develop or publicize his ideas. His technical articles, whether published by the museum or elsewhere, reported his extended series of purposeful collections and measurements as if this was merely part of the taxonomist's task of increasing our fund of accurate information. His first frank discussion of the implications of this kind of research was not published until 1877.[33] So although Allen himself testified that his research had been encouraged by his employer, and although Louis Agassiz trumpeted his wish that systematists address themselves to the great questions of the day rather than "merely adding to the enumeration of objects," it would seem that Agassiz did not press Allen to push forward his theoretical inquiries as part of his museum duties.

Even more suspicious than Allen's slowness to publish a full discussion of his ideas is the peculiar forum in which his article, "The influence of physical conditions in the genesis of species," did appear. In that article Allen argued explicitly that, although natural selection most probably does take place, the physical environment has a larger "share in the work of the production of new species than Darwinists usually admitted." Biologists who now honor him for "Allen's law" go to a Smithsonian Institution publication of 1905 to read his 1877 article, because turn-of-the-century debates about geographical factors in speciation created a demand for a reprint of Allen's early work. Why did he not publish it in a professional journal like the *American Journal of Science* or the semipopular but scientifically respectable *American Naturalist?* Instead, it appeared in a new magazine of politics and opinion called the *Radical Review.* Only one volume of this periodical, edited by the anarchist Benjamin Ricketson Tucker, ever appeared, but among the promised contributors was former Agassiz student E. S. Morse.[34] Whatever induced Allen to compose this article, his choice of medium clearly demarcated it from his scientific output as an officer of Agassiz's museum.

4

"An Object Worthy of a Life's Devotion"

No one who understood Louis Agassiz's pronouncements expected him to embrace evolution. As his colleague Jeffries Wyman confided to Richard Owen: "Having committed himself so fully to a belief in the permanence of species, & of the unalterable nature of Divisions, Classes, Orders, &c., Agassiz cannot be supposed to meet with patience an hypothesis which dashes his whole structure to the ground."[1]

His theoretical structure was in ruins but what of this other structure, the museum; was it not also in jeopardy as Darwinism gradually gained sway? Agassiz was politically astute enough to make sure that evolution would be seen to make the Museum of Comparative Zoology (and its funding) all the more important. He said in 1863,

> The scientific world is divided upon the most vital question ever approached by science. . . . The vast collection of fossils accumulated at the Museum during the past few years may, if properly arranged and critically compared with their living representatives, contribute something towards the solution of this great problem.[2]

And again in 1865,

> . . . work . . . in the Museum, has already extended to comparisons . . . with the view of ascertaining whether there is any probability of tracing a genetic connection between the animals of . . . different geographical areas, and how far geographical distribution and specific distinction are primary facts in the plan of creation. It must be obvious that the question of the origin of species is not likely to be discussed successfully before the laws of the geographical distribution of organized beings have been satisfactorily ascertained.[3]

A perfect fund-raising claim, since endless years of inventory work would have to be done before one could declare the facts all in. One could rest assured that the museum was accumulating relevant evidence, whether one were on the side of fixity of species or evolution.

Agassiz showed great care in the museum's *Annual Reports* not to link the fortune of his museum with the fixity of species. The direct message to legislators and other supporters was that the museum could lead in the fight against evolution, but the indirect message was that collections assembled to serve "comparative zoology" would remain useful whatever the verdict on Darwin's theory. It was no secret what Agassiz's beliefs were, but he knew that the survival of what he had built depended upon the M.C.Z. acting, and being seen to act, in relation to the latest theories rather than becoming merely a memorial to his own.

These statements, and others in his lectures, show that Agassiz subscribed to the model of scientific method called "Baconian" induction, which says that facts (like the position of the planets) are collected, inspection of the facts uncovers regularities (like Kepler's laws); the cause behind these laws is suggested by a theory (like Newton's inertia and gravity); and the theory is proven by testing with new facts (like the predicted return of Halley's comet). The inaccuracy of this image had been identified by Francis Bacon himself, who knew that one could assemble facts only in the light of some theory, and by William Whewell and John Stewart Mill, but it remained commonplace, and it was a very congenial model indeed for the interests of a museum founder.

In spite of any sophisticated quibbles about the philosophy of science, Agassiz's point about the value of a museum appears fair enough. After all, most of the topics that constituted "comparative zoology" were indeed also relevant to the investigation of evolution. Besides the study of geographic distribution, these included the comparison of embryos to fossils, the search for morphological homologies by which to demarcate families, orders, and classes, and even the description of the range of variability within species. Even though Agassiz's former students one by one became evolutionists, they did continue to use the comparative approach they had learned at Harvard in their subsequent studies.[4]

In truth, though, this rhetoric about the museum's collections contributing to the solution of a hotly debated issue was so much guff. The lively research program unifying the students of 1859–63 was exceptional. The daily business of operating any of the great natural history museums in the nineteenth century had scant connection to such fundamental theories as evolution or the thought processes of the Creator. What was really driving the engine at Agassiz's museum, as at the others, cannot be called a theory but a mission. It was the dream of Carl Linnaeus (and encyclopedists before him back to Adam) to catalogue the whole variety of the world. As long as there were species "unknown to science," many of them distinctive enough to require a new genus or higher category to contain them, there was work to be done in institutions like this. Neither Agassiz nor his contemporaries really demanded more. As long as no one questioned that goal, which had become deeply entrenched among scientists and accepted by society at large, whatever doctrinal blizzards may swirl outside, within the walls of a museum the daily routine need not be altered.

A close look at one episode in the curatorial activity of the M.C.Z. will illustrate both the strengths and weaknesses of the Linnaean program as it marched onward through an era increasingly accepting of evolution. Hermann Hagen's 1868 study of North American crayfishes, and its revision by Walter Faxon in 1885, will serve us nicely, for both were produced in the line of duty.

Hiring a Professional

After the painful battle with Clark and the rebellious students, Agassiz came round to his son's view that it was better to employ professionals to look after the collections than to make curatorial tasks part of a student's education. Alexander Agassiz had long felt, and had argued with his father to no avail, that it would be better to pay a few permanent employees full salaries than to dribble the same money out to students who would not do the work properly.[5] By March of 1864 Louis Agassiz was ready to drop his great experiment. He wrote to the American entomologist John Lawrence LeConte (first cousin of Agassiz's student Joseph LeConte),

> I believe I have made a mistake in confiding too much to the young men I have had as pupils in the Museum. At all events I find now that after having had the benefit of all I could do for them, instead of being devoted to me and our collections they show a degree of conceit and egotism which I had not expected and which has already induced me to part with some of them and will probably end in my making a clean sweep, to begin anew. As it is, I have nobody left to take care of the immense mass of Insects which have accumulated for several years past and I come to you for advice and help. Do you know a young entomologist, of trustworthy character, good education and pleasant manners, who would consider the opportunity of building up a great entomological collection an object worthy of a life's devotion. . . . While in Washington I heard that Dr. Hagen wished to come to this country, but I have learned to my cost that it is a dangerous experiment to import Europeans, as they are easily spoiled, not undertaking to make good use of their unlimited freedom or bringing with them an intolerable amount of overbearing confidence [presumably

Edward Desor]. Do you know anything of the man? I would much prefer an American. I think also that for a Museum a man who had a passion for collecting, and enlarging the collection, making exchanges, and keeping everything in good order, with haste and neatness would be preferable to a learned man who lacked some of those qualifications.[6]

Consistent with his preference, Agassiz soon offered the job of caring for the museum's insects to a twenty-nine-year-old American from Baltimore named Philip Uhler (fig. 10).[7] Ironically, Uhler's qualifications in biology largely consisted of the fact that he had recently translated from Latin Hermann Hagen's important study of North American Neuroptera.[8]

Uhler began work at the M.C.Z. in late spring or early summer of 1864, and struck Alexander Agassiz as a "capital man, devoted to his business and faithful, and has done more work in a month, than all the other Entomologists we had [Packard and Scudder] in a year."[9] Alexander might not have thought so highly of Uhler had he known of the scheming going on between him and Albert Bickmore.[10] After Bickmore's dismissal from the museum, Uhler nervously confided to Spencer F. Baird that the two of them were secretly planning to start a museum in New York City, adding that he, Uhler, would be "a ruined individual" if Agassiz were to hear of it.[11] But it was not after all to be a rival museum that drew Uhler away from Cambridge; he accepted a job in his home city as an assistant librarian, with a salary of $1,500.[12] He was then making only $900 for his work in the Agassiz museum.[13]

After Uhler's resignation in January 1867,[14] Agassiz once more thought of Hagen. A fifty-year-old practicing physician, trained by some of Europe's leading zoologists, well respected for his entomological publications,[15] Hagen would cost three times what Uhler had. Asked whether this was wise use of the museum's limited resources, Agassiz replied,

I have been turning the care of the Insects in my mind, with a heavy heart, ever since Uhler left us; and I think I have decided upon the right course in inviting Dr. Hagen to come. It is true as you say, he will be a rather dear man of science. But I am confident from the tone of his letters that he will fit; unless all human signs fail & I never resolved a plan more carefully than this. . . . He is the only living entomologist in Europe who is equally interested in all branches of entomology. . . . Moreover familiar as he is with all entomologists & all collections of Europe I shall direct him first to divide our specimens into what we must keep & what we may give away and at once dispatch as many collections as may repay us for this work . . . in the absence of any special care our whole collection of Insects may be lost.[16]

To his friend Lyman, Alexander Agassiz grumbled, "It strikes me you are getting a pretty good batch of foreigners into the Museum; do you believe it

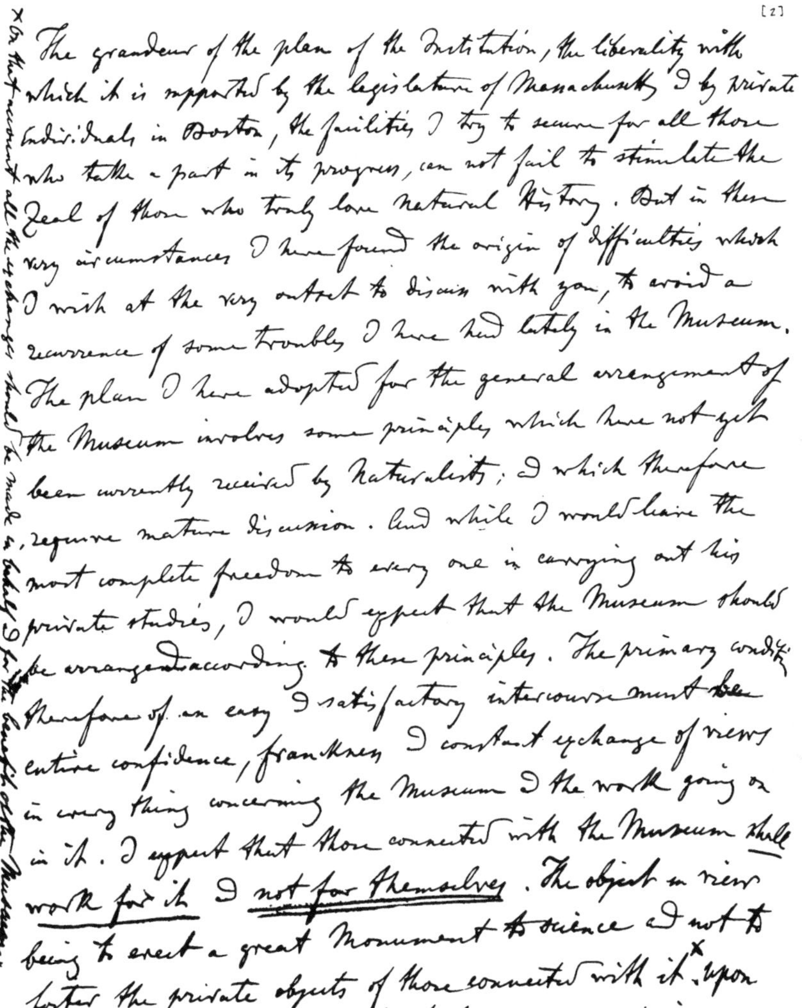

[2]

The grandeur of the plan of the Institution, the liberality with
which it is supported by the legislature of Massachusetts & by private
individuals in Boston, the facilities I try to secure for all those
who take a part in its progress, can not fail to stimulate the
zeal of those who truly love natural History. But in these
very circumstances I have found the origin of difficulties which
I wish at the very outset to discuss with you, to avoid a
recurrence of some troubles I have had lately in the Museum.
The plan I have adopted for the general arrangement of
the Museum involves some principles which have not yet
been currently received by Naturalists; & which therefore
require mature discussion. And while I would leave the
most complete freedom to every one in carrying out his
private studies, I would expect that the Museum should
be arranged according to these principles. The primary condition
therefore of an easy & satisfactory intercourse must be
entire confidence, frankness & constant exchange of views
in every thing concerning the Museum & the work going on
in it. I expect that those connected with the Museum shall
work for it & not for themselves. The object in view
being to erect a great Monument to science & not to
foster the private objects of those connected with it.× Upon
this point I want especially to hear your opinion.

× On that account all the exchanges should be made in behalf & for the benefit of the Museum.

Figure 10. Letter of Louis Agassiz to Philip Reese Uhler, 6 April 1864. "I expect that those connected with the Museum shall *work for it* and ***not for themselves.*** The object in view being to erect a great Monument to science and not to foster the private objects of those connected with it." (By permission of the Museum of Comparative Zoology Archives, Harvard University)

good policy?"* But in Hagen's case, the museum got solid value for its money.

Agassiz had wooed Hagen in the spring of 1867, demonstrating his confidence that the German would like the M.C.Z. by promising to pay all expenses should Hagen wish to leave after the first year. Autumn would be a fine time for him to come, Agassiz suggested, being "the most beautiful season for this country."[17] At 10 A.M. on Saturday the twelfth of October, Dr. Hagen and his wife arrived in Cambridge.[18] Theodore Lyman, who was then serving as the museum's financial manager, dropped by that morning. He "found Hagen, just come: a tall, good humored, intelligent looking German with round eyes" (fig. 11). His diary entry continues, "Agassiz delighted that he should be so promising, and especially that his wife was a cultivated lady."[19]

The museum that greeted Hagen can hardly have impressed him as favorably as he had impressed Lyman and Agassiz. The accumulation of material had for years far outstripped the manpower needed to arrange it, or even the space to contain it. Confident that quantities of valuable specimens would attract their own support, Agassiz had purchased European collections, had sent his students on excursions, had exchanged duplicates with other museums, and gone himself on collecting expeditions. His helpers could never keep up with the material that flooded in. Some of the exhibition rooms were "for the present transformed into storerooms," and most of the specimens were "packed away in barrels and boxes," uncatalogued and inaccessible.[20] Now that Agassiz was back from his expedition to the Amazon and more assistants had been hired, a beginning was being made to sort the Thayer material.† One worker described the scene: "Everywhere in all the laboratories jars of fishes, mainly of the Brazilian collections, cover the tables and occupy all the available space."[21]

As if these physical problems were not enough, the curatorial job as defined by Agassiz had been inflated by his ambitions. Agassiz insisted to Hagen, as he had explained to other new assistants, that specimens must be distributed into at least four and possibly as many as ten subsets, each serv-

*A. Agassiz to Lyman, 8 September 1867, M.C.Z. Archives. Three other foreigners had recently joined the museum (*Annual Report* for 1867, pp. 8–9). Besides Paul Roetter, there was Léo Lesquereux, an expert on fossil plants who had been part of Agassiz's Boston household from 1848 to 1850, who spent several summers at the M.C.Z. beginning in 1867 (Lesquereux to Peter Lesley, 28 July 1867, Lesley Papers, American Philosophical Society, Philadelphia), and Francis R. Staehli, who had been hired to work on mammals and birds and arrived in September 1867 (L. Agassiz to Staehli, 2 September 1867, M.C.Z. Archives).

†James Henry Blake (age twenty-three) and Richard Bliss (age twenty-five) were hired with funds supplied by the benefactor of the Brazilian expedition, Nathaniel Thayer (Lyman, Private Notebooks, 31 May 1867).

Figure 11. Hermann Hagen, entomologist, whom Lyman described as an "intelligent looking German with round eyes." (By permission of the Museum of Comparative Zoology Archives, Harvard University)

ing a different purpose. Agassiz boasted, "When I had Dr. Hagen here to show him how I wanted the Crustacea put up, he was appalled, so different are my notions of a Museum from those prevailing even among the most advanced students."[22] First of all, there would have to be a systematic collection, in which representative members of every species owned by the museum would be arranged to reflect the accepted natural hierarchy of genera, families, and orders; this was to serve as a "dictionary," for identification. In addition, there must be a faunal collection, arranged by geographic region, in order to reveal the natural associations of animals. There also would be separate collections to illustrate certain phenomena, such as one showing all the stages of individual growth. Specimens not needed in one or another of those series must be identified and separated for exchange with other institutions.[23]

In reality, the entomological department had no resemblance to Agassiz's vision, but consisted mostly of boxes of unsorted specimens. Not even the primary arrangement by families and genera could be undertaken until suitable cabinets were obtained. Dried insects are very liable to destruction by dermestid beetles, seemingly created for the special punishment of curators. Hagen insisted on special protective boxes with tight-fitting lids, but these took ten months to have built (at a cost of $4,000).

Crayfishes Accumulate in Cambridge and Washington

While waiting for the insect cases, Hagen busied himself with arthropods other than insects, especially the Crustacea. The approximately 20,000 specimens of lobsters, shrimps, and their allies had been sitting for years in jars of alcohol in the cellar. Every year hundreds of species were added to the collection of Crustacea, but it was mostly unsorted.* Any single container, received from some distant collector, might hold species belonging to a number of different families, so the first step involved distributing the specimens into smaller glass jars with fresh alcohol, each labeled as to collector and locality. Such was the labor that fell upon Dr. Hagen.

Hagen worked thus on the crustacea for three months, and then the museum ran out of alcohol.[24] Agassiz suggested he use this enforced pause in the mechanical part of his curatorial duties to arrange and revise some group of crustaceans, or "whatever else seems to you useful."[25] Apparently

*Agassiz's student Albert Ordway had been working on a family of crabs when he left to fight in the Civil War in 1861; a former student, William Stimpson, had visited from the Smithsonian and done some cataloguing of crustaceans in 1863 (*Annual Report* for 1860, p. 34; for 1863, p. 44; for 1868, pp. 27–28; Hermann Hagen, "Monograph of the North American Astacidae," pp. 1–2).

in response to this suggestion, Hagen applied his professional skills to the American crayfishes, creatures he had never seen alive but which were especially well represented in the museum's collection.[26] His "Monograph of the North American Astacidae" was ready for the printer eight months later, in October 1868.[27] Published in the *Illustrated Catalogue* of the museum two years later, it exemplified both the strength and limitations of current taxonomic method. Though Hagen soon turned his attention back to the insects, the museum would retain its authoritative lead in crayfish classification until after the First World War.

The museum's crayfish were rewarding material for a taxonomic study because they had been acquired for a purpose. Having long been interested in the detailed pattern of fish distribution, Agassiz in the 1850s had been hoping to discover natural faunal regions by tracing animal associations, and so he had proposed to the zoologist Spencer Fullerton Baird a monumental survey of the freshwater fishes of the United States.[28] Though little publication resulted, Agassiz's ideas had shaped his collection. He had let his correspondents know that he was interested not only in fishes but all freshwater animals, and he had called for specimens of both sexes and all ages. Thus although naturalists had named only eight species of North American crayfishes by 1850, Agassiz had accumulated hundreds of specimens from scores of localities.[29] In the early 1850s he was planning, along with countless other projects, to publish something based on this material himself.[30] Yet, at the same time, Agassiz's reason for soliciting and preserving large numbers of specimens of all kinds was institutional as well as intellectual. He was determined to build a museum of international stature, and he knew how to use his growing collections to attract support.[31]

During the 1850s Baird was laying the basis for a museum too, within the Smithsonian Institution in Washington (collections to be called, after about 1880, the United States National Museum, now named the National Museum of Natural History). Agassiz's letters to Baird proclaim his eagerness to cooperate for their mutual benefit and his scorn of rivalry and competition between Cambridge and Washington. He lectured his young colleague, "We have truly a serious task before us, that of establishing the right spirit and giving the right direction to the studies of Natural History in this country."[32] The dominant interest of American naturalists, in naming new species, must be discouraged, Agassiz insisted. At least these "species-mongers" were American, an improvement over the European scientists who had previously dominated the field of describing American specimens, insuring that the permanent reference types got deposited in European museums. But still, Agassiz argued, the mere registration of new species would never win respect for Americans in the scientific world. Any attentive amateur could all too easily discover and name a new species, even though he knew nothing of its variation, geographic distribution, or natural history. To better the respectability of American biology in the eyes of Europeans, said Agassiz, the naming of new species should be done only in the course of an analysis of genera or families.[33]

In the "Essay on Classification" of 1857, Agassiz announced a demanding new ideal for the recognition and description of a species, amounting to a complete knowledge of all its characteristics: its geographical range; its ecological requirements and tolerances; features of individual life cycle like growth, sexual maturity, and longevity; the social relations linking individual members to one another; and their dependence upon members of other species. One must still state all the physical details usually regarded as specific characters, such as size, shape, and ornamentation, but Agassiz added the variability of those characters as a species characteristic: "Well digested descriptions of species ought, therefore, to be comparative; they ought to assume the character of biographies and attempt to trace the origin and follow the development of a species during its whole existence."[34]

Even before the "Essay" was printed, Agassiz himself admitted to Baird that one cannot wait for such a complete state of knowledge before naming a species. "I find that to do it as I think it ought to be done it takes more time than I had expected."[35] In fact, species descriptions he did after the "Essay" were less thorough than those he did before.[36] Traveling across Brazil in 1865, he would declare that he was seeing hundreds of new species of fish, even though he knew nothing about them but their form and color (and sometimes the names given them by local people).[37] Yet, though the concept in the "Essay" was not put into practice, it nevertheless refutes the notion that a believer in divinely conceived forms must believe that a few diagnostic marks can establish a species.

In practice, there was a great temptation for naturalists to work with careless haste, since the rules of nomenclature specified that historical priority, not quality of description, determined what name would be used. Every later worker citing a known species must print with it the name of the first describer, regardless of whether the first description had been found inadequate and had later been improved upon by a more professional worker. Agassiz campaigned to end this situation, proposing that the authority named after the genus and species name should be the author who first recognized the genus to which that species naturally belonged. The naturalist who had failed to perceive correctly the true taxonomic position of his new species would lose his assurance of immortality. The systematic assessment of the degrees of relatedness of species to one another, expressed in monographs and revisions of genera or families, would become more rewarding than the naming of single species. But Agassiz's suggestion was not adopted, except briefly by his own students. The only incentive for caution and thoroughness in species description was one's fear of having a later author revise one's species and "sink" it, by showing that it is identical to, or merely a variety of, a previously named form.

In 1864 Agassiz vigorously opposed Baird's election to the National Academy of Science.[38] Although his complaint failed to prevent Baird's election and was attributed by many to professional jealousy, Agassiz himself felt that Baird had ignored his pleas to "take your stand like a man and frown down upon" the hasty description of species with no understanding of families and genera. One case among many influencing Agassiz's judgment of Baird must have been a classification of crayfishes done at the Smithsonian in 1852. Baird, too, had been accumulating specimens, many belonging to species new to science, but like Agassiz he had found no leisure to describe them. When Agassiz's former student and longtime assistant Charles Girard suddenly left Cambridge and turned up in Washington as an employee of the Smithsonian, one of his first assignments was to publish a study of crayfishes. Besides adding twelve new species to the eight already in the literature, Girard declared them all generically distinct from *Astacus,* the European crayfishes, and gave characters subdividing his new genus, *Cambarus,* into three groups.[39]

Agassiz had responded to Girard's article with devastating criticism. The eminent German zoologist W. F. Erichson had created *Cambarus* as a subgenus of *Astacus* on the basis of a single anatomical difference; having one less pair of gills than the European.[40] Girard's new species were probably indeed previously unknown, but he had not looked at their gills, nor did he offer other characters to explain why Erichson's subgenus now deserved generic status. Agassiz borrowed the type specimens on which Girard had based his new species *Cambarus Gambelli* from the Academy of Natural Sciences in Philadelphia, and then he published a note pointing out that its gills conformed to the European pattern.[41] Instead of accepting Agassiz's correction graciously, Girard made the artless defense that his definition of *Cambarus* was not based on the gill number.[42] What it was based upon, he did not say. No wonder Major John Eatton LeConte, a man of considerable stature in American zoology (and father of John Lawrence LeConte), explicitly rejected Girard's opinion as based on characters "of little moment" and referred to the American crayfishes as *Astacus.*[43] Still, the taxonomic custom Agassiz was trying, with scant success, to alter would leave the credit of first describer to Girard whichever generic name one chose.

For all its pretensions as a contribution to taxonomy, Girard's article had supplied rather meager characterizations of each species, "deferring to another opportunity more full description, accompanied with necessary graphic illustrations."[44] Because this promise remained unfulfilled, in order to do anything with his collection Agassiz would have to divine which of his many unidentified specimens belonged to Girard's new species and which were still unnamed. With some exasperation, Agassiz wrote Baird to ask whether Girard would continue to describe whatever new crayfish came into the Smithsonian. "I have made a thorough revision of the whole family and I should not like to have nothing but criticisms left to publish."[45] No such assurance was forthcoming (apparently the ever-tactful Baird ignored the question). Agassiz's crayfish collection gathered dust for the next fourteen years, until the day when the conscientious Hagen needed a project.

Biological Peculiarities of Crayfish

In his 1868 analysis of crayfishes, Hagen was given a valuable head start. When sorting the M.C.Z.'s collection, Agassiz (or his assistant Henry James Clark, or both men together) had discovered a curious fact which they neither published nor mentioned to their colleagues in Washington. We expect adults of any species to be of two forms, male and female, but American crayfish fall into three: female, one form of male, and a second form of male. The "form II" male has its copulatory limbs less developed.* The scores of specimens of each species which Agassiz had accumulated were, by museum standards of the time, excessive; Hagen had before him about 2,000 specimens, or an average of eighty per species, which he himself felt could "serve to obstruct the judgment of the worker."[46] But these numbers proved to be the crucial element enabling him to establish convincingly the peculiar phenomenon of having two kinds of male, called dimorphism, which, since it does not exist in the European crayfish, would otherwise probably have been received with skepticism:

> The objection that these second-form males may be individuals shortly before or shortly after the casting of the skin I can surely refute, as I have seen many specimens

*Hagen gave Louis Agassiz the credit for having first noticed the two forms of male (Hagen, "Monograph," p. 22n.), but Walter Faxon would later say it was Clark who first recognized the phenomenon ("On the so-called dimorphism," p. 42). Although Faxon did not join the museum until 1873, his report is plausible, for Clark was working closely with Agassiz when Agassiz was building the crayfish collection (Herber, *Correspondence between Baird and Agassiz,* pp. 50, 53, 63, 64). After Clark's "Claim for scientific property," Agassiz may have been in no mood to credit Clark.

at this stage of growth; the Museum collections exhibiting the animal in all the different phases of its existence.

Another objection, that the males of the second form, or perhaps those of the first form, are abnormally developed individuals, is refuted by the great number of the two forms existing and living together.[47]

Agassiz did lecture his students on the value of having large series of specimens, particularly as revealing variation due to age differences.[48] At the same time, his collecting policy was related to his practical plans as a museum builder, for he would need several representatives of each species to arrange into his various special collections, and others to exchange as "duplicates" with other museums. Hagen's material would not have been so rich had Agassiz managed to carry out these ambitious plans. Nine years later Hagen would write in answer to a request from Uhler, "After the publication of my monograph Prof. L. Agassiz asked me so often to put together a set of duplicates to send abroad, that the last time were left only duplicates for 6 species."[49]

It was, of course, Hagen's labor that gave these crayfish specimens extra value to other museums, for they had been identified by himself, author of the definitive "Monograph."

However rich the museum's collection, it could contain no clue to the biological significance of crayfish dimorphism. The reduced copulatory limbs suggested that the form II males might be sterile, so Hagen dissected a few. Finding the testes present but somewhat smaller left the question unanswered; Hagen thought a fertility difference very likely, and suggested as well that a sterile form of female may yet be discovered. He knew, of course, of the sterility of worker ants and bees, and knew also that certain crabs are sometimes found with no reproductive tissue.* This, he said, is an area worth investigating.

As a naturalist, Hagen was curious to know what purpose dimorphism served, but as a museum employee his task centered on the classification of the specimens before him, and he was sure that his utter ignorance of the role of dimorphism in the life of crayfish did not prevent him from establishing its importance to the taxonomist. An author ignorant of dimorphism would think the males were more variable than they are, or he might even describe males of the two forms as two different species. Besides, Hagen could add dimorphism to the list of characteristics distinguishing *Cambarus* from *Astacus*. Girard's premature elevation of subgenus to genus Hagen now certified.

*Hagen, "Monograph," pp. 25–26. That the crabs were in a diseased state caused by a parasite was not yet known.

The Nature of the Species Category, and the Recognition of Species of Crayfish

The hallmark of Hagen's professional style was the care he took in describing species. Here again he was conscious of the advantage of working with the large number of specimens assembled in Cambridge. "The exceedingly rich mass of material before me, thus far unrivalled for such a labor, has permitted a very extensive and careful examination of the constancy of characters."[50] He provided page-long Latin diagnoses of previously known species, when possible using the type specimens, that is, the very individuals used by the first describer, specimens lent to him by the Academy of Natural Sciences, the Peabody Museum in Salem, and the Boston Society of Natural History. He included full illustrations of eight species. He accepted twenty-seven species already in the literature, rejected thirteen names as synonyms, and he added eleven new species. The result was a picture of a continent with thirty-eight distinct species of crayfish, consisting of six species of *Astacus* west and thirty-two species of *Cambarus* east of the Rocky Mountains. No wonder a French critic declared he had overdone things—"dépassé toute mesure."[51] Surely this many names could not represent nature; surely these new "species" must result from pedantic splitting on the basis of unimportant differences? Indeed, many of Hagen's species are so similar to one another that they can be distinguished only by an expert even today; and becoming an expert is at first a distinctly bewildering experience.[52] Yet the species Hagen recognized have stood the test of time. Subsequent specialists have admired Hagen's judgment.[53] He was a master taxonomist of the classical style.

Agassiz knew the kind of man he was getting when he hired Hagen, a man dedicated to "keeping everything in good order, with haste and neatness." This was not a man who cared for philosophy, theory, or speculation. A letter from Hagen discussing evolution, written from Germany to Benjamin Dann Walsh in 1864, was later destroyed in the Great Chicago Fire, but we can infer the tenor of Hagen's comments from Walsh's reply:

> By the way, have you ever read Darwin's book carefully through? I was once as strongly prejudiced as any man could be against his views, but the first perusal of his book staggered me and the second convinced me. . . . You say there is no philosophical necessity for the existence of Darwinism, and that it is just as difficult to understand how the most simple organic being is created as how the most complex is created.[54]

Hagen's point had been one often brought against Darwin, namely, that he had not improved upon the creationist explanation because his theory still required, but did not explain, the coming into existence of the original simple forms of life from which the others evolved. Walsh appealed to Hagen to recognize the irrelevance of this complaint:

> Granted. But if Darwinism is *true,* that is no reason why we should reject it. I object to asking whether any discovery will explain or not the mystery of life before we consent to accept it as true. Follow truth whithersoever she leads us, & don't be afraid that one truth will contradict another, or will not decipher all the hieroglyphs of Nature.[55]

When Hagen did not reply to this letter, Walsh assumed he had been dropped, but Hagen reopened the correspondence when he took up his new duties at Harvard in October 1867.

Whether or not Hagen ever did "read Darwin's book carefully through," certainly he studiously ignored the *Origin* while composing his monograph on crayfishes. He could not ignore, however, another work published in Russia in 1859, "Ueber die Flusskrebse Europa's." The author, Georg Gerstfeldt, suspicious as to whether there really could be in Europe as many as eight different species of *Astacus,* had gathered new material to test the existing definitions. Besides comparing specimens in a few museums, he examined "very many" freshly collected crayfish and was able to show that all the characteristics that had been used to describe these species, such as the length of antennae, shape of rostrum, shape of great claw, and so on, varied sufficiently to obliterate most of the claimed distinctions cited by taxonomists. He argued that there was justification for only two full species of European crayfishes, though four other forms were recognizable enough for Gerstfeldt to allow them the status of varieties.

The American species already named must have seemed very ripe for a similar consolidation. Major LeConte, while describing eight new species in 1855, had innocently confessed his doubts:

> Their extreme similarity renders it difficult to distinguish the species from each other, and this difficulty will undoubtedly increase as the number of species is multiplied. To what extent this number may reach cannot be guessed at; it appears however that when these animals come to be more closely studied, this number will be very great, and the passage of one species into another become almost imperceptible.[56]

Thus the task facing Hagen in 1868 was not merely to recognize species in Agassiz's collection but to convince zoologists of their validity. If, indeed, the passage of one crayfish species into the next was imperceptible, as LeConte thought, Hagen should sink them all into a single populous, widely varying species. But he did not find himself to be as overwhelmed as Le-

Conte had. His goal was to distinguish the species, and he found that this was possible. The situation with American *Cambarus* was not parallel to the European *Astacus,* for though the species were more numerous than anyone would have expected, they were in most cases perfectly definable.

Gerstfeldt's assault on the European species meant that it would not be enough for Hagen to inspect his material, come to his own decisions, and pronounce the distinctions he judged significant. Certainly he had to learn to recognize the species, by close familiarity with the specimens, but he did not have to start with a chaotic mass. All his specimens had labels telling him where they had been collected. Often he had scores of specimens from a single locality and could use them, as Gerstfeldt had used his freshly collected ones, to give him an indication of the range of variation within one species, revealing differences due to sex, age, or individual variation. Sometimes two species did live together in the same body of water, rarely three, but in those cases nature conveniently made them easy to distinguish.

We can recognize instantly individual faces we would be hard put to describe. Hagen's job was to reduce to a written formula—the taxonomists' "diagnosis"—what the experience of looking at the specimens had taught him to recognize. The structural characters which constituted the general appearance of American crayfishes were elusive and open to the same critique Gerstfeldt had made of diagnoses of European forms. So Hagen looked for other characters, ones sufficiently constant within his perceived species to withstand scrutiny. Nature cooperated. It so happened that a remarkable feature of dragonflies Hagen had uncovered earlier gave him the solution to these crayfishes as well. The organ with which the male deposits sperm, the end of the dragonfly's tail or the first pair of abdominal limbs of *Cambarus,* was in both groups remarkably unchanging within each species, yet distinctive from one species to another.

Hagen reported that the tip of the cambarid copulatory limbs is not simple, as it is in *Astacus,*

> but transformed into more or less corneous hooks and teeth. We find also in Cambarus the tip of the first abdominal legs bifid, and the two branches more or less elongated, equal or not in length and breadth, straight or curved, and very well adapted to form specific characters. Having examined a very great number of specimens, I am able to state that these different forms are very constant in the same species.[57]

Hagen made careful drawings of the first abdominal appendage for almost every species of *Cambarus* he had, both first and second form male, from front and side view. Paul Roetter, who had left Switzerland in 1845 and was hired by Agassiz when Burkhardt died, transformed these hundred draw-

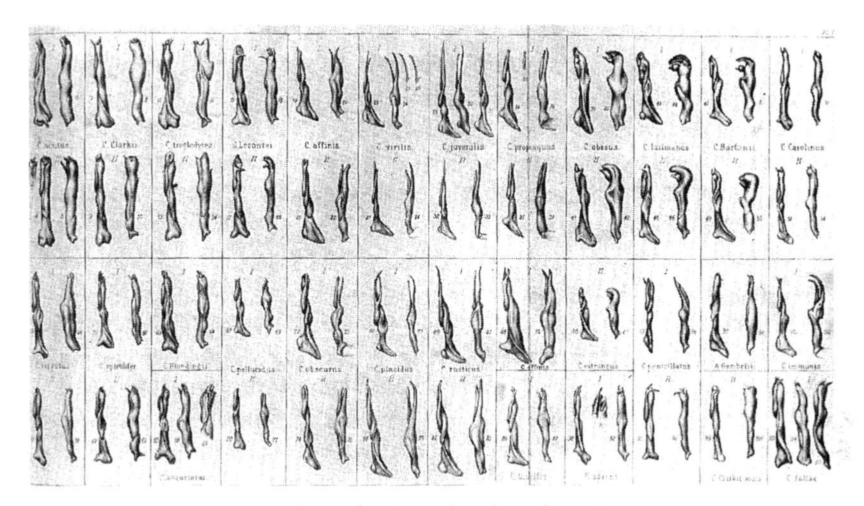

Figure 12. The first abdominal limb of male cambarid crayfish, lithographed by Paul Roetter. Having discovered a taxonomically effective feature, Hagen made careful drawings, front and side view, for each of twenty-nine species, two forms of male per species. He offered no explanation for the special value of this character, but later specialists agree with his choice.

ings into one remarkable lithograph (fig. 12). In 1852 Girard had briefly mentioned in some of his descriptions the shape of the first abdominal appendage of the male, but it was Hagen who established the reliability of this character, which remains the sine qua non of the study of *Cambarus*.*

Hagen also noted, though more briefly and without illustration, the constancy and taxonomic usefulness of a feature of female *Cambarus* that he knew was absent in *Astacus:* a small sculptured button, which he called the "annulus ventralis," located on the middle of the underside. As with male dimorphism, Hagen's ignorance of the biological function of this organ was quite irrelevant to his job as a taxonomist. He pointed out the existence of the structure, explaining that his material did not reveal its function; then he moved on to the task of identifying species. He was, in fact, not even close to guessing the function of the annulus ventralis, for he was assuming that during mating the male's first abdominal appendage deposited sperm in the female's oviduct. It was to be a quarter-century after the publication of Hagen's "Monograph" before any zoologist bothered to look closely at how crayfish mate (simple enough to see the pair in their

*For the North American crayfishes, it is "the most useful taxonomic and evolutionary character yet discovered. It is usually stable and, insofar as is known, is little affected by ecological factors" (Fitzpatrick, "The Propinquus group of the Crawfish Genus Orconectes," p. 137).

quiet long embrace, tricky to see precisely what is going on).[58] In fact, the male deposits a viscid package of sperm into the pocket of the annulus ventralis. Fertilization does not take place until the female produces eggs, which may be several months after mating. Knowing none of this, Hagen nevertheless could tell from his series of specimens that the curves of the annulus ventralis can help us identify a species, just as we use the curves of a fingerprint to identify an individual.

Subsequent taxonomists have agreed with Hagen, that in cambarid crayfishes the copulatory limbs of the male and the female annulus ventralis are remarkably constant within a series and sharply distinct between species. Here again, as with his discovery of crayfish dimorphism, Hagen limited his interest to the taxonomic usefulness of the phenomenon. Yet within the covers of Darwin's book he could have found a dozen suggestions for relating these discoveries to questions of interest to biologists not concerned with identifying crayfishes. Dimorphism is a striking characteristic of female ants which Darwin admitted posed a major difficulty for his theory: How could a useful variation in the sterile worker be preserved and accumulated by natural selection, since she leaves no offspring? Darwin's answer depends on her aiding her nieces and would not apply to nonsocial animals like crayfish, so Hagen's guess that the second-form male is sterile could have given ammunition to an antiselectionist. Indeed, sex was at the very center of Darwin's theory, since survival means nothing without success in leaving offspring, so to find features of the sexual organs distinguishing one species from another should have been particularly thought provoking for a Darwinian. Or the lack of variation in those features could have been invoked as a problem for Darwinism.

Hagen did say that crayfish dimorphism was interesting and should be investigated, but no one noticed. Further information on the subject came along entirely by accident. In 1875 Walter Faxon, Hagen's successor in charge of the M.C.Z.'s crustacea, received some live crayfish from Kentucky.* Probably these had been requested in connection with Faxon's graduate studies in embryology; he kept them in an aquarium long enough to mate. "After pairing, three of the males moulted and were thrown, while in the soft-shelled state, into alcohol together with their exuviae." There they sat, unexamined, for eight years. Then, in the course of a routine taxonomic review, Faxon took a look at these bottled animals and realized

*Faxon, "So-called dimorphism," p. 43. These were probably lots 3442 (seventy-five specimens) and 3443 (three specimens) which the MS catalogue of M.C.Z. Crustacea records as collected and donated by W. M. Linney of Perryville, Kentucky, and received 30 October 1875. The "remarks" column, likely entered by Faxon, says of lot number 3443, "Form II, and exuviae form I!"

that they differed from their own cast-off shells. Thus in 1884 it finally became known that "form I" and "form II" males are stages in the life of the same individual. So crayfish are not after all dimorphic in the proper sense of the word, as ants are—some individuals having one form and some another; each male alternates from form I to form II and back to form I.

Historians will be inclined to accuse me of the professional equivalent of Monday-morning quarterbacking, if I imply that it was obtuse of Hagen not to learn from Darwin why external sexual characters may be constant within a species and distinct between species. After all, the concept of isolating mechanisms was not clear in the *Origin,* would later be a source of confusion and disagreement between Darwin and Wallace, and was not well developed until the middle of this century. The "lock-and-key" theory to explain dragonfly genitalia is still a subject of confused controversy, and we have yet to explain the peculiarities of crayfish very well.[59] We can hardly blame Hagen for not solving these difficult questions. Nevertheless, his decision not even to ask them I think signals a source of trouble for the scientific status of taxonomy.

Hagen's correspondent Walsh, who had read the *Origin* with such excitement on a secluded farm in Illinois, stands in sharp contrast to Hagen. The explanatory potential of Darwin's theory struck Walsh forcibly. With no formal training in science, complaining often of his lack of books and colleagues, Walsh repeatedly invoked Darwinian concepts in his entomological contributions. He was, of course, familiar with Hagen's demonstration of the specificity of the male organs of dragonflies, and he allowed himself to speculate on its cause, suggesting that

> when a variation useful to a male in his sexual operations does take place in the male reproductive organs, it is often seized hold of by Natural Selection to originate a new species, the less favored males being beaten in the struggle for females.[60]

Walsh was suggesting that a new species might bud off from a parent species with no improved adaptation in its way of life, with scarcely even a change in morphology beyond that of the sexual organ itself. To most nineteenth-century evolutionists, such a thing would have been quite as inconceivable as it would be to creationists, for it follows no principle of progress, nor law of development, nor temporal unfolding of a divine plan. Had Hagen thought about it, though, he might have noticed that Walsh's idea does accord with the general similarity and species distinctions of American crayfishes. Walsh's notion was far from a mature theory; whether a new species can thus originate, without a geographic barrier to protect it from breeding with unaltered forms, is still a subject of debate.[61] But Walsh's idea could have been a fruitful speculation, suggesting new questions such as how the overlapping ranges of species relate to different patterns of variation in their genitals, or whether members of similar species will attempt to mate.

In contrast to Walsh, biologists with more professional training, even the ones who heartily adopted evolution, were rarely willing to credit natural selection with such power. Thus suggestions like his were ignored, and mechanisms of speciation, indeed, sexual selection more generally, got very little serious attention until well into this century.

Hagen made not the slightest allusion to Darwin's theory in his "Monograph of the North American Astacidae," nor did he make any use of Agassiz's ideas. He would doubtless have denied having any theory at all, but we may infer from his work that he did have a strong set of working assumptions virtually invisible to him because held in common with almost all of his fellow zoologists. He assumed that each organism does belong to a species, even though some species are highly variable and others not, and even though sometimes a species is so similar to another that specimens are easy to confuse. Hagen subscribed to the norm that there is always a correct answer to the question, Does this animal belong to a different species than that one?—even though in some cases he might not yet have enough information to do more than guess at the answer and point the way for future research. To describe species, and to associate them into genera, he exercised the practical arts already highly developed among entomologists and other taxonomists.

The concept of species implicit in the work of men like Hagen was mainly derived, not from any philosophical school like idealism, but from the practical art of taxonomy as it had evolved over the previous two centuries. Although a few leaders had set forth discussions of their principles, and a few authors introduced their lists of mammals or butterflies or snails with explanatory sections, for the most part the identification and classification of species was carried out according to unwritten guidelines. There was much trial and error, following rules of thumb passed on by informal apprenticeship from expert to young enthusiast. To write descriptions of species and genera that would prove useful to later workers, there was no substitute for intimate familiarity with one group, because each group of animals had its own peculiarities. Features such as color, hairs, and teeth that are dependable marks of a species in one group may vary so much in another group as to be taxonomically useless. Thus taxonomists were divided into many small communities of specialists, modifying their respective predecessor's catalogues over the years.

Hagen and taxonomists like him were comfortable with the Linnaean program, not because they held a particular picture of creation or had given deep thought to what sort of thing a species is, but because, having begun with the assumption that constant forms could be identified, described, and grouped, their own experience confirmed their expectations. At any one point in time and space, the majority of living things, or at least of those most noticed by humans, are indeed very like their nearest relatives, and distinct from their distant cousins, so that assuming the existence of species is a highly rewarding approach.

When a taxonomist declares a form to be a species, he is proposing a hypothesis, venturing a prediction that further research will not fill in the gap between this form and any previously known form. The fact that such predictions were often borne out by subsequent investigation reinforced the feeling that the species is a natural category rather than an artificial device imposed by us on the living world. For example, in his crayfish monograph Hagen compared four species already in the literature, *Cambarus Blandingii, acutus, Clarkii,* and *troglodytes,* and noted that they "are very similar in size and form."[62] The species *Blandingii* had first been described and figured by Richard Harlan in 1830 on the basis of one specimen from Camden, South Carolina; Girard in 1852 had christened the species *acutus* and *Clarkii* on the basis of a few specimens sent to the Smithsonian from Mississippi and Texas, respectively; and John LeConte had described *troglodytes* from the highlands of Georgia.* Each author had had a severely limited sample, with no information about local or geographic variability. Hagen had before him similar crayfishes from more than thirteen additional localities, over 150 individual specimens as well as type specimens for all but *acutus.* Yet the effect of this additional material was not to blur the distinctions between the four forms; on the contrary, Hagen was able now to document many more differences between them than had been cited in the original descriptions. Individually these differences were indeed slight, but their combination was recognizable, and he found no specimens of an indeterminate nature to blur the boundaries between any of these four species.

In a number of other cases, however, Hagen could not decide, on the basis of material before him, whether certain forms were varieties or species. His new species *placidus,* for example, resembled Girard's *rusticus;* "the identity is not impossible," he confessed.[63] LeConte's species *spiculifer* and *angustatus* were separated by differences "perhaps rather too

*Harlan, "Description of a new species of the Genus Astacus"; J. LeConte, "Descriptions of new species of Astacus from Georgia." Harlan and LeConte put their new species in the genus *Astacus,* but all four of these are now called *Procambarus.* I depart from the proper taxonomic style by dropping the generic portion of the species names, because their changes are complicated and not helpful to the issue at hand.

minute," but Hagen let them stand with the comment that "a further examination may perhaps bring out more strongly the differences of the two nearly allied species."[64] (Modern experts consider them quite distinct.) Borderline cases like this did not seem to cause him any anxiety, nor make him doubt whether nature might be playing tricks on him by allowing forms to exist that really occupied a no-man's-land between "good species" and "mere varieties." He was confident that problematical cases would be decided in time by his successors, who would use the same sort of museum-based methods he used. Distinctions might be sharpened, as in these cases, or recognized to be only variants within one species, but Hagen assumed that any doubt would be caused by imperfect information, not by ambiguity in nature.

Darwin on Species

Hagen's attitude in the late 1860s fits very well Darwin's description written a decade earlier. In the *Origin* he noted that

> naturalists have no golden rule by which to distinguish species and varieties; they grant some little variability to each species, but when they meet with a somewhat greater amount of differences between any two forms, they rank both as species, unless they are enabled to connect them together by close intermediate gradations.[65]

The question Darwin tackled, and Hagen ignored, was, Did the absence of such a golden rule indicate that the distinction between species and varieties might be purely arbitrary?

The fact that nature does present the systematist with problem cases was critical to Darwin's argument. Reviewing the practice of contemporary taxonomists, he showed that neither degree of morphological difference nor inability to interbreed was an infallible criterion. There was no definition of the terms "species" and "variety" that was universally followed. What naturalists did in practice was to rank as varieties, rather than species, forms which were "closely similar to some other forms," or "closely linked to them by intermediate gradation," or assumed to be so linked by forms yet to be discovered.[66] Usually restrained in his language, Darwin scolded,

> It must be admitted that many forms, considered by highly-competent judges as varieties, have so perfectly the character of species that they are ranked by other highly competent judges as good and true species. But to discuss whether they are rightly called species or varieties, before any definition of these terms has been generally accepted, is vainly to beat the air.[67]

Darwin's theory of the gradual modification and divergence of species insisted that the difference between species and varieties is one of degree, not of kind.

To adopt his theory, Darwin claimed, would free biologists "from the vain search for the undiscovered and undiscoverable essence of the term species."[68] Naturalists would then

> not be incessantly haunted by the shadowy doubt whether this or that form be in essence a species. This I feel sure, and I speak after experience [his eight years revising the barnacles], will be no slight relief. The endless disputes whether or not some fifty species of British brambles are true species will cease.[69]

The bramble or blackberry, genus *Rubus,* turns out to have been a particularly poor example for Darwin to have chosen. They were, indeed, notoriously difficult, with a great number of barely distinguishable forms, but the reason for this is not as simple as he was implying. Their seeds are produced asexually, so the various forms, whether called "species" or "varieties," are not comparable to either species or varieties of most animals, which are interbreeding populations. Giving names to British brambles remains a complex and contentious business. Botanists have been faced with the fact that there exist several distinct kinds of variation, speciation, and species far more often than zoologists have.

Quite aside from these complexities, Darwin was sadly mistaken in his prophecy. Even those zoologists who thought of themselves as evolutionists could not shake off the traditional feeling that they ought to be able to distinguish species in every case. After all, it was their business to do so.

All this being said, it is important to note that Darwin was not arguing that there was not really any meaningful distinction to be drawn between varieties and species, nor that "species" was a man-made concept arbitrarily applied to infinitely variable living things. He stated clearly the crucial difference between species and varieties: that varieties "are . . . connected at the present day by intermediate gradation, whereas species were formerly thus connected."[70] Whether any given form is linked to its nearest relatives by living intermediates can be discovered by close investigation, and in zoology the overwhelming majority of cases are not borderline. For that reason, Darwin explicitly predicted that the overall methods of taxonomists would not have to be altered by his theory:

> Systematists will be able to pursue their labours as at present. . . . [They] will have only to decide (not that this will be easy) whether any form be sufficiently constant and distinct from other forms, to be capable of definition and if definable, whether the differences be sufficiently important to deserve a specific name. This latter point

> will become a far more essential consideration than it is at present . . . we shall be led to weigh more carefully and to value higher the actual amount of difference between them.[71]

Again, Darwin's prediction was mistaken. A modern evolutionary taxonomist decides this question not so much by the "actual amount of difference," that is, morphological distinctness, as by the kind of difference, namely, how much interbreeding occurs between the two forms. Although this criterion may seem to have been implicit in Darwin's thought, it was not explicit in the *Origin,* and his followers went off in rather the opposite direction from the one he expected. In the nineteenth century, evolutionary taxonomists' only criterion of whether a difference was "important" was whether they could detect it, no matter how tiny it was.[72]

Yet as Darwin said, systematists could if they chose carry on their work much as they had been doing, since the general principles inherited from Linnaeus, including the existence of species and their arrangement in a hierarchy of categories, represent the phenomena very well. In the generation after the *Origin,* evolutionary skeptics and agnostics like Hagen formed one continuous community of practice with followers of Darwin. The Linnaean program generated plenty of rewarding questions, especially for those animals whose inventory was in the early stages, as was true for most non-European forms and most invertebrates. Thus, in every museum, there was work to be done that would be acknowledged as valuable, even if one did neither "comparative zoology" nor research into the origin of species.

"Apparently Capricious Aberrations" of Crayfishes

Darwin's theory included a clear discussion of the distinction between affinity—similarities due to common descent—and analogy—similarities due to independently acquired adaptations. If our classifications are to represent degrees of genetic closeness, as our family trees represent relationship by "blood," we must take care not to mistake analogies for affinities and not to use them as taxonomic characters, however interesting they may be in the natural history of the organism.

The distinction between analogy and affinity was well established before 1859. William Sharp MacLeay stressed it in 1823, using analogies to form the cross-connections between his five-member circles of affinity; Darwin and others credited MacLeay with heightening awareness of the difference.[73] When the eminent anatomist Richard Owen offered formal definition of the term "homologue" in 1843, he meant merely to codify an already accepted concept.[74]

In practice, analogies would spring to view between forms whose relationships of affinity were otherwise clear, that is, where the number and kind of resemblance were so strong that zoologists easily agreed on the natural taxa; in such cases a feature resembling some other type stood out as anomalous. For example, when we first see, in a quick glimpse from shipboard, what seem to be swallows darting along above the waves, we casually class the fliers as birds, but when one lands on deck, we can make a proper study. Immediately we recognize that our first impression was an error. It does not require the hours Agassiz's students spent on their first assignment for us to be positive this birdlike creature is a flying fish. What it flies with are certainly real fins, analogous, not homologous, to the wings of birds. It is self-evident that the reason these fins are the same general shape as wings is that their function is so like that of wings. What is not self-evident is what kind of treatment the taxonomist ought to give to the birdlike character of this fish, for the resemblances certainly seem to belong to the organism, not merely to our perceptions. W. S. MacLeay, Louis Agassiz, and many others viewed analogies as an extra dimension of pattern woven across the network of affinities by the imaginative Creator. Yet everyone agreed that the Linnaean categories should express affinities, not analogies.

As long as the two forms being compared were as unlike as birds and flying fish, and as long as the features held in common were adaptive, the phenomenon of analogy presented no confusion to taxonomists, even if a few wanted to give it a cross-linking role in their classifications. But when the similarities were striking, or seemed to serve no function, or occurred in closely related groups, difficulties did arise. Echinoderms (sea urchins) and coelenterates (jellyfish) were united by their radial symmetry for over half a century, and even after knowledge of their anatomical differences had accumulated to the point where they were divorced, it still took a study of their embryology to establish the idea that this symmetry was analogous rather than homologous. The mimetic resemblance of some flies to bees and of certain butterflies to others did not cause them to be misclassified, because entomologists used structural characters. Still, it was hard to understand why such elaborate and detailed resemblances existed, seemingly contrived not so much for benefit of the insects as for the amusement or confusion of the naturalist.

Darwin's theory promised a powerful new understanding. Similar features in related organisms had been inherited from a common ancestor which had had that feature, while in unrelated organisms similar features had been evolved independently because of analogous adaptive needs. If classification is to reflect genealogy, analogies must not be admitted as taxonomic characters. The idea is simple, but its application often is not, because evolutionary relationship is inferred in exactly the same way as natural affinity had been determined. There is no problem when the analogous features are not too much alike, or when the organisms differ greatly otherwise. But when closely related forms have met parallel circumstances by evolving parallel adaptations, perhaps by selecting the same genetic variation, what ought to be called analogy can perfectly mimic homology. Bees that specialize in stealing honey from other bees all lack pollen-collecting apparatus, which caused them to be classified together, though other points of structure reveal this thieving life-style to have been invented by different ancestral kinds over and over again. The crayfishes confronted Hagen with two such cases.

Among the many characters Hagen carefully included in his description of species was a set of thornlike hooks or spurs near the base of the walking legs of the male. In some species such a hook is present only on the third pair of legs, in some only on the fourth, in others on the second and third, in still others on the third and fourth. For any given species, the presence and location of these hooks is constant enough to make them very useful taxonomically. Hagen seems not to have inquired into the function of these hooks; he would have to have watched a mating pair very carefully to see how these hooks help the male keep his grip during the long hours of coupling.

Hagen did notice that some species of *Cambarus* seemed to have a closer affinity to one another than to other species. He did not think the differences important enough to split the genus into several genera, as modern workers, who have hundreds of species to deal with, have done, but he did want to express these relationships by forming three subdivisions of the genus. He had a feeling for where the natural associations lay, but he ran into trouble when he tried to write down diagnoses of the three groups. He confessed that no matter what characters he chose, a few species would always spoil the rule. He defined the groups anyway, with the disclaimer, "Nature never agrees with the strict principles of a particular scheme, so that apparently capricious aberrations are to be found everywhere the stumbling-blocks of the naturalist who wishes to arrange everything in a regular series."[75]

He did discuss the ill-fitting species. To one new species that would not fit his rule he gave the appropriate name *extraneus*. If he paid attention to the hooks on its walking legs, he would have to put it in the group typified by *affinis*, but the shape of its first abdominal appendage, its copulatory stylets, were obviously of the *Bartonii* type. In his third group, he signaled the uncertain membership of the last four species, whose copulatory stylets resembled those of the first group. Had he followed his perceptions of relationship, instead of keeping strictly to his definitions, he would have successfully distributed his species among the three largest genera of cambarids recognized today.[76] It is not surprising that it did not occur to Hagen to regard the occurrence of these hooks as mere analogies, for they are as nearly identical in detail as any homologous character. Current speculation envisions an ancestor with spurs on several walking legs, making the hooks themselves homologous; it is imagined that the pattern of their disappearance or persistence on different legs has been produced independently several times by a parallel process.[77] Probably Hagen would have found this conjectural narrative no more scientific than his own admission that nature refuses to be orderly.

Before we decide on the basis of this example that evolution will automatically give its adherents a deeper insight into the principles of taxonomy, especially by illuminating the distinction between analogy and homology, we should pay attention to the case of the cave animals, in which crayfishes figured prominently.

The Blind Crayfishes of Kentucky and Indiana

In the early 1840s naturalists began to notice the existence of animals living deep in the Mammoth Cave of Kentucky—spiders, beetles, crickets, shrimp, fish, and crayfish—differing from their aboveground relations by being small, colorless, and blind. The power of environmental conditions to modify animals directly, just as Lamarck had claimed, seemed perfectly demonstrated by these creatures. In 1851 Agassiz's opinion was asked,[78] and he replied,

> This is one of the most important questions to settle in Natural History, and I have several years ago proposed a plan for its investigation, which, if well conducted, would lead to as important results as any series of investigations, which can be conceived, for it might settle, once for ever, the question, in what condition and where the animals now living on the earth, were first called into existence. . . . If physical circumstances ever modified organized beings, it should be easily ascertained here.[79]

Agassiz's plan called for "thorough anatomical study of the species found in the caves, with extensive comparison of allied species, found elsewhere—next, an investigation of the embryology of all of them, and when fully prepared by such researches, an attempt to raise embryos, of the species found in the cave" under different amounts of light. While stating his own belief that species originated as we see them, Agassiz declared, like the inspiring teacher he was, that his own views were mere inference, and whoever would settle the question by direct experiment, might be sure to earn the everlasting gratitude of men of science. And here is a great aim for the young American naturalist who would not shrink from the idea of devoting his life to the solution of one great question.[80]

Harvard's other great zoology teacher, Jeffries Wyman, was also interested in cave animals, and in the 1850s he carried out dissections of the eyes, ears, and brain of the blind fish and crayfish.[81] In the *Origin* Darwin attributed the atrophy of the eyes of cave animals to disuse, not to natural selection, because he found it "difficult to imagine that eyes, though useless, could be in any way injurious to animals living in darkness."[82]

Among Agassiz's students, Hyatt and Shaler had already visited the Mammoth Cave in 1859, but for a number of their colleagues the first opportunity came when the American Association for the Advancement of Science met in 1871 in Indianapolis, about 180 miles north of the cave.[83] As guests of the Louisville and Nashville Railroad Company, a number of former students of Agassiz and Wyman made the trip, including Putnam, Packard, and Caleb Cooke.[84] Edward Drinker Cope was with them. Putnam and Packard published reports of their collections there, both explicitly discussing the great relevance of the cave fauna to evolutionary hypotheses.

The blind crayfish, *Cambarus pellucidus,* had first been described briefly in 1844 by Theodor G. Tellkampf, member of the Lyceum of Natural History of New York.[85] Hagen, true to form, provided in his "Monograph" a much fuller description based on thirty-eight specimens.[86] Packard added no new observations; indeed, the line drawing accompanying his article was copied from Roetter's illustration in Hagen's "Monograph." But Packard made much of Tellkampf's comment, which Hagen had passed over, that the eyes are not so rudimentary in the young *C. pellucidus* as in the adult. This, to Packard, pointed "back to ancestors unlike the species now existing," and offered partial proof that the causes producing atrophy of the eyes acted first "on the adults and [was] transmitted to their young, until the production of offspring that become blind becomes a habit."[87]

Taking advantage of the influx of East Coast scientists for the Indianapolis meeting, the state geologist of Indiana asked Cope for a professional report on the animals of the Wyandotte Cave and took him there for a two-day collecting trip.[88] Cope's initial reports called the blind crayfish *Astacus pellucidus* instead of *Cambarus pellucidus*[89] (presumably because he had not yet seen Hagen's "Monograph"). After comparing his specimen to four specimens in the Academy of Natural Sciences in his home base of Philadelphia, Cope decided that his specimens were less spiny and deserved a new species name, *inermis*. Furthermore, the undeveloped eyes of both should be sufficient structural ground, Cope argued, for placing both *pellucidus* and *inermis* in a new genus, which he named *Orconectes,* meaning "swimmer of the underworld." His creation of a new genus made no mention of the many differences Hagen had used in making the generic distinction between *Cambarus* and *Astacus*—such as number of gills, relative length of antennae, male dimorphism, annulus ventralis, and copulatory stylet. Cope simply alluded to Hagen's own description of how aberrant *pellucidus* is:

Dr. Hagen, in his monograph of the American Astacidae, suspects that some will be disposed to separate the *Cambarus pellucidus* as the type of a special genus, but thinks such a course would be the result of erroneous reasoning. Dr. Hagen's views may be the result of the objection which formerly prevailed against distinguishing either species or genera whose characters might be suspected of having been derived from others by modification, or assumed in descent. The prevailing views in favor of evolution will remove this objection.[90]

Cope was alluding to the old taxonomic rule for deciding whether a form should be ranked as a variety or species: if descended from another species, it was by definition merely a variety within that species. Evolution had of course rejected that definition and thus rendered the rule meaningless. But there had been no such rule at the genus level, so it was sophistical of Cope to suggest that evolution had removed it.

Hagen's reaction to Cope's article was immediate and vigorous. The implication that being an avid evolutionist endowed Cope, who had compared his one specimen of *inermis* to four specimens of *pellucidus,* with better taxonomic judgment than a man who had examined over 2,000 specimens of crayfish, had to be exposed as nonsense if classical taxonomic methods were to retain respect. In the very next number of the same journal, Hagen demolished Cope's claims. With respect to the Wyandotte form being a new species, he said,

the description of the single specimen does not give any character by which to separate it from the old species, *C. pellucidus.* I have not seen Prof. Cope's type, and though he states that his specimen is a male, he omits to inform us to which of the two forms of males it belongs, but his description applies perhaps to the second form of the male, the characters of which are always less marked than in the first.[91]

Without examining Cope's specimen, Hagen could not conclusively prove that the Wyandotte form was not a new species, but he could greatly diminish Cope's credibility by hinting that Cope was not even aware of the phenomenon of male dimorphism in crayfish. (Hagen's guess that what Cope had was a second-form male was correct.)[92]

As to blind crayfishes constituting a genus distinct from *Cambarus,* Hagen cited as evidence the case, "well known by those who have studied the cave insects," of beetles blind only in the female, "while the males have well developed eyes." The case was European, not American, and probably news to Cope, but Hagen pressed his advantage mercilessly:

> Would Prof. Cope have the cruelty to separate husband and wife so far as to put them in different genera because one of them is blind and the other not? If the prevailing views in favor of evolution demand such a separation, would it not be more human [humane], and perhaps more courteous to the feminine sex, to wait a little while until the poor males shall be able to follow their more advanced wives? It is rather hard for Nature to follow, or even compete, with the fast driving of the evolutionary disciples, but as she is after all a very good natured old lady I have no doubt she will do her best not to stay too far behind the prevailing views of evolution.[93]

Hagen carefully directed his sarcasm, not against the idea that evolution may have occurred, but against Cope's implication that anyone who did not share the current enthusiasm for it must be old-fashioned, incompetent, or blinded by prejudice.

Cope had to yield to the species judgment of the expert at the Agassiz museum, but he held fast to his opinion on genera. He dropped the name *inermis,* afterward referring to both blind crayfishes as *Orconectes pellucidus.*[94]

The question whether its cave adaptations entitled a crayfish to its own genus did remain a matter of opinion. Just as Louis Agassiz had complained in his "Essay," there was no objective or rational criterion one could refer to; each case was decided by the specialist. If authors differed, there was no court of appeal. Darwin's theory had not altered this state of affairs at all; on the contrary, Darwin had declared explicitly that one of the virtues of his theory was that it could explain why a rational ground for categories had not been found, and never would. For him the category levels (though not the taxa) were truly arbitrary.

Crayfish from the Nickajack Cave

When an even more interesting new form, from the Nickajack Cave of Tennessee, came to Cope in 1881, he called on his entomologist friend Packard to help describe it. Presumably he wanted to avoid another devastating review by Hagen. Their joint article strove to meet the standards of description set by Hagen's "Monograph," including an illustration of one view of the first appendage of the male; this specimen, like the one from the Wyandotte Cave, turned out to be a second-form male. There could be no doubt this time that the species was new, for the copulatory stylet was entirely different from that of *pellucidus*, curved rather than straight at the tip.[95] The name they chose for the new species denoted that taxonomically important character: *hamulatus*, from the Latin for "broad little hook."

The fact that the new blind crayfish and a number of other creatures in Nickajack Cave differed from the inhabitants of caves anywhere else in the Southern United States was, Cope and Packard declared, "a matter of considerable interest from an evolutional point of view," because

> it shows that these cave forms are the descendants of different out-of-door species from those of the caves to the northward. The Nickajack cave may be in a different faunal region from the Mammoth or Wyandotte caves, and thus the blind crawfish has perhaps originated from a different species of Cambarus than that which gave origin to *Orconectes pellucidus*. Thus while the conditions . . . of cave life are much the same throughout the United States, the ancestors of the different cave animals were, in most cases, distinct.[96]

Packard repeated this interpretation in *Popular Science Monthly* in 1890.[97] His use of geographical reasoning is parallel to a point Darwin had made in 1859. Citing the fact that American and European cave animals, though living under almost identical conditions, are more closely allied as judged by normal taxonomic characters) to the aboveground species of their own region than to cave dwellers on the other continent, Darwin had stated, "It would be most difficult to give any rational explanation of the affinities of the blind cave-animals to the other inhabitants of the two continents on the ordinary view of their independent creation."[98]

Yet even though Cope and Packard so clearly imagined an evolutionary history of the Nickajack *hamulatus* quite independent of the history of the Mammoth Cave *pellucidus*, they insisted on classing both together in the genus *Orconectes* (fig. 13). Presumably Cope liked the idea of a new genus for *pellucidus* because he wanted to show that the modifying force of the environment was powerful enough to create generic as well as specific changes. The theme of his article "On the Origin of Genera" was that Darwin had only explained the forces that create species, leaving room for Cope to explain the other forces that create genera. No one could deny, Hagen's earlier sarcasm notwithstanding, that *pellucidus* might "deserve" generic designation. But when Cope added to the genus *Orconectes* another species, *hamulatus*, which he believed to be descended from a different ancestor, he was proposing a relationship between taxonomy and evolution radically unlike the one Darwin understood. Darwin had said that "all true classification is genealogical."[99] Naturalists seeking to uncover the "natural order" had already learned to value homologies and beware of analogies, and Darwin saw his revolution as leaving their methods intact:

On my view of characters being of real importance for classification, only in so far as they reveal descent, we can clearly understand why analogical or adaptive character[s], although of the utmost importance to the welfare of the being, are almost valueless to the systematist. For animals, belonging to two most distinct lines of descent, may readily become adapted to similar conditions, and thus assume a close external resemblance; but such resemblances will not reveal—will rather tend to conceal their blood-relationship to their proper lines of descent.[100]

The blindness of cave crayfish was obviously adaptive, and while important to the animal, it did not give clues to ancestry. No clearer instance of analogical characters could be imagined. But if it surprises us to see a fervent evolutionist like Cope proposing to ignore genealogy in his classifica-

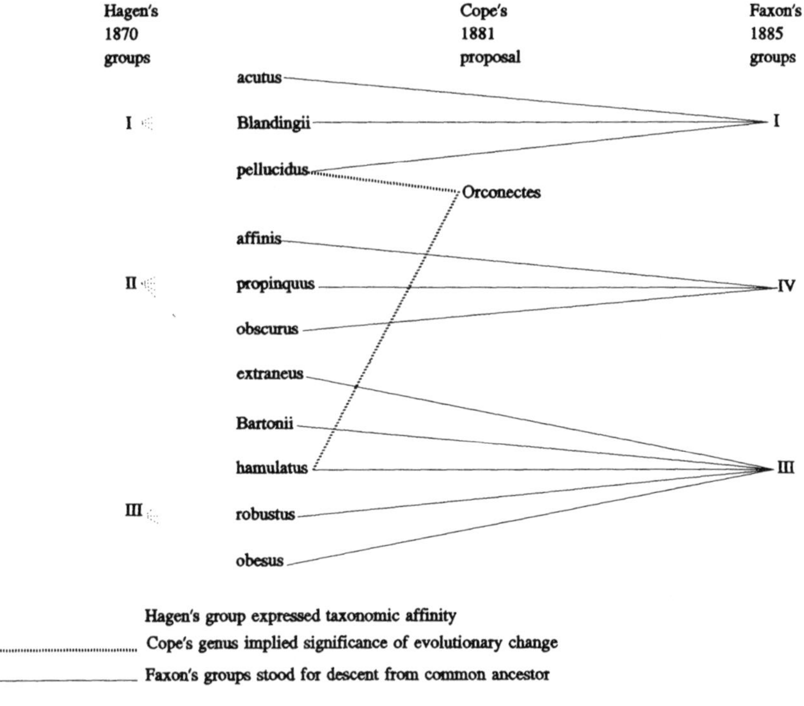

Figure 13. Taxonomic relationships of crayfishes as seen by the traditionalist Hermann Hagen, his successor Walter Faxon, and the enthusiast of evolution Edward Drinker Cope.

tion, we are forgetting that there was and is no law making Darwin the final authority on evolution.

Faxon Inherits a Problem

The duty of passing expert judgment on Cope and Packard's *Orconectes hamulatus* did not fall to Hagen, who was by then fully occupied with the museum's insect collection, but to Walter Faxon. The crustaceans had passed into Faxon's care in the fall of 1873, and although for a number of years he had ignored the crayfishes, in 1883 he was hard at work on the family.* In February Packard sent him a box "containing a male and female Cambarus—or rather *Orconectes* hamulatus, *Cope's types* and Cope's type of *O. inermis,* which I regard as synonymous with *O. pellucidus.*"[101] In his "Revision" published two years later, Faxon declared that *inermis* was within the normal range of variation of *pellucidus.*†

Faxon agreed that the ancestors of *hamulatus* were neither *pellucidus* nor the sighted ancestors of *pellucidus;* indeed, he strengthened Cope and Packard's point by showing that, according to the details of its sexual characters, *hamulatus* should be classed at a far remove from *pellucidus.* Within the main subgroups of *Cambarus,* typified respectively by *Blandingii, affinis,* and *Bartonii,* Hagen had thought *pellucidus* was an aberrant member of the *Blandingii* group, and Faxon called it an aberrant member of the *affinis* group; the position of *hamulatus,* in contrast, was unmistakably close to *Bartonii* (see fig. 13).

The taxonomic location Faxon assigned to *hamulatus,* adjacent to species with similar copulatory stylets rather than adjacent to a species with parallel adaptations to cave life, conformed to Darwin's idea that taxonomy should be genealogical, but, of course, it was also exactly what the evolutionary skeptic Hagen would have done. Faxon did allude favorably to evolution and seemed eager to incorporate it into his taxonomic work,

**Annual Report* for 1874, p. 17. It is not surprising that Faxon gave his curatorial attention to those portions of the museum's Crustacea collection which were less well arranged than the crayfishes. In 1877 Hagen encouraged Uhler to take up work on crayfishes (Hagen to Uhler, 29 September 1877, M.C.Z. Archives), which strongly suggests that Faxon then had no plans to work on them himself.

†Today *Orconectes inermis* is recognized as a species, and is even divided into subspecies, but Faxon's 1885 judgment was not mere conservatism. At that time only one specimen was under consideration, a second-form male, in which the differences from *pellucidus* were much less than the first-form is now seen to be. Experts in this century with better material at hand still disagreed for several years over whether *inermis* was a variety of *pellucidus* or a distinct species.

but he overlooked his nice opportunity to describe the difference between Cope's views and Darwin's principles. Instead of explaining that if one chooses to consider eyes of generic importance, then an additional new genus would be required for *hamulatus,* Faxon simply ignored the proposed genus *Orconectes.*

Lacking such an explanation, the evolutionists remained confused. Packard accepted Faxon's authority as monographer but wondered why *Orconectes* should not be recognized as a subgenus to contain the cave crayfishes:

> I do not see how we can in our system put nature in to a straight-jacket, and regard as simply of specific value the loss of eyes. . . . It is a convenience not to lump such forms with a mass of other species, where they will lose their individuality, but rather to emphasize them by giving them distinct generic or at least subgeneric names.[102]

Today the group that was *Cambarus* has grown to hundreds of species, so the major subgroups have been raised to the status of genera, but still no extra taxonomic weight is given to the cave adaptations. Strict application of the rule of priority, however, has revived the name *Orconectes* for the genus holding *inermis* and *pellucidus,* even though most of its members are not denizens of the underworld at all, and even though Cope's definition of the genus is still regarded as foolish.

Faxon did allow himself to speculate on the evolution of the blind crayfishes. Perhaps, he suggested, the reason *pellucidus* is so hard to locate among the subdivisions of *Cambarus* is that it has stayed closer to the ancestral type:

> The simple form of the male appendages, and the combination of characters belonging to different groups, seen in *C. pellucidus,* indicate, to my mind, that it is a very ancient form, which has been preserved in the seclusion of the cave, while its nearest kin succumbed in the sharper struggle incident to life outside or were replaced by modified descendants.[103]

This speculation, though still reasonable by today's standards, may have helped tempt Faxon into uncritical acceptance of a "fact" of which he should have been suspicious. In 1879 it was reported that the limestone caverns of Carniola, Austria (now part of Yugoslavia), contained a blind crayfish very similar to *Cambarus pellucidus.* In fact, there is no evidence of a blind crayfish ever having lived in those caves. Had any other species of *Cambarus* supposedly been collected from the Old World, it would certainly have met with some skepticism, because otherwise all European crayfishes are *Astacus,* while *Cambarus* is limited to one continuous geographical realm in the Western Hemisphere. One of the duties of a taxonomist undertaking a revision is to pass judgment on supposedly new species named since the previous expert monograph: this report from the caves of Carniola would have withered beneath any critical attention. The report was based upon one dried specimen which was not even found in the caves but found labeled in a collection; the report's author admitted that his own search in those caves had yielded no crayfish at all.[104] But Faxon, instead of suspecting a mix-up, was eager. "It is to be lamented," he wrote, "that a fuller account of this animal has not been published, on account of its important bearings on the subject of the geographical distribution of these animals."[105] A careful reader of Darwin should have regarded egregious anomalies of distribution with special caution, but Faxon welcomed a report supporting his own historical speculation. Not until 1914 did Faxon relegate the Carniola *Cambarus* to his list of doubtful species.[106]

Taxonomy's Ultimate Goal and Conservative Practice

The most remarkable thing about Walter Faxon's "Revision of the Astacidae," published fifteen years after Hagen's "Monograph" and twenty-six years after *The Origin of Species,* was how unremarkable it was. He did make use of material from the United States National Museum, whose crayfish collection, like the M.C.Z.'s, had grown. Faxon, therefore, could report on a score of new species, including a couple that did not fit the definitions of any of Hagen's three subdivisions of the genus *Cambarus*. Besides creating a new group for these species, Faxon split up Hagen's third group, fulfilling Hagen's prediction that that group "perhaps unites two groups of co-ordinate value."[107] Of course, he also was able to add greatly to the number of localities at which particular species had been collected. But he presented this information exactly as Hagen had done: by giving faunal lists state by state, not by plotting collection records on a map. The investigation of "the laws of the geographical distribution of organized beings" which Louis Agassiz had seemed to promise in his *Annual Reports* got only raw material from Faxon. His work was mechanical and uninspired. Faxon remained in charge of the M.C.Z.'s Crustacea (and for many years the Mollusca as well) until his death in 1920, producing two additional reports on new crayfish species.

The idea that systematics need not change course was endorsed by Darwin's lively champion T. H. Huxley, apparently motivated by the same "Baconian" model Agassiz had taken for granted. Before 1859 Huxley had argued that comparative zoology was an objective science of observed form, independent of theories of the origin of species, and he held to this position, more or less, ever after.[108] His 1879 textbook, *An Introduction to the Study of Zoology,* used the crayfish as its model animal. There he described the classification and geographical distribution of *Cambarus, Astacus,* and other genera of crayfish, emphasizing that taxonomic families, genera, and species are abstractions, having no real existence in nature; they are merely our summing up of all the common characters of their members. This was not pure nominalism, for he also asserted in a technical article that the "ultimate goal . . . is a genetic classification,—a classification, that is, which shall express the manner in which living beings have been evolved one from the other." I suspect that his failure to state this goal in his textbook was deliberate. Taxonomy would be in a better position to serve as evidence for evolution if it was logically independent of evolution. Besides, having been trained himself in the era of MacLeay, Forbes, and Owen, Huxley always maintained that relationships were there to be discovered just as if species had sprung into existence in the Garden of Eden.[109]

Huxley saved the question of how all these kinds of crayfish had come to exist until the very end of his textbook. The only two explanations imaginable, he said, are some form of creation or of transformation. As to the first,

> apart from the philosophical worthlessness of the hypothesis of creation, it would be a waste of time to discuss a view which no one upholds. And, unless I am greatly mistaken, at the present day, no one possessed of knowledge sufficient to give his opinion importance is prepared to maintain that the ancestors of the various species of crayfish were fabricated out of inorganic matter, or brought from nothingness into being, by a creative fiat . . . the hypothesis of transformism remains in possession of the field.[110]

So does might make right in scientific discourse, too.

Huxley then proceeded to speculate on what sequence of migration, geophysical change, and transformation could explain the present world distribution of crayfishes. In order to do this, he had to do a considerable amount of original research, because although his method was that of classic comparative morphology and biogeography—simply the "comparative zoology" Agassiz had advocated—no such review of the crayfishes had been done. Using specimens in the British Museum, Huxley closely examined the gill structure of representatives of eight different genera, expressing the result in tabular "branchial formulae."[111] Their distribution, especially the disjunction of *Cambarus* standing between the *Astacus* of the Western United States and Europe, as well as the total absence of crayfish from South America, Africa, and India, could not be easily explained, but nothing in those facts, he claimed, was inconsistent with "the hypothesis that they have been gradually evolved." His textbook concluded that

> it is well to reflect that the only alternative supposition is, that these numerous successive and coexistent forms of insignificant animals, the differences of which require careful study for their discrimination, have been separately and independently fabricated, and put into the localities in which we find them. By whatever verbal fog the question at issue may be hidden, this is the real nature of the dilemma presented to us not only by the crayfish, but by every animal and by every plant; from man to the humblest animalcule; from the spreading beech and towering pine to the *Micrococci* which lie at the limit of microscopic visibility.[112]

This wonderfully ringing polemic was the closest Huxley came to suggesting that the careful study of the taxonomist who discriminates between similar forms could be a means of investigating evolution. No wonder that Faxon, except for imagining that caves may protect their inhabitants from "the sharper struggle incident to life outside," had no idea how to use museum crayfish collections for any theoretical inquiry.

But Faxon, like Hagen before him, had not been motivated by any theoretical question. His work was a contribution to the Linnaean dream of a world catalogue, not to the biological ideas of the 1880s nor even to Agassiz's original vision of comparative zoology. Nor were new specimens pursued with the ambition that had driven Agassiz in the 1850s. Even the new blind species Faxon described in 1889 was acquired at the instigation not of Faxon but of Samuel Garman, the assistant in charge of the M.C.Z. fish collection, who had asked a woman in Missouri to collect in a cave.[113]

There is evidence to suggest that Faxon undertook his crayfish work exactly as Hagen had done, as a museum employee doing his job. Faxon recorded in his *Annual Report* for 1882–83: "Professor Baird being desirous of sending a collection of United States Crayfishes to the London Fisheries Exposition, I identified for the National Museum a set made up from the material in both museums, containing nearly every species described from this country."[114] Spencer Fullerton Baird, who by now was not only director of the USNM but was also Commissioner of Fish and Fisheries, was in charge of the official United States contribution to the Great International Fisheries Exhibition in London in 1883. When Baird had asked the British organizers for suggestions, he had received in reply a printed list of sixty-one official divisions, with some categories ticked off and some annotated. Next to "Division 50. Crustacea of all kinds," was noted, "I should be disposed to limit this to the Freshwater Crustacea (crayfishes!) to the economic marine forms & to crustacean fish parasites."

These suggestions to Baird were in the handwriting of the chairman of the exhibition's executive committee, none other than Thomas Henry Huxley.[115]

If it is amusing to find the high priest of Darwinism thus commanding, even indirectly, activities in the Agassiz museum, what is actually more ironic is to realize the extent to which Louis Agassiz's own ideas had left no stamp there. This was certainly not because the crayfishes had already yielded up their secrets to the masterful Hermann Hagen. Arnold Ortmann's virtuoso "Crawfishes of the State of Pennsylvania" of 1906, a foreshadowing of the "new systematics" of a later era, is proof enough of what sort of riches remained. Instead, the explanation involves the momentum of taxonomic assumptions and practice, which the objects of study obligingly reinforced.

Agassiz's hope to raise the status of taxonomy by infusing it with theoretical issues had failed. Instead, Darwin's prophecy, "Systematists will be able to pursue their labours as at present," was dismally fulfilled. The widely held faith in Baconian induction allowed the majority of museum workers to do exactly as Faxon did and hold their course.

5

"The Many Plans Started by My Father"

A principle of multiple use had always been central to the founding ideas and practical operations of the M.C.Z. Not only the students, the ones who doubled as curatorial staff, but other staff members, the specimens, and even the building itself, all had to serve a confusing array of overlapping functions. Sometimes multiple use was talked about explicitly, but often it manifested itself unbidden. If the overlapping functions were complementary, then the museum would operate with special effectiveness because of them. Unfortunately, this was not always the case. Louis Agassiz had already begun to discover that several of these concurrent uses were inconsistent or even in conflict. During the fourteen years of his directorship, Agassiz occasionally struggled with various forms of the multiple-use issue, but much remained for his successors to sort out. What was not clear was whether the activities central to Agassiz's dream of the M.C.Z.—a place where students would learn systematics by working with specimens, where the research of professionals would advance the science of zoology, and where the arrangement of specimens would enlighten the general public—would survive intact, or whether its special qualities would be lost.

Agassiz's plans were grandiose in scope. He wanted the M.C.Z. to rival the national museums of Britain and France, an ambition whose very audacity was doubtless part of its appeal. Agassiz made sure that enough land was allotted to the M.C.Z. to allow for enormous future growth (see figs. 20, 21, 23, 25, below). He began with only enough money for a modest building, yet had the architects draw up plans for one four times as large and had the builders leave rough the west wall, in anticipation of an addition. That blank wall, and the distance the little new M.C.Z. stood from the rest of Harvard in the 1860s, were visible expressions of Agassiz's hope for a glorious future.[1] Agassiz did not hesitate to appeal to the legislature of the Commonwealth of Massachusetts with his grand plans, encouraged by their policy that income from the sale of newly created land from the filling of the Back Bay was to be used to promote education. Agassiz legitimized his request by promising that he would use his museum to give special lectures to schoolteachers and future schoolteachers.

Specimens for Public View

Agassiz was demanding several functions from the specimens with his decision that the museum should be for the public rather than just a teaching institution. In every publicly supported museum, and even in most of the great private collections, it was taken for granted that any respectable visitor was entitled to inspect all its holdings. The job of a curator or keeper was to have all his specimens laid out for examination by the ordinary visitor; in the early nineteenth century, there was no distinction between specimens to be used by researchers or students and those open to public view.

In the 1850s, Agassiz's barrels and boxes and jars of preserved animals were really just a private collection, used for his own teaching and research; occasionally a visitor might be shown some of his treasures, but the specimens were not arranged for public view. Everyone seems to have assumed that with the founding of the Museum of Comparative Zoology, this deplorable situation would change, and that as soon as it could be managed all of the specimens would be on display for the edification of the public. For a year after the transfer of the specimens in November 1859, most of the assistants' work was devoted to arranging displays of the material in preparation for the public opening. During the first dozen years of the M.C.Z.'s existence, the small proportion of the collection on exhibit was cause for apologetic explanation; there was not enough space, or display cases, or helpers. The fact that so many specimens were still "packed away in boxes and casks," "crowded into the cellars and attic," was used to argue to the state trustees that Agassiz needed more money. Though this institution was supposed to provide research material for advanced students and to enable Agassiz and visiting scientists to push back the frontiers of knowledge, that was no reason to change the standard by which a museum's richness was judged, namely, the number of mounted specimens exposed to view. Before long, however, circumstances would conspire against this traditional attitude.

The old standard of proper museum arrangement was beginning to be questioned by the middle of the nineteenth century. Two opposing forces were working against it: the ever-increasing numbers of specimens, and the democratic demand for comprehensible exhibits. Government surveyors and private collectors measured their success by the numbers of species they could send home, and at the same time professional naturalists wanting to study geographic distribution and variation required many specimens and not just a few representatives for each species. Museums were thus overwhelmed by masses of material, and attempting to display them resulted in a less edifying experience for the ordinary visitor. At the same time, there was a growing popular interest in natural history and an increasing expectation that state-supported institutions should serve the public at large. John Edward Gray, keeper of zoology at the British Museum, had already decided before 1858 "that it would be desirable to form a *study-series* as distinct from the *exhibition-series*."[2] The goal of advancing knowledge ought to be clearly distinguished, he argued, from the goal of diffusing knowledge. Explaining his ideas to the British Association for the Advancement of Science, meeting in Bath in 1864, Gray said,

> What the largest class of visitors, the general public, want, is a collection of the more interesting objects so arranged as to afford the greatest possible amount of information in a moderate space, and to be obtained, as it were, at a glance. On the other hand, the scientific student requires to have under his eyes and in his hands the most complete collection of specimens that can be brought together, and in such a condition as to admit of the most minute examination of their differences, whether of age, or sex, or state. . . . In the futile attempt to combine these two purposes in one consecutive arrangement, the modern museum entirely fails in both particulars.[3]

This address, published in the widely read *Annals and Magazine of Natural History*, stimulated much discussion.[4] In the twentieth century, we are so accustomed to research museums that display a moderate number of specially selected objects while storing a larger number behind the scenes, that it is difficult for us to imagine that such a rational policy was not always obvious. Presumably what had long been obvious to any serious zoologist was that for research one needs a collection that is stored so as to be safe from light and dust, arranged and catalogued so that parts can be retrieved for close study. Yet none dared articulate this before Gray, since hiding specimens away conflicted with the accepted definition of a public museum.* Gray's idea met "considerable opposition by persons who referred to the great Continental Museums, in which, at that time, every specimen was on exhibition."[5]

Gray's superior at the British Museum, Richard Owen, adamantly op-

*My "none dared" is hyperbole. I hope my claim is taken as a challenge, for the history of museum theory and practice is greatly in need of fresh investigation.

posed the idea of a study series distinct from the exhibition series. Owen insisted that it would be absurd to choose a "typical" elephant or rhinoceros, and that the national museum of a great empire must show God's creative energy to the full, including every variety of elephant, seven species of rhinoceros, and all the whales. As a matter of fact, Owen was using Agassiz's rhetoric and citing Agassiz's success to shame the British into equaling the grand plans of the Commonwealth of Massachusetts.[6] Of course, Owen recognized the practical considerations preventing every species from being open to view, but certainly Gray's idea that only a selected set of specimens would be on exhibit was not official policy in London in the 1860s. (One should note, however, that Owen's views may have been based more on financial need than personal opinion, since he was using calculations of the space needed to display the entire collection as his main argument to convince Parliament to enlarge the British Museum.)

The idea of having separate collections for study and for the public was faring no better in Washington than it was in London. In the United States National Museum, circumstances had always forced Baird to put only a small portion of its material on display, making it, he explained, not so much a museum as a warehouse, saving up "the rough materials from which science is to be evolved."[7] Twenty years later, there was much discussion among museum administrators in Europe and North America of the "new museum idea" of "dual arrangement."[8] This was nothing more than Gray's idea finally becoming orthodoxy. Alexander Agassiz made this the explicit policy for the M.C.Z. in 1878, and the USNM followed in the same course in 1881.[9] In 1887 Alfred Russel Wallace, describing his visit to the United States, hailed the M.C.Z. as an example of what the best museum of the future would be like. He quoted Alexander Agassiz's exposition of the dual arrangement idea and praised his implementation of it.[10] Wallace's report became useful ammunition in the hands of William H. Flower, who had begun to implement Gray's idea in the British Museum from the time of his appointment in 1884.

Several museum administrators of this later period, including Alexander Agassiz, assigned credit for the dual arrangement idea to Louis Agassiz; later writers on museums, possibly allured by the image of a founding father, have tended to enlarge upon his originality and influence.[11] Strictly speaking, Gray certainly had priority, having talked along these lines for several years and then having published a clear and full discussion of the dual collection idea in 1864. Louis Agassiz had met Gray, and he cited Gray's address in his *Annual Report* for 1864.[12] Agassiz was then concerned to record a claim to his own priority, "since I see that the directors of other Museums begin to feel the imperfections of the present arrangement of their collections, and are proposing as new, schemes identical with those which for many years have been in active operation with us."[13] This ought not to be understood, however, as a reference to the dual arrangement idea, that is, the separation of public from research collections, for the context shows clearly that Agassiz was referring to Gray's other suggestion—that the arrangement of typical specimens could exemplify each class, order, family, and genus, with special series to show the variations of age and sex; Agassiz went on to say that Gray's suggestion "coincides, though on a limited scale, with what we have been doing upon a much more extended plan for several years past." The plan which Agassiz had been announcing for the preceding years had to do with the intelligible arrangement of related specimens and not with reserving a large part of the collection away from exhibition entirely, solely for the use of specialists.[14] Experience would force Louis Agassiz to that situation by 1873, but the intent of the M.C.Z. in the 1860s, as in similar institutions of its day, was to put everything on exhibition. For example, in the *Annual Report* for 1863, Packard submitted for his department: "Good progress has been made during the past year in the final arrangement of the Insects. Mr. Scudder has placed on exhibition in the cases *all the Orthoptera in the collection,* having completed their arrangement, begun last year."[15] (My emphasis)

The fact that the M.C.Z. always differed in practice from its European models by having the bulk of its material not on exhibit, was not so much the result of any innovative policy as an unintended consequence of the director's greediness to acquire more specimens than could be dealt with properly. The opposing concepts and motives of Gray and Owen were intertwined in Agassiz's rhetoric and action. He could see the educational value of having only selected types on exhibit, but he knew the fund-raising advantage of complaining that he had treasures hidden away for want of exhibition space. Thus throughout his directorship, from 1859 through 1873, Louis Agassiz projected several conflicting messages about his museum.

Agassiz did have an idea for the arrangement of specimens which would make his museum distinctive and, he argued, more educational than other major museums. Visiting Europe in 1859, he had wandered through the British Museum, undoubtedly looking for tips as to what to adopt and avoid in his new establishment. Contemplating those well-stocked halls with the ideas of his own "Essay on Classification" fresh in his mind, Agassiz decided that the very magnificence of a collection could interfere with the ordinary person's ability to perceive the profound principles that underlie classification. Such a crowd of mounted birds, for example, each differing only slightly from the one next to it, "tends rather to make one dizzy than to instruct one in basic principles."[16] Thus began, he said, his plan for the educational arrangement of representative specimens. Within one room he would assemble a few typical birds, not every bird in his collection, along with some typical reptiles, mammals, and fish, so that the major outline of the Vertebrata could be seen. Agassiz planned four "synoptic" rooms, one for each of Cuvier's *embranchements* (Vertebrata, Articulata, Mollusca, Radiata). Other rooms would house selected specimens arranged according to their homeland, or alongside their fossil relations. In the "systematic" rooms, the bulk of the collection would be arranged in accordance with the natural system of classification. And throughout, scientific principle, not aesthetic considerations, would rule.

This would indeed make a display radically different from those in any existing museum, though it is not clear that Agassiz was utterly original in thinking it up. He had spent a day visiting Richard Owen in the summer of 1859,[17] and in that same year Owen suggested to the trustees of the British Museum an "index museum," that is, a space within the British Museum "devoted to specimens selected to show the type-characters of the principal groups of organised beings."* In the face of uncooperative subordinates and an already-arranged collection, Owen was unable to implement his plan for an "index" area within his museum. Nor did Agassiz manage to create in reality the exhibitions he could describe so persuasively. Implemented or not, the "index" or "synoptic" displays should not be confused with the dual arrangement idea. In all Agassiz's speeches, annual reports, and private directions to the museum assistants in the first decade of the M.C.Z., there is no hint that the synoptic and faunal rooms would be open to the public while the systematic collection would not.

Throughout the 1860s, the disparity between the growing bulk of undisplayed material, much of it unattractive, and the ideals of orderly or even inspirational exhibition, posed a problem for Agassiz when he annually sat down to explain his enterprise to his supporters. After his return from Brazil in 1866, with countless new barrels of Amazon fishes to multiply his embarrassment, he tried to get his supporters to understand that his museum was to serve researchers, not just the public. He declared in 1868, "There are really two very distinct objects to be kept in view in organising a great Museum—the public exhibitions and the scientific use of the specimens."[18] But these were still multiple purposes, not separate purposes, for it was at this same time that the man in charge of Mollusca in the M.C.Z.,

*It was to accommodate Owen's idea that the design for the new British Museum building in South Kensington featured side bays around the central hall, although the space was never thus used (Günther, *A Century of Zoology at the British Museum*, p. 352).

John Gould Anthony, was securely gluing each clamshell and snail to a glass base or cardboard tablet, and Agassiz proudly reported that 67,749 specimens had been mounted so far. There is no inkling of the notion of a permanently separate study collection in Agassiz's complaint that many of those tablets "for want of room, are hidden in drawers and cases."[19]

While Agassiz did give a few cryptic warnings that "the object of our institution is not that of popular exhibition,"[20] he also continued to plead for building funds on the grounds that the public had a right to see for themselves that the collections were as vast as he said they were.[21] Ten years after the museum's founding, he was still suggesting in his *Annual Reports* that the day would come when, except for duplicates awaiting exchange with other museums or distribution to public schools, the collections in their entirety would be on display. With adequate funding, he promised,

> we shall be able to put our immense collection in complete order, and to enlarge the building sufficiently to exhibit all our specimens in their true relations. I hope that in three years any intelligent observer will be able to say that a mere walk through our Museum teaches him something of the geographical distribution of animals, of their history in past ages, of the laws controlling their growth as they now exist, and of their mutual affinities; in short, that the whole will be so combined as to illustrate all that science has thus far deciphered of the plan of creation.[22]

When that day came, the coherent arrangement would speak for itself. Meanwhile, the actual state of the museum fell far short of this; visitors could see only the four rooms representing the four branches, and those rooms were extremely crowded.

Some workers in the M.C.Z., facing daily reality and not burdened with the job of explaining policy to legislators, did begin to think in terms of the kind of dual arrangement Gray had advocated. Both segments of a collection could be seen by the public, but one would be on display in the exhibition halls and the other accessible only by special request. Practical circumstances, not an image of the ideal museum, prompted this shift. Alexander Agassiz, left in charge while his father was in Brazil in 1865, reported: "The collection of insects has been packed away, as, owing to their exposure to light, they were rapidly fading, and hereafter only a small typical collection of specimens which can easily be replaced will be left on exhibition."[23] A few years later Hermann Hagen, progressing in his monumental task of arranging the material entrusted to him into several series, endorsed this policy.

> Since most of these collections require to be kept as much excluded from light and air as possible, there will be a collection for public exhibition, containing species remarkable for their beauty, or as being either useful or obnoxious; besides, a col-

lection representing types of families, and genera for entomological students; the other collections will always be accessible on special applications, or for purposes of study.[24]

That same year, 1868, the curators of the new Peabody Academy of Science in Salem, Massachusetts, Agassiz's mutinous students Putnam, Packard, Hyatt, and Morse, announced their intention to exhibit only a part of the academy's collection, storing the other specimens "in drawers or other repositories, where they should be thoroughly arranged and available for study, but not for general exhibition."[25] They did not credit either Agassiz or Gray as their inspiration, but they knew very well the realities of the M.C.Z. and undoubtedly would have known of Gray's article as well.

Finally, about 1870, Agassiz himself began to come around to the idea of separate collections for study and for exhibition. His gradual change of mind can be seen in a series of letters he wrote to a dealer in natural history objects in Rochester, New York, his former student Henry Augustus Ward, whom Agassiz was commissioning to procure and prepare skeletons of different ages and breeds of cattle. Promising future orders of all breeds of horses, pigs, sheep, and dogs, Agassiz explained his scheme to Ward:

I propose to put up a perfect skeleton of a male and a female, a half grown and a young *of each breed* of all the different kinds of domesticated animals raised not only in the country, but in every part of the World, as a monument to the progress of civilisation in that direction.[26]

It was a wonderful idea, characteristically ambitious to the point of absurdity, having both practical appeal, from its relevance to agriculture, and theoretical importance, since Darwin begins the *Origin* by discussing domestic variation. Before long, however, it became evident that Agassiz had not thought very clearly about the multiple functions these specimens were expected to serve. He complained to Ward about the attractively mounted specimens that were being sent, because the varnish hid some details, and the varying postures would interfere with rigorous comparison:

You may say that skeletons arranged in rows all looking the same way, with legs turned in the same position, must look very monotonous. So they will; but that is the only position in which they are really serviceable for study. It is the only kind of mounted skeletons I care to have in the Museum of C.Z.[27]

But a few months later he realized that specimens suitable for exhibition were not as useful for close examination and comparison as were disarticulated bones: "The fact is the more I think of it, the less do I care to have mounted skeletons in the museum for study. They are only useless show things."[28]

Agassiz's next few *Annual Reports* contain hints of his shift in policy. Once again he had added to his great stock of unsorted material the product of a voyage around the hemisphere to California aboard the *Hassler* in 1872. The specimens acquired on that expedition, at a cost of $17,000, would require several thousand more dollars for sorting and labeling. In his pleas for more funds, Agassiz offered a new definition of a great museum. Emphasis should be placed on the scientific potential of its collection and not on its public displays. He promised that

> if I am allowed to make these collections available by the proper means of preservations, we have now the greatest working Museum in the world; the one, that is, which supplies the most extensive and varied material for special and comprehensive zoölogical research. I do not exclude the oldest and largest Museums of Europe from this statement, believing as I do that the time has passed when the value of a Museum is to be measured by the number of its stuffed birds and empty shells.[29]

Seashells and stuffed birds were what the general public found most attractive, but the museum's benefactors must learn how to take pride in a "working" museum where specialists did their research.

> The only question now is whether a Museum of first order is needed in Massachusetts, or not. If the legislature will favor us with a visit I would gladly submit our institution to the most critical examination of its organization. I think I can satisfy any competent visitor, that by her liberal support of the Museum, our State has earned the right to say that among civilized communities there is not a purely scientific establishment of higher character or distinguished by more active, unremitting, original research in various departments of knowledge.[30]

The general public might not be impressed by pickled fishes, but the legislators ought to recognize that civilization's highest expression was the original research of experts. In other words, it was by research behind the scenes, based upon hidden collections whose true worth required a competent judge, that the greatness of this museum must now be measured.

Louis Agassiz had attracted extraordinary financing for his dream, on the strength of a wide range of promises, which the esoteric specimens and highly specialized scientific researches based upon them did little to fulfill. In various ways Agassiz attempted to assure the general public, benefactors, and the Commonwealth of Massachusetts that the M.C.Z. was an institution that did serve their needs, even if its meager public exhibits fell short of their idea of what they had been promised. And so in the early 1870s, when even the few rooms open to the public were used partly for storage of cans and barrels, Agassiz developed a new use for specimens that would help satisfy the legislators' expectation that their expenditures were buying an educational benefit. The assistants were instructed to prepare small collections of duplicate specimens, well labeled, to send to the Massachusetts normal schools for the use of students there. Anthony reported that in 1871

> I have commenced the arrangement with the normal school in West Newton Street, Boston, under the management of Mr. Hunt, and, in order to make this more useful to that institution, have required him to detail a few of his best scholars to come at stated times for many weeks in order to learn the proper method of handling and mounting shells, and this work has been done by them very neatly. Upwards of 200 generic types illustrated by a large number of species, have been thus prepared and mounted ready for delivery whenever the cases are ready to receive them.[31]

At the Smithsonian in Washington, Baird was applying the same remedy to the same embarrassment. Yet all such efforts, of course, drained manpower away from the basic needs for arrangement and preservation of the collections.

If Agassiz had been immortal, how much longer could he have gotten away with this policy? The United States in the nineteenth century was notoriously neglectful of pure science while enthusiastic about technology. The philanthropist Abbott Lawrence had intended to support practical science and the mechanic arts at Harvard, until he was dazzled by Agassiz. Nathaniel Thayer was willing to support the study of entomology, but only on the condition that the "curator would make the collection one especially devoted to Insects Injurious to Vegetation." (He also underwrote the cost of the Brazil expedition, but this was likely out of fondness for Agassiz personally rather than from interest in the scientific goals of the trip.) Baird, who had neither mandate nor money for exhibits, was continually justifying his collections to skeptical congressmen; the title "Commissioner of Fish and Fisheries" was very useful to him even though it carried no salary. The American Museum of Natural History kept the interest of donors by careful attention to techniques of attractive display; it pioneered new fossil mounts, lifelike taxidermy, and habitat groups of animals, and the trustees clearly counted such activities more important than purely scientific research and publication.[32]

Anyone contemplating the M.C.Z.'s future in 1870 would have been hard pressed to guess what would be left of Agassiz's dream a generation after his death. In spite of the yearly attempt to put the situation in its best light in the *Annual Reports*, all those closest to the museum felt grave concerns about its achievements. Alexander Agassiz, visiting European museums, wrote to Lyman, "I must say frankly that for means we have had compared to European Institutions we have not much to show."[33] Perhaps not even in comparison to an institution closer to home, across the Charles River, half an hour's walk or a few minute's ride on the horsecar. After the opening of the new building of the Boston Society of Natural History in 1864, its collections were displayed in a building far more elegant than Agassiz's (it later became the home of Bonwit Teller's department store), and, until the 1871 addition to the M.C.Z., newer and larger too.[34] Benefiting from the leadership of the much-loved anatomist Jeffries Wyman and the energetic curatorship of Alpheus Hyatt, the Boston museum was attracting benefactors, possibly including some who were weary of Agassiz's bombast; it had public exhibits and collections in storage for the use of specialists. In 1871, it also had a coherent plan for the future logical arrangement of the public exhibits (fig. 14).[35]

Agassiz might claim some moral priority, since the Boston plan was the brainchild of his student Alpheus Hyatt, but there was not much visible superiority in Cambridge in 1870. In that year the BSNH museum received an average of 400 visitors a day, on the 104 days it was open to the public; it is not clear from Agassiz's *Annual Report*s whether his museum was in a state to receive visitors at all; the only mentions of exhibits are promises for the future. For example, in 1873 Agassiz reports,

> The exhibition rooms have been more than doubled, owing to the addition of one story to our building; unfortunately, they cannot yet be thrown open to the public, our means being insufficient for the present to provide the necessary wall cases and other appliances to protect the specimens from injury by ignorant or careless visitors.[36]

We may be tempted to suggest it would have been an efficient specialization for the museum at Harvard to concentrate on maintaining research collections and teaching qualified students, leaving the Boston Society to concentrate on public exhibits, but neither institution could afford to limit itself.

The BSNH museum, like Agassiz's, acted as the biology department of a college (the Massachusetts Institute of Technology, which adjoined it), while Agassiz had promised the taxpayers of Massachusetts displays to uplift even the casual visitor. Hyatt wanted the Boston museum to specialize in the fauna of New England, but Louis Agassiz was in the early '70s regarding his museum's weakness in New England insects as a serious lack that must be remedied. Both museums boasted the skeleton of a large whale hanging from the ceiling. The survival at Harvard of a museum, combining the many different activities Agassiz had at first intended, was far from guaranteed.

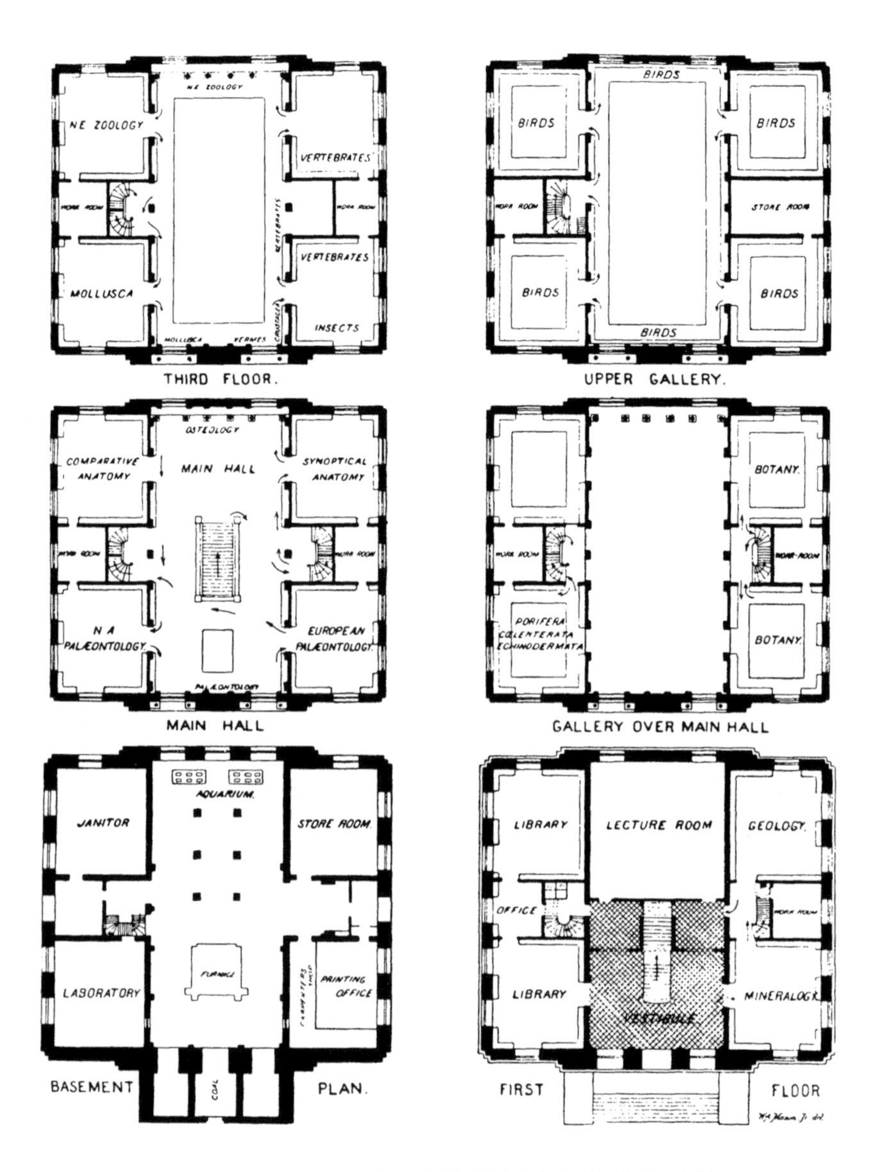

Figure 14. Plan of the museum of the Boston Society of Natural History, from Bouvé, "Historical Sketch," p. 144. Agassiz's colleague Jeffries Wyman and several of his former students, including Samuel Scudder and Alpheus Hyatt, helped the BSNH to flourish. The building represented here was completed in 1864; this is Hyatt's 1871 proposal for future arrangement of exhibits rather than an indication of the actual layout.

The Problem of a Successor

The future of the M.C.Z. was often discussed within Agassiz's close circle. One of its central problems was the multiple roles knotted together in the person of Louis Agassiz: teacher, administrator, keeper of collections, researcher, and fund-raiser. It was a strain just to watch him attempt this superhuman balancing act. As early as 1862 his son Alexander was already complaining to Theodore Lyman:

> He is killing himself by inches with the Museum, his book [the unfinished *Contributions*], the lectures he has to give [he was Lowell Lecturer again] to get money to keep the machine going . . . this has become too serious a question, his health cannot stand the amount of work. The Labours at the Museum ought to be divided so that Father should not have anything to do with the Direction of the Museum. . . . All I say only serves to irritate father and to make me fume and boil with rage.[37]

Lyman wrote his friend in reply,

> I agree about the Museum woes. But what is bred in the bone won't go out of the meat, and what a man is at 30 he will be at 50 [Louis Agassiz was 54]. Sometimes I think, why don't Professor answer letters? why don't he save money? why don't he carry out one thing before he starts another? why don't he do this, that, & the other? But then I turn round and say, how few of the best men accomplish 1/10 as much?!!—You must serve him like a trotting horse—hang on to the reins, and, when he runs away (which he will, periodically) "saw his mouth" till you bring him down.[38]

Think of a Standardbred charging past you on the racetrack! Not a very practical means of transportation, but a powerful source of pride, as a world-class museum would be to intellectual Bostonians.

What was perfectly clear to everyone involved, including Agassiz himself, was that no other single person could replace him. Besides holding the Lawrence professorship, Louis Agassiz was designated both curator and director of the museum. It was always assumed that after him these three offices would be held by at least two men. Back at the founding of the museum, when he had recommended to Harvard's President Thomas Hill that Henry James Clark would make a suitable curator if he himself were lost at sea, he had suggested Lyman as director.[39] The titles "curator" and "director" had been created when Agassiz first persuaded the Massachusetts legislature to contribute to his project, promising to open his collections to the public and to provide lectures to teachers and other members of the public. Agassiz had been made responsible to the state-appointed trustees as well as to the museum's Faculty, the five-member governing board appointed by Harvard. The curator, specified to be whoever was professor of zoology in the Lawrence Scientific School, was the scientific head, and the director was responsible for making the museum serve the educational needs of the general public. On closer examination, the two posts overlapped inextricably; they had to do not with separable function or authority but with the obligation to report to different funding agencies. The museum was like a ship trying to serve as a private merchant carrier, a naval vessel and a yacht simultaneously, a situation requiring a captain skilled at writing tactfully worded dispatches.

The museum's statutes stated that the director was responsible for public access to specimens, while the curator would acquire specimens and lecture to teachers and would-be teachers.[40] Alexander Agassiz wrote to Lyman,

> It is evident from the experience I have had during this year [managing the museum during Louis Agassiz's Brazil trip] that it will be hereafter impossible for one man to attend to the care of the collection, the money matters and the teaching and scientific direction; the sooner the attributes of the Director are properly fixed the better, and the Director ought to have the whole charge of the collections, the work of the Assistants and money department, while the Curator should say what is needed, direct the Assistants and teach the Students. The Curator ought to be powerless in the executive except as far as recommending what should be done, while the Director should do what he can with the means at his command. That is the way I look at it. As things stand now the Curator has the spending account.[41]

This peculiar suggestion for a separation between the authority over budget and collection on the one hand, and over the students and the assistants who worked on the collections on the other hand, undoubtedly arose from Alexander's perception of how his father had kept the museum on the brink of financial ruin. Of course, a scientifically qualified curator could *say* what purchase or collecting of specimens would be desirable, but he should be powerless to tie up the future income of the museum by committing it to purchases beyond its means. The moment the senior Agassiz departed for Brazil, leaving his son to mind the store, Alexander fired most of the museum staff and warned away others whom his father had invited to Cambridge, in an attempt to reduce the annual expenses down to the level of annual income, a management principle unknown to Louis Agassiz.

The question of who could fill Louis Agassiz's various roles was at times more than a hypothetical one, for on several occasions he threatened to resign, and at other times he was traveling or ill. It was never imagined that his son should or could step into his shoes; Lyman knew that "it is not in reason to hope for a *Successor* to such a steam-engine."[42] Alexander was willing to help, but he harbored a certain distaste for the vast enterprise of preservation a great museum entails.

> As far as I am concerned personally, the Museum is of very little use to me, as I believe in study ex natura, and have but little fancy for closet investigations where you get long Memoirs about animals which have never been seen living or in state of nature by the author.[43]

This was written just as Hagen was completing his "Monograph," in the preface of which he admitted to never having handled a live crayfish.

Who else was there to give scientific direction? There were no clearly outstanding candidates among Agassiz's students. Harvard's President Eliot did not see any, and thought the lack part of a wider phenomenon caused by the rigid patterns in American education:

> To illustrate the failure of the system of the last 40 years to breed scholars, let us take the most unpleasant fact which I know for those who have the future of this University to care for—Gray, Peirce, Wyman & Agassiz are all going off the stage & their places cannot be filled with Harvard men, or any other Americans that I am acquainted with. This generation cannot match them. These men have not trained their successors. This is a very grievous fact which had better not be talked about—it is altogether too significant.[44]

Lyman perceived the same problem but blamed it on the passion for money which diverted the best minds into business.

What, indeed, was wrong with the next generation? Verrill was narrow, and Bickmore a lightweight, but Putnam, Scudder, Hyatt, and Morse would later demonstrate in their own careers both scientific and administrative competence. But both Agassizs bore such grudges that the list of candidates would have to exclude the rebels of 1864. It was one thing to deal with them professionally when museum business called for it, but none who had seemed disloyal in that crisis were ever fully trusted again. The name of Edward Drinker Cope, a paleontologist of growing fame, did not come up; his pro-evolution evangelism may have counted against him, but more likely it was his reputation for being less than scrupulous with borrowed specimens (and intemperate with drink and women) that ruled him out.[45]

Whatever the reason for the lack of impressive potential curators, it was also the case that no one tried very hard to find one. Thomas Henry Huxley was approached, but not until 1882, when Alexander Agassiz knew very well the English scientist was not moveable.[46] Anyway, another steam engine might not have worked, even if one had existed, because there was no longer an open arena where such a person could exercise his full ego. The people concerned with the future of the museum cared too much to give over real control to anyone but themselves.

Financing the Directorship

Louis Agassiz had brought no personal wealth to the museum but had created it out of nothing, as it were, by donating his services. The M.C.Z.'s statutes stipulated that the professor of zoology in the Lawrence Scientific School would be the curator, but would receive no additional salary from the museum even though Agassiz had actually added to his duties by his promise to the state that his activities in the museum would include public lectures.[47] In so doing he was, of course, using the Lawrence gift for unintended purposes, though neither the donor nor Harvard apparently complained. More important, his successors would inherit a budget that included no provision for the administration of the museum.

In the middle of the nineteenth century many branches of science chiefly consisted of amateur efforts, that is, unpaid work. Agassiz himself must rank as a principal actor in the professionalization of biology in America, instilling in his students not only high standards but expectations for a paying career in science. But change was slow and irregular, so it remained true, as it had been when Louis Agassiz arrived, that a prime qualification of a potential museum administrator was that he require no salary.

In the early 1860s, Alexander Agassiz had seen that he could be useful as an administrator and financial manager of the museum, but he could not afford to take on an unsalaried position. Of the people he knew who did make their living doing science, several besides his father—Jeffries Wyman, James Dwight Dana, and Asa Gray—were employed as professors. This did not encourage Agassiz, as he had learned from experience (he had given a course during his father's trip to Brazil)[48] that he was neither skillful nor happy as a teacher.[49] He had thought in 1860 that he might make the museum his career, but his father's policy of spreading the cash among many students, and the recognition of how uncertain the museum's finances were, quashed that idea:

> About my taking the Directorship: if I take it I ought to do it without receiving a cent from the Museum; that is impossible placed as I am [his wife was expecting their second child] and I know the finances of the establishment too well to have the impudence to ask for a Salary. So that this is out of the question as matters stand now.[50]

Alexander knew that the small salary he was receiving as "agent" in the museum had been donated just for that purpose by Theodore Lyman. He could not be comfortable with that arrangement indefinitely, though Lyman assured him "the annual sum I have given, so far as encouraging science goes, has been the best possible investment—dirt cheap!"[51] But Alexander Agassiz saw that he could not earn his living as a scientist or scientific administrator, and he determined to find work elsewhere:

> As soon as father returns [from Brazil] I shall look about town and endeavor to obtain a place which will keep me in town during the morning and give me time enough after that to look after the Museum generally; and what I receive now from the Museum can go to keep some one to do the manual work I am doing there now. This is the only way in which I could be the Director, by having some occupation to give me my pay and the rest of my time I can then give to the Museum.[52]

In spite of Lyman's warning that he would be taking on too much and working too hard, trying to support himself while volunteering in the museum, Alexander could not imagine abandoning his father when he so evidently needed help, nor could he picture supporting himself as a scientist. Besides helping his father with the administration of the museum, he wanted to do his own scientific research, but he would have to buy the freedom to do those two things with money earned elsewhere. Thus in spite of the difference in personality and scientific style of father and son, and in spite of Alexander's acute perception of the strain his father was under, Alexander would also, like his father, take on the burden of multiple roles. He didn't lecture or stump for contributions, but he became a fund-raiser on a grand scale.

The Circumstances of Fortune

Of course, there were plenty of other young men in the 1860s who longed to make enough money to free some of their time to pursue their nobler dreams. All the young naturalists who were Alexander's colleagues in the museum wondered how they could do science and support themselves at the same time. But Alexander Agassiz had several factors in his favor, once he set his mind on making his fortune. He had studied mathematics and chemistry as a Harvard College undergraduate as well as engineering and geology for two years in the Lawrence Scientific School, with an eye on mining. Alexander was bright and willing to work hard. He had in addition an extra advantage that enabled him to convert his intelligence and character into vast profits: family connections. Not through his mother (though it was helpful to him to consult with her brother Maximilian Braun, who was a successful mining engineer), nor through his father (though the fame of his name never hurt), but through his stepmother, Elizabeth Cary Agassiz. It was her membership in the mercantile aristocracy of Boston that placed young Alexander, and his sisters Pauline and Ida, comfortably within the right circles of wealthy and influential Bostonians. Alexander Agassiz's special bit of luck was in having the right in-laws.

In the fall of 1860 Alexander married Anna Russell, and his sister Pauline married Quincy Adams Shaw. Shaw's elder sister was the wife of George R. Russell, Anna's father, so that Alex's wife's "Uncle Quin" was also his sister's husband.[53] His wife's sister, née Elizabeth Russell, was married to Theodore Lyman, whose sister Cora was the wife of Quin's brother Gardner Howland Shaw (fig. 15). Alexander was now related to the wealthy Lyman family, and related several times over to Quincy Adams Shaw, who along with several siblings would inherit the considerable fortune of the mercantile families of Parkman and Shaw.[54] Alexander Agassiz and Theodore Lyman had already been friends as undergraduates, were now both studying echinoderms (Lyman the brittle stars and Agassiz the sea urchins and starfishes), and both wanted to help Louis Agassiz succeed

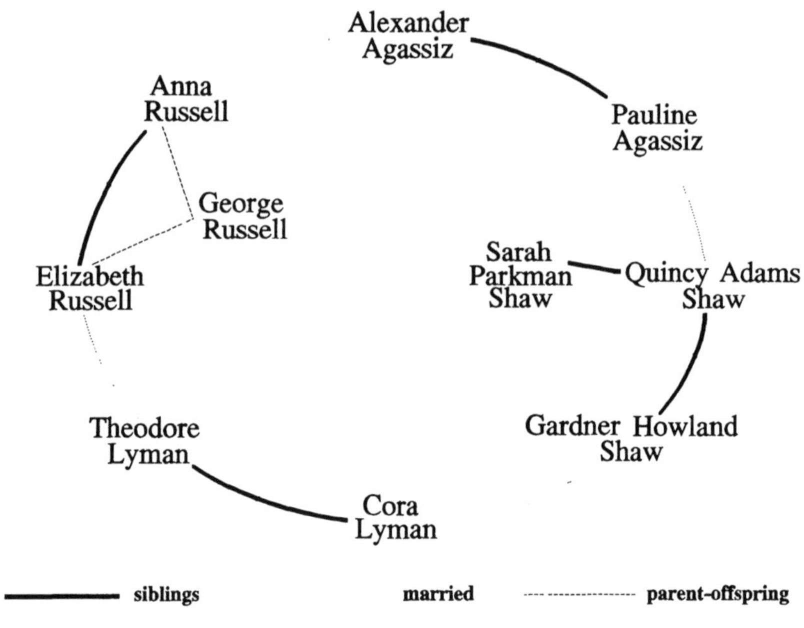

Figure 15. Alexander Agassiz's relatedness, by blood and marriage, to wealthy Bostonians. Louis Agassiz's first wife, the mother of Alexander, Ida, and Pauline, died in 1848. In 1850 he married Elizabeth Cabot Cary, who remained childless. Agassiz's three children all married into the Boston aristocracy.

with the museum. After 1860 they not only thought of each other as brothers but they could be called brothers-in-law.

Young Quincy Adams Shaw and his cousin Francis Parkman had gone West in 1846 to seek adventure together; the narrative of their travels, *The Oregon Trail,* which Parkman dictated to Shaw, became an American classic. Their trip took place during the period of the first American mining boom, "the copper fever," so they must have heard of the Keweenaw Peninsula of northernmost Michigan.[55] Alexander Agassiz had undoubtedy heard of it as well, for his father, after an excursion to Lake Superior in 1848, had published a speculation on the geological cause of its famous copper deposits.[56] In 1865 a surveyor, who had discovered a rich lode on land he could not afford to buy, brought some sample rocks to Boston, looking for investors. His secret and Shaw's money began the Calumet and Hecla mining companies.* Scores of little mines in that region were selling stock, but because the price of copper was low and the distances to transport equipment and ore was great, few returned a profit. It may have been the excitement of speculation that attracted Shaw, for he had several safer investments closer to home. He visited the company's land and returned with the idea that his appropriately educated brother-in-law Alex might go up and have a look.

This was a boom period for mining; Alexander Agassiz was expecting to work part-time as engineer-manager of some mining company, and he did work in April 1866 for a coal concern in Pennsylvania.† But he could not immediately take on such a big project as Shaw proposed because he was needed at the museum, Louis Agassiz having sailed off with much fanfare on the Thayer Expedition to Brazil. After Louis Agassiz returned, Alex traveled up to Michigan, in late August or early September of 1866. He was so excited by the quantities of pure copper he could see that he borrowed $10,000 from Lyman to buy stock in the mining company, and led Shaw to increase his land holdings by acquiring the nearby Hecla mine. All that would be required for great profits was able management, so far lacking, and Quin was glad to hire Alex to get the mine properly established. Alex

*The Calumet Mine and the Hecla Mine became the Calumet and Hecla Mining Company in 1871 (G. R. Agassiz, *The Founding of the Calumet & Hecla Mine*).

†G. R. Agassiz, *Letters and Recollections of Alexander Agassiz,* p. 53. "You know that since I left college I have spent nearly all my spare time in preparing myself to be a mining engineer. I took mining engineering as it seemed the only branch of Engineering which was not glutted with engineers" (A. Agassiz to Lyman, 1 September 1859, M.C.Z. Archives). "Alex has got the Presidency of the Gilberton Coal Co. which will do to start in, though the salary will be no more than say $900" (Lyman, Private Notebooks, 7 April 1866, Massachusetts Historical Society, Boston).

moved to Michigan in March 1867, joined that same summer by his wife, five-year-old George Russell Agassiz, and infant Maximilian.

There was copper in the ground, but converting it into a profitable business turned out to be much harder than anyone had foreseen. Alexander Agassiz's original estimates of richness of ore proved exaggerated, local workers were uncooperative, and the expenses of operating railroads and mills in that region kept multiplying. It began to be clear why so few mining speculations ever paid off. Alexander exerted himself to the utmost, only to report finally that much more capital would have to be poured into equipment to produce a profitable quantity of metal; the stockholders, far from receiving dividends on their investments, received notice of assessment, that is, that they would have to contribute more cash to the venture or lose their stock. Alexander had purchased his own holdings with borrowed funds; his wife's brother lent him the extra amount he needed to meet this assessment. All of Alex's abilities and ambitions could not turn Michigan rock into profit without still more capital from Boston.[57]

That the capital did come, after Shaw's own credit was exhausted, was not so much the result of a prudent investment decision as of a crisis of family honor. Quincy Adams Shaw was embarrassed by the failure of the investment firm of Ward, Mellen & Co. At first (in March 1867) it seemed he might weather the storm by mortgaging his house in Jamaica Plain and his three stores on Commercial Wharf to his brothers. He even sold his horses, and for a while he and Pauline paid their social calls on foot. Finally, the day came when he had to confess to Theodore Lyman that he had borrowed heavily from a trust fund for his late brother Robert's widow and from a charitable trust set up by their father. This involved Lyman, because his sister Cora's recently deceased husband Gardner Howland Shaw had been legally responsible for the funds Quin had used. If Quin were to go bankrupt, and the charity's deficit thus be uncovered, Howland's good name would be compromised. Since the Calumet and Hecla stocks were Quin's only property, yet were just worthless paper unless the mines could be made profitable, Theodore Lyman advised his sister Cora to lend $100,000 of her own to the mines. Thus "she would help Quin, his brother [Samuel] Park[man Shaw], & Alex and the Fund; but would run a risk. If she lost the loan she could show the paper to prove she had done all for Howland's name. Cora was much moved by Quin's gambling, bad faith, and folly." But she made the loan.[58]

Soon after this infusion of capital, Calumet began to produce copper in satisfactory amounts. Also, in 1868 a tariff on imported ore, though harmful to Eastern seaboard smelters, was passed by Congress and increased the value of the Michigan product. Calumet paid its first dividend in December 1869, and by the end of the next year it was clear that major shareholders like Shaw, Lyman, Alexander Agassiz, and Ida Agassiz's husband Henry Lee Higginson would reap millions.[59] Quin and Alex divided the labor of running the mining companies, Alex spending every summer and fall on site, keeping on top of technical and management problems.[60] Quincy Adams Shaw's near disaster turned to riches so quickly that he not only repaid Cora Lyman Shaw with interest in September 1870 but he could also spare another hundred thousand in May 1873 to give Louis Agassiz as a sixty-sixth birthday present from his wife and himself, and another hundred thousand or so the next year, after his father-in-law's death, for the Agassiz Memorial Fund. Shaw would make one more major donation in 1900 to help build the last segment of the museum. It is usually assumed that these benefactions reflected his wife's devotion to her father, for Shaw otherwise betrayed no interest in natural history. No doubt they did, but perhaps Quin never forgot how Theodore Lyman's advice, and Alex's labor at the mine, had rescued him from imminent disgrace.

The Problem of Museum Leadership

While Alexander was in Michigan struggling with the mines between March 1867 and September 1868, Theodore Lyman took over his role as bookkeeper and "refrigerator" at the museum. Alexander took thoughtful note that in his absence his father seemed to be behaving more reasonably and museum affairs proceeding more smoothly. Returning to Cambridge, Alex did not resume his position as agent but limited his help to the arrangement of the group he was studying, the echinoderms. He used the museum—specimens, library, artists, and workspace—as his base for the taxonomic work he was doing on the sea urchins, but otherwise he made a conscious attempt "to steer clear of all Museum interference."[61] Louis Agassiz's brother-in-law Tom Cary filled the role of volunteer business manager of the "baby" instead of Alexander. With Alexander's wealth multiplying, though, he assured his father that he would look after the well-being of the museum.* He likely hoped to do this somewhat from the sidelines, as financial benefactor and advisor.

Though Louis Agassiz's health did not cease to be a worry to his family, especially after a stroke which kept him bedridden for much of 1870, he seemed to have recovered his old self in 1872, when he accepted an invita-

*"It was always agreed between Professor Agassiz and myself . . . that, as his scientific executor, the care of the museum should be my special charge, and that toward its support alone whatever I could contribute personally or command from others would be applied" (*Boston Daily Advertiser*, 26 October 1875).

tion to sail on the Coast Survey steamer *Hassler* to San Francisco by way of the Galapagos Islands. In 1873 he made arrangements for a summer school on Penikese Island, southwest of Cape Cod, where in his inimitable style he preached to schoolteachers that they and their pupils should "Study Nature, Not Books." But whether through retirement or death, he knew that his museum would soon need other managers.

The man Agassiz chose to succeed him in the job of curator was Count Louis François de Pourtalès (fig. 16). In his teens a student of Agassiz's in Neuchâtel, Pourtalès was among the many who emigrated with him or soon after him (others were Charles Girard, Jules Marcou, Auguste Sonrel, Jacques Burkhardt, Edouard Desor, Arnold Guyot, and Léo Lesquereux). Hired to do oceanographic work for the Coast Survey, Pourtalès published respectably, raising questions of geographic distribution. Upon the death of his father in 1870, he became financially independent. He accompanied Agassiz on the *Hassler* cruise and stayed on thereafter to help in the museum.[62] Since the curator was supposed to be in charge of teaching in the museum, Louis Agassiz then tried (without success) to arrange a professorship for Pourtalès at Harvard.[63]

Early in December of 1873, after a remarkably successful summer on Penikese and autumn of lecturing and fund-raising, Agassiz, aged sixty-six, lost consciousness. Death released him on December 14. He was sincerely mourned by thousands, from schoolchildren who had been taught the legend of a simple naturalist to colleagues who had been driven to exasperation by his bluster. No one could, and no one person ever did, replace him in all his roles. In line with the division of labor that had already been evolving, Professors McCrady and Shaler continued to teach zoology and geology, and Mrs. Agassiz's brother Tom continued to donate his time managing business affairs as superintendent. Count Pourtalès, quiet, "modest to a fault," took on more paperwork to do with the collections but was not appointed curator, nor was anyone appointed director.

With the museum's founder and main cog gone, both Harvard and the state's trustees agreed to change the regulations to reflect the new reality rather than try to fit the existing cast of characters into this awkward mold. Massachusetts essentially transferred its interests to the care of the Harvard Corporation (without altering the museum's responsibilities to the public), legally eliminating its board of trustees in 1876. The regulation stipulating that the curator be professor of zoology at Harvard was removed. Pourtalès was called "keeper," a title that did not exist in the statutes; his volunteer administrative labors were gratefully acknowledged every year, but his position was distinctly that of an assistant to the curator.

The year 1874 was a hard one for Alexander Agassiz. On the twenty-

Figure 16. Louis François de Pourtalès and Louis Agassiz with dried sea urchins. Agassiz wanted Harvard to make Pourtalès a professor; perhaps this photograph was taken to forward that plan. Pourtalès served as volunteer keeper in the museum from 1876 until his death in 1880. (By permission of the Museum of Comparative Zoology Archives, Harvard University)

third of December, 1873, just eight days after the death of his father, Alexander's wife, Anna, caught cold and died. Theodore Lyman, who was still mourning the death of his only child, Cora,* recorded in his diary,

> Alex stood in the library, his sisters with their arms round his neck—my near friend from our Freshman days, and my brother for these thirteen years. I took him in my arms and said, "Alex, when we die, you will see Annie, and I shall see my little Cora." He sat on the sofa with his head on my shoulder for more than an hour; and I cried and cried over him. He said, "When father died, I could cry, but now I cannot." . . . We were fit company, two broken-hearted men together. . . . A woman—only thirty-three—strong, healthy, brimming with life and motherly beauty—gone in one week—and for no cause![64]

Alexander suffered profound grief and desolation. For many years thereafter he could not bear to hear mention of his wife.

Soon after that awful December, Alexander Agassiz resolved to finish the projects his father had begun. His pledge is remarkable for its tone of self-immolation, a sort of longing for meaning without any hope of finding it.

> I shall try and carry out, to the best of my abilities, the many plans regarding Penikese and the Museum which were started by my father, and I shall at least have the melancholy satisfaction of knowing that in his case at least his views, whether right or wrong, and his dearest wishes, will be faithfully executed, and that I may raise a monument to him expressing what he hoped to be able to show, better perhaps than he himself would have done, because I shall not be constantly drawn aside by new plans and shall not have the incessant temptation of remodelling as I go along.[65]

Of his father he could speak and take action. But it was the death of Anna Russell Agassiz that cast him into the depths of misery in which he formed this cold resolution to memorialize his father. Striving to carry on the summer school on Penikese, he collapsed from emotional exhaustion and had to leave the island.[66] Funding for a third summer did not exist, yet people blamed him, Alexander bitterly imagined, for the school's demise.[67]

At the museum, Theodore Lyman worked closely with Alexander Agassiz, as he had done off and on since its founding. The two were designated "the Committee on the Museum" until Alexander was ready to take on the curatorship in 1875. In subsequent years Pourtalès, Lyman, and Cary continued to carry some of the burden of bureaucratic work for Alex-

*Though they had lost their only child when Cora died, Lyman and his wife would later have two sons: Theodore, after whom the Lyman Laboratory of Physics at Harvard is named, and Henry, whose son, the mammalogist Charles P. Lyman, kindly made available to me the wonderful diary kept by his grandfather.

ander Agassiz. They quickly found that as always, and in spite of the recent infusions of copper money, expenses were far outstripping income.

Thousands of people, those who had been touched by the lectures, ideas, or legend of Louis Agassiz, grieved at his death and saw the museum as the obvious monument to his spirit. "Dr. Oliver Wendell Holmes compared the unfinished work of Agassiz to a cathedral left incomplete at the death of the architect. In each case the noblest memorial is the completion of the work."[68] As one eulogist put it, "Shall this great work be allowed to fail? Let every person who honors the name of Agassiz, say No! . . . it shall go on and be built and filled, and stand firm, a glorious temple of science forever."[69] A fund drive to this end was announced. Some observers expected a million dollars to be raised, but the public drive netted only $64,000, plus $50,000 from the Commonwealth and $190,000 from Quincy Adams Shaw and Alexander Agassiz. Of course, this was invested to guarantee the future income of the museum (something Louis Agassiz never willingly did with any money he had ever raised), more than doubling its annual income.

Alexander Agassiz Takes Charge

When Louis Agassiz had used the image of cathedrals and temples in relation to his museum, he was not referring to the building's structure, for its architecture was plain, but to the inspired selfless zeal impelling people to devote their money and labor to building institutions such as these. A natural history museum, housing the works of God, should inspire support no less generous and pious than a church does, he claimed.[70] He had insisted on selfless devotion from his staff, on the grounds that the "object in view" was "to erect a great Monument to science."[71] But some wise instinct had made him recoil from the image of museum as memorial. He had refused to let the new museum be named after him, on the grounds that "personalities must be banished from science."[72]

Another aspect of his museum demanded a different metaphor; Agassiz declared that "it is with museums as with all living things; what has vitality must grow. When museums cease to grow, and consequently to demand ever-increasing means, their usefulness is on the decline."[73] Alexander Agassiz and Theodore Lyman in their letters had been calling the museum "the baby," because of all the caretaking it demanded.* Lyman showed considerable insight in his eulogy of Louis Agassiz:

*A. Agassiz to Lyman, 8 December 1871, 15 and 20 November 1872, M.C.Z. Archives. Alexander Agassiz used the image of a troublesome and demanding infant for the copper mine as well (A. Agassiz to Lyman, 7 December 1873, M.C.Z. Archives).

But this child [the museum] kept outgrowing its clothes. We could never get enough jars, or drawers, or alcohol enough! In a museum of natural history everything pours in, and nothing pours out, except money. Nature has no beginning or middle or end; the process of increase and arrangement is an everlasting one.[74]

But the dominant metaphor in 1874 was not organic, it was that of the cathedral left unfinished by its great architect. Alexander Agassiz felt duty bound to see the last stone laid in place, literally as well as metaphorically. He announced that he would "erect to him a monument . . . a valuable historic record of the interpretation of nature by one of its most enthusiastic worshippers."[75] After the Memorial Fund drive of 1874 had ended, neither the state nor the general public was again asked to support the M.C.Z., perhaps because the millionaire Agassiz thought the role of beggar did not suit him, or because he feared that strings would be attached, or believed that an endowed institution properly managed should be self-sufficient. In any case, it was largely to erect a public monument that Alexander set about the task of creating the exhibition halls envisioned by his father.

Now for the first time the policy of dual arrangement was made explicit and unambiguous.[76] The staff would concentrate on creating the public exhibits until the job was done. Alexander ordered the repainting of the bases on which stuffed birds and mammals stood, had new labels written, and personally bought display cases. The themes his father had trumpeted, that there should be a synoptic room with representative forms, plus separate rooms for the fauna of different regions, Alexander adopted as the formal plan for the exhibition halls.[77] Not only would each specimen be labeled, the arrangement itself would also be labeled in large letters on the wall announcing the faunal area or systematic order. But the several "new plans," like the skeletons of domestic animals, which had constantly drawn his father aside, Alexander dropped.[78] For Louis Agassiz, in spite of his fund-raising rhetoric, the education of the casual visitor was really just a side effect, arising out of the fact that the specimens had been placed in a natural and thus intelligible arrangement. Now that the museum had become a memorial, the specimens set out and labeled for the public must be attractive. Louis Agassiz had been so determined to keep science and not popular amusement paramount that he had insisted that the curator must never prefer beauty to substance. Alexander made a sharp division between the specimens on display and the "collections," the vast series of dried or pickled objects arranged so that an expert could study them. The assistants in each department were instructed to pick out those specimens suitable for display. Whatever was lacking to complete the exhibition would be purchased. Soon the M.C.Z. became the single best customer of Ward's Natural Science Establishment in Rochester, New York.

Henry Augustus Ward had been one of those eager young men in love with natural history whom Louis Agassiz, at the height of his rhetorical powers, had attracted to Cambridge. Ward spent only a few months there in 1854, but it was long enough for him to observe how a collection could be used to attract financial support. Acquiring material on credit from European museums and dealers, Ward soon became a dealer himself. By the 1870s Ward's Natural Science Establishment was not only supplying existing museums and collectors with specimens, it was convincing cities and colleges that they needed to found their own museums (and become his clients). He encouraged the taxidermists who worked for him to avoid the stiff look of stuffed skins and to aim at life-like poses. Besides selling individual specimens, Ward offered to cities and colleges collections he described as complete, or at least perfectly representative of the whole of nature. As his catalogue boasted, Noah had probably not had such a complete collection with him on the ark.[79]

Alexander Agassiz sent skins already in the M.C.Z.'s collections to Ward for mounting, including a llama, alpaca, and monkeys, and told Ward what species the museum needed to complete its exhibition series. Agassiz personally purchased and donated to the museum a rhinoceros, hippopotomus, gorilla, antelopes, and giraffes, solely for the purpose of completing the public rooms. He had a frank relationship with Ward.

> I could not believe that you would send me such a specimen except as a joke. I am well aware that the Dromedary is not a handsome animal but where that skin has been and who the artist was who placed him in that most unnatural attitude I cannot imagine . . . certainly no one would expect it to hail from your establishment. I really cannot give it houseroom . . . I should have nightmares any time I went to the African room, to have to face it.[80]

By 1878 five refurbished rooms were open to the public, and by 1885 six more had been added, with another seven in preparation. Museum policy had moved so far away from the multiple purposes of specimens that it was now forbidden even for a staff member to remove an exhibition specimen from its case for closer study.[81] Yet in the years when the public halls were much visited and admired, little if any money was spent on enlarging the research collection.

A decade after the death of Louis Agassiz the museum building had more than doubled in size and the exhibition rooms were nearing completion. By now, however, it felt to Alexander Agassiz like a very thankless task. The death of Pourtalès in 1880 had increased his load of routine management, and he began to think of who might take over the administrative load. "I had an elephant left to me; my successors can feed him or kill him as they like,"[82] he told President Eliot, and wrote asking Huxley to come.[83] But this feeble attempt, like a later interest in hiring H. F. Osborn, was ineffectual because of Alexander Agassiz's deep ambiguity about managing his father's project, an ambiguity he would never escape.

> Now I have been slowly coming to the conclusion that . . . I've been a good deal of a fool. . . . This is simply insane. I have also allowed myself from sentimental reasons to carry out plans which are not my own and to which I had but little interest, practically sacrificing any views or intuitions of my own. I have thus spent the ten best years of my life in being simply an executive officer and have lost all chance to do the work I was interested in. . . . It is rather late to wake up, as I have done at 48, to the lesson that nobody should undertake another man's work if he has any he can do himself.[84]

6

"Shall We Say '*Ignorabimus*,' or Chase a Phantom?"

For the second head of the M.C.Z., the surname "Agassiz," like the museum itself, was a mixed blessing. He was exasperated when people attributed his father's ideas to him, and rightly so, for the two men thought and acted very differently in most respects. People in the know were careful to call him Mr. Agassiz, to show that they did not confuse him with Professor Agassiz. At the same time he was fiercely loyal to his father and ready to defend his memory (fig. 17).

The great watershed marked by the appearance of Darwin's *Origin* in 1859, one of the greatest revolutions in the history of thought, left Louis and Alexander on opposite sides. Real watersheds, though, wander erratically across an irregular landscape, and real revolutions leave the populace still concerned with cooking their meals, earning their pay, and talking with friends. Likewise in science, even the most fundamental intellectual transformation is complicated by local circumstances, takes time, and leaves considerable room for continuity. Scientists are for the most part more committed to the gathering of facts than to any particular explanation of those facts, and are more committed to instruments and techniques of observation than to any higher abstract truth. Biology has seen no greater shift in its fundamental explanations than that symbolized by the opposition of Louis Agassiz to Darwinism, yet many workers, especially in museums, carried on after 1859 much as they had before, defining their problems much as their predecessors had defined them, just as if Darwin's work was not so earthshaking after all. Not only did specialists trained before 1859, like Hermann Hagen, see no reason to change any detail of their activities, but younger taxonomists who fully accepted evolution, like Walter Faxon, could also continue the tradition of their predecessors with no significant change of method. The enthusiasts would have it that there were only two options, to be for evolution or against it, but there was a third option—to duck your head and carry on. Alexander Agassiz and most of those in his museum believed that good science required less enthusiasm and more patience, less hasty judgment and more information.

The Theory His Father Rejected

Darwin's momentous book had appeared when Alexander Agassiz was just turning twenty-four. Like his fellow students in the museum, he had not doubted that this new theory must at least be taken seriously, in spite of his father's skepticism. At first Alexander, too, had been skeptical, since his own experiences with embryology impressed him with the power of inheritance to keep the offspring essentially true to the parental type. Even so, he approached the question from a set of assumptions very different from those of his father. The similarities of form that Louis Agassiz attributed to the ordering intelligence of the divine mind Alexander could imagine would someday be explained as the product of necessity, like the laws of geometry. Even while trying to withhold judgment, he had acknowledged that evolution could explain the phenomena of embryological resemblance.[1] The resemblance between the embryos of starfish, sea urchins, and sea cucumbers, which to his father were just another wonderful example of the Creator's love of pattern, were to Alexander Agassiz suggestive of descent from a common ancestor. His father knew the direction Alexander's thoughts were taking. In 1869, Louis Agassiz gave his son a letter of introduction to Darwin which said, "You will find Alex more ready to accept your views than I shall ever be."[2] In January of 1870, the younger Agassiz let his closest friend know that he was leaning toward belief in a theory of development, though not necessarily Darwin's.[3]

Even though Alexander Agassiz was moving steadily to an acceptance of evolution, several aspects of his personality prevented him from embracing it with any enthusiasm. By nature the very opposite of his father's romantic, visionary temperament, he thought of himself as an engineer, a student of mathematics, a practical man. The zealous tone of committed evolutionists was enough to make the dour younger Agassiz distance himself from them. But besides these matters of perception and style, the complex emotions linking him to his father may have helped shape Alexander Agassiz's treatment of evolution in his scientific publications.

It seems to me that Alexander Agassiz was impelled by powerful motives to try to please and impress his father. His attitude of duty, caring, and helpfulness went beyond ordinary filial devotion and may have been his way of dealing with unconscious feelings of anger, feelings for which he must

Figure 17. Alexander Agassiz. He was "Mr. Agassiz" to all but his few intimates. It was a faux pas to call him "Professor Agassiz," his father's title, because he was not a member of the Harvard faculty, though for many years he did serve on its governing board, the Harvard Corporation. (By permission of the Museum of Comparative Zoology Archives, Harvard University)

atone. He had been nine years old when his mother, increasingly unhappy with Louis Agassiz, quit her home.[4] After his father left Switzerland a year later, the boy may well have experienced conflicting feelings upon hearing about the wonders of the New World and his father's reception there. For two years Alexander was in the position of caring for his two younger sisters as well as his mother, while she lay dying of tuberculosis.

> As Mrs. Agassiz became weaker, the thoughtful child [then aged 12] grew more quiet and serious, for he adored her and must have realized that her end was near. He now took charge of the pathetic little household, kept the small accounts, did the errands, went to market every day, and strove like the true mite of a man that he was to relieve his mother of anxiety. In the summer of 1848 she died and the children were taken to their Uncle Alexander's. . . . after he had lived nearly a year with his uncle, he was sent for by his father. . . . Hearing that America was a land of freedom where one could do what one chose, Alexander celebrated his departure from Freiburg by jumping on his violin as he set out for the New World.[5]

Years later Alexander must have amused his own children by telling that story of daring destruction (this version is from the biography by his son George), without consciously recognizing the emotional wellsprings of such an act. Hating music lessons was allowed, but to hate one's father would be too frightening. In America the boy adjusted to his new life with the help of a loving and tactful stepmother. Growing up surrounded by people who lionized his father, he would have observed that Louis Agassiz respected two kinds of men: conscientious scientists and wealthy patrons of science. He was fiercely determined to become both.

In the autumn of 1873, father and son had separately been putting on paper their thoughts on evolution. Louis Agassiz's words were widely read, appearing in the *Atlantic Monthly* shortly after his death, while Alexander's views were buried within a formidable technical monograph, "Revision of the Echini," seen by a very few specialists.[6] Though the intended audiences of the two documents were utterly different, each text contains signs that its author was solicitous of the other's feelings. Louis Agassiz was obviously trying to sound fair and open-minded and to show that he fully understood and respected those who found Darwinism attractive. Alexander Agassiz stressed those aspects of his father's work that could be said to have been in some sense confirmed by evolution, and he cloaked his rejection of the "Essay on Classification" and his acceptance of evolution in obscure and indirect language.

Louis Agassiz emphasized that the word "evolution" originally referred to individual development, and he suggested that the pioneers of embryology (a status most Americans accorded to him, though historians of biology do not) should receive due appreciation. He also declared that no amount of theorizing could substitute for the careful empirical work that was the basis of good science. Exactly such empirical work was Alexander Agassiz's monograph, with very few of its 762 pages given over to interpretation. Alexander declared that it had been Louis Agassiz who had been the first to point out the agreement between "the paleontological and the embryological genesis," and he castigated the speculative style of the evolutionists, who "outdarwin Darwin." While agreeing with Darwin that parallels did represent genetic and not merely ideal connection, Alexander said that he was unable to see how natural selection could apply to sea urchins. His allusions to evolution all seem to have been carefully worded so that no one sentence could be quoted and held up as a trophy, proof that an Agassiz now accepted Darwinism. Alexander's most ringing declaration leaves unclear whether he believed in evolution himself: "No one appreciates more than I do that the explanation of the theory of evolution, as given by Darwin, has opened up new fields of observation in many departments of biology, the importance of which can hardly be overestimated."[7]

Even if it were true that Alexander Agassiz had unresolved childhood feelings predisposing him not to offend his father by embracing Darwinism, still there is no reason to suppose that in expressing his reservations Alexander would have been writing hypocritically. The tightrope he had to walk between the views of his father and those of his father's opponents was not merely a problem of public relations or even of family relations; his problem was to develop a position of his own which would allow him to respect himself as a scientist. To explore the threefold parallelism of embryology, taxonomic rank, and paleontology was an attractive theoretical research program, modern because illustrative of evolution yet faithful to one of his father's favorite ideas. This provided the positive theme for Alexander's researches on sea urchins, in which he compared the anatomy, embryology, homologies, and geological record of the group. He presented himself as a properly skeptical but thoughtful scientist, loyal in sentiment yet modern in substance.

Alexander seems to have taken up the classification of echinoids in the spirit of objective observation, studying nature not books, as his father preached. He certainly must have been familiar with the ideas about natural categories, from species and genus up to classes and *embranchements* his father had expounded in his 1857 "Essay on Classification." But experience taught Alexander that these formulas did not prove useful for echinoids and, furthermore, that the more extensive the material, the more uncertain became any given characters and definitions. Especially when he included the developing young in his scrutiny, all definitions failed. Alexander confided to Darwin in 1872,

The number of young I have been compelled to examine has led me to modify my views of the nature of genera, species, and in fact of all subdivisions. I cannot find anything that *is stable,* the greater the material in space and number (age) the more one is adrift to get a correct diagnosis of a genus or a species, and the gradual passage in Echini of the most widely separated groups leaves in my mind but little doubt that our classification is nothing but the most arbitrary convenient tool, depending upon the material at our command at a special time.[8]

Louis Agassiz's ideas about each taxonomic category, from *embranchement* down to species, expressing distinct categories of the Creator's thought had been quite forgotten by the 1870s; even their author had long since ceased alluding to them. Yet the central assertion of the "Essay on Classification"—that groupings at all levels, from species up to *embranchement,* were chunks of reality and not creations of the observer—this Louis Agassiz could never abandon. Nor did Darwinism make this view out of date. Many evolutionists, perhaps most, still believed in natural taxa—species and groups of species it was the duty of the respectful observer to discover, which he was not free to erect according to his fancy. But Louis Agassiz's son was entertaining a view of classification rarely heard since Buffon's attack on Linnaeus; that "classification is nothing but the most arbitrary convenient tool." This was not the conclusion of a proselyte holding on to his teacher's theories as nearly as the changing climate of contemporary thought would allow. No; more radical than any of Louis Agassiz's other students, Alexander was reaching toward an intellectual position directly antithetical to that of his father.

In his "Revision of the Echini," Alexander insisted that the precise definition of species was an old-fashioned and quixotic goal, and that it was time zoologists acknowledged taxonomic problems for the practical matters of mutual convenience they were. Once we have recognized a reality consisting of countless individual organisms changing during their own lifetimes, differing from one another, covering a wide area across the earth and stretching back in time toward some different ancestral form, then we must admit the impossibility of ever making a perfect list of defining characters.

Are we to attempt to define with mathematical accuracy what we mean by a species because we find it convenient to use a binomial nomenclature to express zoölogical units? As well attempt to solve an equation of an infinite number of unknown quantities by means of an equation of the second degree. . . . The fact is, that we can no longer define species as has been customary.[9]

The taxonomist's job becomes reduced to the recording of whichever structural details he has found useful for distinguishing between the specimens he happens to have before him, in the hope that his work may be useful but in the certainty it will appear inadequate to anyone with access to more extensive collections.

The implication of evolution for any transcendental meaning of species was well known and was stated unequivocally (though gently) in the *Origin*. As to the higher categories, Darwin was less explicit, but the challenge was no less fundamental. If his theory were true, then two species so different that they are properly placed in different families or orders could trace their heritage back in unbroken sequence to a time when their ancestors were varieties within a single species. In that case, the higher divisions have no meaning beyond the historical accidents of divergence and extinction; it would be as absurd to call them "categories of thought" as to look for thoughtful artistry in valley and mountainside, landscapes that are the products of blind forces and not handicraft. So again, the taxonomist's task is merely to record similarities and differences in a useful fashion. In Alexander Agassiz's view, that assignment was important enough, and he did not bemoan the absence of any deeper rationale. He argued that

> . . . in no way do we lessen their value by saying that we have no accurate definition of species, or by saying that species belong to the same categories as genera, differing only in degree; and so in admitting all the most zealous evolutionist could require, it does not lessen the fact of the finite condition of the differences we now notice, and which we call species or genera or families or orders, as we class them in various categories. For their transition, if such a transition does exist, can only take place through an infinite series, which still leaves the problem capable of a definite solution within fixed limits at any special time; and this is all that is needed for our purpose.[10]

Clear prose was not Alexander Agassiz's forte. I take him to mean that the number of steps required to work one's way back from a member of one family, through all the different parental forms, species by species, and then forward from the common ancestor until we reach a member of an entirely different family is so large as to be virtually infinite; and yet this "infinite flexibility" of a species does not alter the taxonomist's ability to describe a species in one time and place.

> We know nature only through individuals, and whatever conclusions we draw are based upon the examination of a number of individuals showing a certain range of variation . . . as long as we confine ourselves to the interpretation of nature, susceptible from such infinite data, we need not trouble ourselves as to the metaphysical existence of species, genera, etc., or because we have no suitable definition of species applying to all classes of the animal kingdom, which, in the present state

of biological science, it is absurd to expect. We are agreed for the present to call certain categories specific, others generic, others ordinal, and it matters only to us that we should distinctly state the limits we assign to these categories in some way readily understood; and this the individuals or groups of individuals themselves belonging to the different categories will supply.[11]

It is a testimony to the tact of both Louis and Alexander Agassiz that the extent of the difference in their beliefs was not generally noticed, for it was vast indeed.

Alexander Agassiz's Bombshell

In 1880 it was the duty of Alexander Agassiz, as vice-president of the American Association for the Advancement of Science, to deliver an address during the August meeting. Since he was just finishing his report on the sea urchins collected by H.M.S. *Challenger,* he used those data for his talk. The scientists gathered on a Thursday morning in Sanders Theater, within sight of the Agassiz museum, but, if thoughts of its founder were not far from anyone's mind, what they heard left no doubt as to the independent scientific credentials of his son. The body of Alexander Agassiz's paper was a detailed comparison of living adult sea urchins with immature stages and with fossils. He cited genera and species of echinoids by the score whose names can hardly have meant a thing to the audience, none of whom were echinoderm specialists. But his talk held their attention nonetheless, for he was discussing evolution, and the notorious opinions of his father created an almost morbid curiosity as to the allegiance of the son.

Facing the assembly that August morning, Alexander was even more loath to comfort the evolutionists than he had been while his father was alive. He chose as his central theme the claim he had made in the "Revision," that his own close studies of a wide variety of species had confirmed the parallels pointed out by Louis Agassiz in his echinoderm work of the mid-forties, and that the idea of evolution only showed how prescient those early studies had been. Alexander proposed the explanation that the different forms in the fossil records may be the result of the same unknown "genetic" forces that shape individual development. Aware that his audience would expect theoretical commentary on such an occasion, and that much of the interest in the hauls of the *Challenger* resulted from the belief that the deep sea harbored primitive creatures, "living fossils," Alexander must have felt under pressure to outline the evolutionary history of this very ancient group of animals. Instead of doing so, he dumbfounded the session by declaring that he had a mathematical proof that this favorite program of the evolutionists was an impossible one.

Anyone who had been listening carefully to Agassiz must have been totally unprepared for this declaration. The speaker expressed no doubt that modern echinoids were the evolved descendents of extinct echinoids. The first part of his lecture must have pleased his audience, as he described new parallels between series of particular fossil genera and the larval stages of living species. Growthlike sequences could be traced, from "regular" spheroidal species to heart-shaped or flattened ones, from those with simple teeth to those with complex teeth, from those with a few long spines to those with many short spines. Evolutionists like Ernst Haeckel and Edward Drinker Cope took pride in sketching out such series, constructing elaborate phylogenetic trees. But now Alexander declared that in this group he knew so intimately, consisting of forms certainly related to one another, whose fossil record was exceptionally long and full, and whose embryology gave such suggestive resemblances to fossil types, in spite of all those advantages no genealogical tree could be defended. Anyone could draw such a tree, but it would be arbitrary and imaginary.

Of course, Agassiz conceded, we could construct series of genera illustrating a gradual transition from one character state to another: from long spine to short, from globular form to elongate, or whatever. We could do this for each character singly, but then we would find that the sequence of species would have to be different depending on which character we followed. Trying to compare each species with all its peculiarities to every other species was not merely difficult, Agassiz claimed, but virtually impossible. He warned that

> We must, on the theory of the independent modifications of special structural features, trace the many and complicated affinities which so constantly strike us in making comparative studies, and which render it impossible for us to express the manifold affinities we notice, without taking up separately each special structure. Any attempt to take up a combination of characters, or a system of combinations, is sure to lead us to indefinite problems far beyond our power to grasp.[12]

Without in the least doubting that later echini are the descendents of earlier forms, we can and should doubt, Agassiz was arguing, our ability to uncover their actual lines of descent. "It certainly has been shown to be an impossibility to trace in the paleontological succession of the Echini anything like a sequence of genera. No direct filiation can be shown to exist."[13] He compared this paradoxical situation to the history of the human population of North America: "bold indeed would he be who would attempt even in a single State to trace the genealogy of the inhabitants from those of ten years before."[14] But his fellow biologists had been oblivious to such difficulties, he went on, and had given us genealogical trees where we may, in the twigs and branches and main limbs and trunk, trace the complete filiation of a group . . . while we cannot but admire the boldness and ingenuity of these speculations . . . we are building in the air. Ordinarily, the twigs of any genealogical tree have only a semblance of truth; they lead us to branchelets having but a slight trace of probability, to branches where the imagination plays an important part, to main limbs where it is finally allowed full play, in order to solve with the trunk, to the satisfaction of the writer at least, the riddle of the origin of the group.[15]

"The time for genealogical trees is passed," he declared.

To clinch his argument, to demonstrate the true futility and full hopelessness of reconstructing phylogeny, Agassiz offered "a simple calculation." We are acquainted, he said, with about 2,300 species of sea urchins, living and fossil. In order to construct a picture of the extinct progenitors and transitional forms which link them into one family tree, we must consider, he seems to be saying, all other imaginable sea urchins, and their number is beyond our control.

Let us take, for instance, the ten most characteristic features of Echini. The number of possible combinations which can be produced from them is so great that it would take no less than twenty years, at the rate of one new combination a minute for ten hours a day, to pass them in review.[16]

Reflecting on this fact should convince us, in Agassiz's view, that there exists no hope of solving "the problem of derivation."

Reconstructing Alexander Agassiz's Calculation

Agassiz's calculation was confused, and his reasoning deeply flawed, but the essence of his argument contains an important and original point. How did he come up with twenty years? To play the game of constructing an imaginary echinoid with ten different characters (a = shape, b = teeth present or absent, c = length of spines, d = pattern of tubefeet, e = position of anus, and so forth), we would need first to know how many different states are possible for each character. In my example, character b has just two states, but the others are indeterminate; before we can proceed we must decide whether our imaginary organisms will have only two shapes (regular or not), three (regular, heart-shaped, or flattened), or some other number. In practice, within a limited group, taxonomists try to find characters that exist in just a few different states, and when they construct a key they carefully limit themselves to characters that are either present or absent, long or short, green or not green. Playing the imaginary-echinoid game with just two states for each character gives 2^{10} different forms, or

1,024 species. Following Agassiz's postulate of one per minute for ten hours a day would take not twenty years but less than two days. Allowing four states for each character would bring us up to around five years, but if we move to five states per character the time jumps to forty-five or fifty-two years (depending on whether we count Sundays as working days). We have not yet reproduced Agassiz's reasoning.

College textbooks of the period included the formula for computing simple combinations, which is $n!$, the factorial of n. The factorial of ten, or $10 \times 9 \times 8 \times 7 \times 6 \times 5 \times 4 \times 3 \times 2 \times 1$, works out to 3,628,800.[17] Looking at each combination for one minute, ten hours a day, but never on Sunday, you would be at this task for 19.32 years. It is my guess that this was the "simple calculation" Agassiz had performed (saying "no less than twenty years" when he meant "somewhat less than twenty years").* That calculation is the correct solution to the following kind of problem: given ten different cards or tokens, how many different ways can we arrange them? Given a set of ten different letters, how many ten-letter "words" can we form? The same letters occur in each word; it is only their order that is different. This is not the correct answer to the question, How many different "species" can I make if each of ten characters varies independently?

Can we define an echinoid game to which the factorial of the number of characters is the correct solution? If the object of the game is to trace the evolutionary history of a given echinoid, and each of ten characters has been acquired by its ancestors at some point in the past, then the number of different sequences in which those events could have occurred is the factorial of n. If the ordered steps by which it acquired its present form is its phylogenetic history, there are $n!$ different histories possible for this one species.† To have defined the problem thus and calculated its solution would have been a remarkable anticipation of ideas which have attracted serious attention only recently, but I think it likely that this is exactly what Alexander Agassiz did do.

Yet, if he did have some such idea in mind, it quickly slipped away from him. By the time he set down the same argument a little later that year in his *Challenger Report,* he had decided his calculation was wrong. Now he upped the number of characters to twenty and stated that the number of combinations works out to 2^{19}. This looks like the first version of our echinoid game, with two states for each character (except that it should be 2^{20}); he was apparently confused by the formula for the total number of possible outcomes where the null case must be excluded, which is $2^n - 1$, not 2^{n-1}).

*My research assistants Stephen Bocking and Gordon McOuat helped me to construct this solution.

†I am indebted to my colleague Roger Hansell for leading me to see this possibility.

He has dropped his reference to how long it would take to study them at the rate of one per minute (which would be less than three years), simply claiming that their number is enormous in proportion to the few hundred genera known.[18]

The switch between these two versions of his argument confirms what his characteristically obscure language suggests, that Agassiz had failed to work out a rigorous statement of the problem. I think he fell short of posing to himself either of the two echinoid games quite as baldly as I have put them. Sometimes he put his case as though the problem was the vast number of possible forms (species or genera), vastly more than we actually have, and vastly more than we could keep in mind if we did have them; if we mean to postulate imaginary ancestors, we ought to pass in review such a staggering number of forms that the exercise could not reasonably be performed (until computers were allowed to play). But the concrete examples he gives makes clear that his central concern was not the number of possible imaginary echinoids but the number of possible systems of relationship, that is, phylogenetic trees, linking those forms together. Early in his AAAS address he had said, "On taking in succession the modifications undergone by the different parts of the test [shell], we can trace each one singly, without the endless complication of combinations which any attempt to trace the whole of any special generic combination would imply."[19] That is, if we limit our attention to a single character in all its various states, we can easily trace a gradual sequence of modifications to form a hypothetical genealogy, but if we combine this with the evidence offered by a second character, and a third, up to the many constituting the definition of the genus, the options are too complicated to follow. At the end of that address, he seems to be using both of our echinoid games at the same time, with a complaint about inadequate fossils thrown in for good measure.

> Let us take, for instance, the ten most characteristic features of [genera of] Echini . . . each one of these points of structure is itself undergoing constant modification [many character states per character] . . . in spite of the millions of possible combinations which these ten characters may assume when affecting not simply a single combination, but all the combinations which might arise from their extending over several hundred species . . . we yet find that the combinations which actually . . . leave their traces as fossils—fall immensely short of the possible number.[20]

At no point did Alexander Agassiz cast doubt on the belief that similar forms are, in fact, genetically related; what he was criticizing was the assumption that we can reconstruct the long sequence of forms constituting the precise evolutionary history of living species, using fossils and taxonomic inference. He was attacking the evolutionists of his day who wanted to make fossil species into ancestors.

Although it was not the basis of his argument, Agassiz also explicitly pointed out that what various combinations of characters we do have give strong evidence that separate lineages can give rise independently to the "same" character (now called "convergence"). Also, in a single long lineage, a character may transmute into an alternate state and then back into its original state.

> The more recent the genus, the greater is the difficulty of tracing in a direct manner the origin of any one structural feature, owing to the difficulty of disassociating structural elements characteristic of genera which may be derived from totally different sources. . . . [Many genera] show affinities [implying descent] with genera following them in time, to be explained at present only on the supposition that, when a structural feature has once made its appearance, it may reappear subsequently, apparently as a new creation, while in reality it is only its peculiar combination with structural features with which it had not before been associated (a new genus), which conceals in that instance the fact of its previous existence.[21]

In an age of triumphant and optimistic science, when faith in progress was pervasive and an unrolling history of life was vivid to the imagination, no one wanted to be told that their hypotheses were vain speculation. Some voices of caution, however, were being raised. The eminent German physiologist Emil Du Bois-Reymond was about to proclaim that there exist natural boundaries that will always restrict our scientific knowledge. In his youth he and other biophysicists had scorned the vitalism of their predecessors by promising that no mystery of life would be able to withstand the techniques of physical or chemical investigation; Du Bois-Reymond had once encouraged biologists to believe that every question to which their current answer was "ignoramus" (we do not know) would in due time yield up its solution. But in a lecture to the Congress of German Naturalists and Physiologists in Leipzig in August 1872, he added a critical restriction. The very nature of our own understanding, he said, made it impossible for science ever to explain either the cause of matter and force or the cause of consciousness. The scientist must acknowledge that there are limits to what can be known, and in those realms "must resign himself once for all to the far more difficult confession—'IGNORABIMUS!' [we shall not know]"[22] Alexander Agassiz was reminding his audience of that solemn proscription when he concluded his 1880 address with the rhetorical question, "Shall we say 'ignorabimus,' or 'impavidi progrediamus' [let us press on undaunted] and valiantly chase a phantom we can never hope to seize?"[23] The mining magnate was not in favor of chasing phantoms.

Agassiz's Latin was as sloppy as his calculations were obscure, but no one missed his central intent.* He was challenging the very foundations of phylogenetic reconstruction.

The Fate of the Ignorabimus Idea

The debate that followed Agassiz's bombshell was reported in *Science* as "one of the most important events of the great Boston meeting."[24] It was not as though he were finding fault with a particular hypothetical ancestor; he was questioning the entire enterprise of seeking to trace ancestries. The next day Edward Drinker Cope, better known for his paleontology than for his scrap with Hagen, was scheduled to speak on the evolution of the cat family, and he closed his paper by proclaiming that his successful use of fossil teeth to reconstruct the history of this group was evidence that refuted Mr. Agassiz's pessimism. Burt Wilder, former student of Louis Agassiz and now professor at Cornell, endorsed the general sentiment of Alexander Agassiz's address by warning against "the fallacy of hasty generalization." The discussion closed with Agassiz "emphatically repeat[ing] his statement of the day before."[25]

In spite of the interest it aroused on the spot, the debate over Alexander Agassiz's challenge to the devotees of genealogical trees soon evaporated. Cope, in his role as editor of the *American Naturalist,* quoted half a page of Agassiz's argument and made the powerful point in reply:

> If Mr. Agassiz had insisted that any or all of the millions of possible combinations he has pictured may have existed as extinct species, he would indeed have presented us with an inextricable genealogical puzzle. But he does not do this, for he admits that the number of the forms which have actually existed is limited. Does Mr. Agassiz mean that there has been no order in this limitation; that there have existed no causes which have rendered some combinations possible and others impossible? [If that were so] we can expect to find fossil centaurs or sphinxes.[26]

Agassiz's complaint, said Cope, was "the objection of a mathematician, and not that of a practical biologist"; the success of his own work on the fossil cat, camel, and rhinoceros families (and O. C. Marsh's on the horse) was evidence enough, he claimed, that something was awry with such ab-

*He should have said "impavidi progrediamur." My Latin being at least as rusty as Agassiz's, I am grateful to my colleague at Victoria College, Wallace MacLeod, for this correction.

stract figuring, for "no question of mathematical probabilities can invalidate the significance of the wonderful closeness of the successive stages which they present."[27] That wonderful closeness, however, was largely a result, as Cope fully acknowledged, of the practice of defining groups according to just one character. "[Mr. Agassiz] admits that genealogies of single characters may be constructed; therefore genealogies of orders, families and genera can be constructed, for they are, or ought to be (for they ultimately must be), defined by single characters."[28] *Nature*'s reviewer of Agassiz's *Challenger* report had reacted the same way, reading it, he said, "with a sort of shudder," and though he could not put his finger on the flaw in the mathematics, he concluded that since a connection between recent and fossil sea urchins must exist, "we cannot see why the attempt to indicate it graphically or descriptively is to be condemned as futile."[29]

Agassiz's arithmetic of permutations may have been a bit confused, and his mode of expressing himself was certainly opaque, but he had perceived a dilemma of fundamental importance in the area of intersection between classification and evolution, one that specialists in both fields have been the poorer for ignoring. Charles Darwin, aged seventy-one, with characteristic sagacity, did see some of what he was getting at. He wrote Agassiz,

> I read your address with much interest. However true your remarks on the genealogies of the several groups may be, I hope and believe that you have overestimated the difficulties to be encountered in the future. A few days after reading your address I interpreted to myself your remarks on one point (I hope in some degree correctly) in the following fashion:—
>
> > "Any character of an ancient generation or intermediate form may, and often does *re*appear in its descendants after countless generations, and this explains the extraordinary complicated affinities of existing groups."
>
> This idea seems to me to throw a flood of light on the lines, sometimes used to represent affinities, which radiate in all directions often to very distant subgroups—a difficulty which has haunted me for half a century. A strong case could be made out in favor of believing in such reversion or atavism after immense intervals of time. I wish the idea had been put into my head in old days, for I shall never again write on difficult subjects.[30]

Darwin had captured some but not all of what was bothering Agassiz. Atavism, reversion to an ancestral character state *A*, is one source of difficulty in trying to infer ancestry, but it is not the only one. By assuming that each character could vary independently of all the others, Agassiz was allowing for those situations in which species that were distant cousins could both independently acquire some new character state A' not present in their common ancestor, the same feature as far as appearances go, distinct only in their histories. If the laws of evolution are such that this kind of thing happens very often, then groups uniting similar forms will not accurately reflect phylogeny.

Alexander Agassiz, cantankerous though he was, did not enjoy debate. He felt isolated in his views. Perhaps the ambitious professionals who were flogging evolution so enthusiastically could not afford, as Darwin could, the luxury of giving a thoughtful and sympathetic reading to awkward prose bearing a disturbing message. For his part, Agassiz had no reason to throw his time and energy at an unwilling audience. To his uncle, the eminent botanist Alexander Braun, he expressed his feelings:

> I hope . . . the younger men who are indulging in such high flights . . . [will] put their noses down to the grindstone again and do a little hard work before they finish unravelling the mystery of creation. It is hard . . . to fight against the crowd, and the mania which seems to have seized all the younger workers makes me often doubt the wisdom of saying anything, it all falls so flat; still I know our time will come, and those who have kept cool, and continued to work quietly during this time of transition, will find themselves some day just so much ahead of their . . . opponents, and for that time I am quietly waiting.[31]

He was right that science does go through periods of fashion, when voices challenging the dominant view are paid no heed. He was wrong to expect that such an unorthodox idea as his would gain currency in his own lifetime, irrespective of his leaving it unchampioned. I know of only one writer who alluded to what he called Alexander Agassiz's "rather obscured thought" in later years.[32]

Perhaps his challenge was a foolish one and received all the consideration it deserved during the 1880 Boston meeting? If so, there is an entire subdiscipline of foolish academics who are wasting time (their own and their computers') with similar games today.* They have constructed rigorous proofs that, if you create a set of imaginary forms and let them vary and diverge step-by-step to "evolve" new ones, you can indeed end up with "descendants" whose relationships cannot be inferred with certainty. The number of genealogical trees that might plausibly have connected them is so enormous that the job of comparing them cannot reasonably be performed, not even by computer, and certainly not by the human mind.[33] Given a moderate number of species, the number of possible genealogical trees connecting them is so large that the puzzle of their relationship is not solvable. We know that only one solution is correct, for they did evolve, yet an enormous number of other solutions can also look correct. The extreme implication of that model is that knowledge of phylogeny is unattainable.

Agassiz was right to see that believing that something really did take

*I am indebted to Roger Hansell and Jim Riosa for helpful discussions.

place does not require us to believe that its history is recoverable. It is an important new stage in our understanding of taxonomy, that there are now serious disagreements about the precise methods proper for constructing phylogenies, even including several schools of theorists who insist that they must always be so speculative that it should not be attempted at all. But so, too, are the possible moves in a game of chess beyond practical calculation, and still the game can be played. We learn something important about the nonmechanical heart of human reason when we teach our chess-playing computers to follow model games, to adopt rules of thumb, to learn from their mistakes, and to act upon likelihood instead of waiting upon certainty. The mathematical model speaks to absolute proof, while the scientist expects only the best likelihood.

Cope was saying, in effect, Why should we care for your numbers when we are building such lovely phylogenetic trees? But Cope's trees, based on very few characters, no longer look so lovely. He would have improved his inferences had he taken Agassiz's challenge seriously. Just as we can play endless new chess games for our amusement, you can indeed construct phylogenies, but why should the Alexander Agassizs of today not accuse you of building high flights of fancy in the air?

Of course, neither Agassiz's calculation nor a modern computer simulation fully describe the world of flesh and blood. Agassiz's model assumes random and independent variation of each character, whereas we are convinced that on this planet very many characters have acted as a conservative background of stability against which a small amount of interesting deviation stands out. We have a claim to be describing the real world only if we can show what powerful extra clues nature gives us for shortcuts, what simplifying assumptions we are justified in adopting, and what care we take to avoid asserting that something is so because we wish it to be so.

7

"The Slender Thread Is Practically Severed"

Late in 1872, two of Louis Agassiz's brightest students were following events in the M.C.Z. with intense interest. Alpheus Hyatt, on leave from his two part-time jobs (custodian [curator] of the Boston Society of Natural History and instructor at the adjacent Massachusetts Institute of Technology) was visiting Germany to investigate a series of fossil snails he thought showed non-Darwinian evolutionary forces. Samuel Scudder, a well-respected but underemployed entomologist, despondent over the death of his wife, was thinking of moving West. Hyatt advised against it, hinting that his own position at the BSNH might fall vacant, if a professorship developed at Harvard:

> I look also for considerable changes in the Museum at Cambridge in the course of a few years. Whether Agassiz lives or not I think the government must pass out of his hands and the institution become more Americanized. There are many things that point that way. The battle between the conservators of the Museum on the one hand and the teachers in the College, who wish to use the specimens, has already begun. In this country it cannot be doubtful who will win, and as Shaler is the leader on the College side, there will be no very merciful use of victory.*

Though the broad Atlantic instead of only the Charles River lay between Hyatt and the Agassiz museum, he was right about the clashing armies. His prediction of a swift and clear victory, however, could not have been more wrong. The battle dragged on for the rest of the century and well into this one. Louis Agassiz, ever the politician, took care to keep the struggle hid-

*Hyatt to Scudder, n.d., Museum of Science, Boston. This letter, though undated, can be assigned to late 1872 with confidence because of Hyatt's reference to Scudder's mourning (Scudder's wife died in June of 1872) and the recent Boston fire (which occurred on November 10 of the same year).

den; although his son took rather less care, few outsiders recognized the manned trenches and barbed wire. This long war was tragic indeed, shattering the ideal institution Louis Agassiz had glimpsed, the one that would cradle a new discipline, and thus also destroying whatever possibility may have existed for the elevation of taxonomy from amateur natural history to a rigorous department of biological science.

The battle Hyatt wrote Scudder about, with their former classmate Nathaniel Southgate Shaler "the leader on the College side," is concealed in the *Annual Report* of the museum that had appeared the previous spring. Describing the activities of 1871, Agassiz gave the impression of wonderful cooperation within the museum.

> Professor Shaler is now chiefly occupied with teaching our under-graduates; and, in order to render his instruction more practical and impressive, every possible effort is now making to prepare suitable collections for the lecture room, for the use of the professor, as well as for the purposes of the students, of which Professor Shaler will have charge. All the heads of the different department[s] are instructed to assist Professor Shaler in the preparation and systematic arrangement of these collections. . . . Besides this, Mr. Shaler continues to take part in the work of the Museum, and is at present devoting his attention to the Silurian Fossils.[1]

Only someone aware of the changes underway at Harvard would notice in the phrase "teaching *our* under-graduates" Agassiz's subtle assertion of his authority over all students in the museum. While promising "every possible effort . . . to assist" the college, Agassiz leaves no doubt that the museum's own curatorial staff would decide the disposition of specimens. In his letters to Shaler, Agassiz was less subtle: "I shall do everything that can be done to accommodate as large classes of working students as are likely to be brought together. . . . But the thing must be done by the administration of the Museum and not by that of the College."[2]

Of all his students, Shaler (fig. 18) was, Agassiz said in 1866, "one . . . whom I love best."[3] He had welcomed him back to the museum after Shaler's service in the Civil War, delegating him to teach in Agassiz's place as well as giving him responsibility for the paleontological collection. For his part, Shaler set about making sure that his teaching would satisfy his employer both administratively and intellectually. He suggested helping the museum meet its obligations to the Commonwealth of Massachusetts by initiating courses of public lectures. Agassiz knew of Shaler's sympathy for evolution, but Shaler assured him that

> I hope to . . . impress upon the student the feeling which I have acquired from your teachings and which I value above any other intellectual result: that there is an intelligence guiding the changes of nature. . . . the leading thought of the lectures on

Figure 18. Nathaniel Southgate Shaler, student of Louis Agassiz, instructor in the museum, initiator of the summer school idea, first dean of Harvard Graduate School. (By permission of the Museum of Comparative Zoology Archives, Harvard University)

paleontology will be the evidences of an intellectual plan in the history of the animal kingdom.[4]

Likewise following Agassiz's rhetoric, Shaler proposed "to begin the formation of type collections of specimens"[5] to illustrate elementary teaching.

The battle began after Shaler's position in the museum had been fundamentally altered from this lieutenantship. The lines of power shifted in 1869, when Harvard University chose a vigorous new president, Charles William Eliot. Shaler was "the leader on the College side" because he was affiliated with Eliot.

Everyone recognized the appointment of Eliot as a bold move, signaling a new era in American education.[6] It was also seen by some as a defeat for Louis Agassiz. This appears paradoxical, for Eliot was expected to enlarge the importance of the sciences in the university. The changes Eliot aimed at, though, could only lessen the independence of Agassiz and his museum. Previous presidents had restricted their attention to Harvard College, allowing various associated foundations, like the medical school, scientific school, and botanical garden, to run themselves. Agassiz, whose museum operated with gifts and endowment given specifically to it by private donors and the state legislature, was indebted to Harvard only for a few hundred dollars of seed money and a wooden building. With Eliot as president, there were sure to be pressures to make the museum devote itself to undergraduate instruction.

Eliot knew that Agassiz had initially opposed his appointment to the presidency, and he had accepted the post only after conferring with his good friend and first cousin, Theodore Lyman.* Lyman had "advised him to accept, promising to stand by the Museum with Alex."[7] Alexander Agassiz's close association with Eliot went back at least to 1858 when they had rowed together in a famous crew race.† Louis Agassiz's two most intimate supporters were resolved to see the M.C.Z. cooperate with Eliot's reforms.

Eliot's inaugural address contained an ominous warning that the price of developing undergraduate science might be high. He declared that

the prime business of American professors in this generation must be regular and assiduous class teaching. With the exception of the endowments of the Observatory, the University does not hold a single fund primarily intended to secure to men of learning the leisure and means to prosecute original researches.[8]

*Lyman's father was the brother of Eliot's mother; the boys were born a year apart.

†This Harvard crew, the first to sport crimson, consisted only partly of undergraduates. Eliot and Agassiz were several years beyond their Harvard B.A.s; Eliot was tutoring chemistry and mathematics, Agassiz was doing postgraduate study in chemistry (G. R. Agassiz, *Letters and Recollections of Alexander Agassiz,* p. 22; James, *Eliot,* 1:79–85).

Eliot's failure to mention the Gray bequest and other funds of the M.C.Z., which were held by the Harvard Corporation, could well have been meant and taken as an attack upon the idea that Agassiz's collections should be devoted to the advancement of scientific zoology. Eliot, of course, had no animosity toward research, and, a chemist himself, he certainly favored science. However, Agassiz's ambition to compete with Europe by fostering original research, at a time when the provision of solid introductory science courses in the regular college curriculum had scarcely begun, Eliot judged premature.

Nevertheless, there seemed every prospect for genuine and successful cooperation between the museum and the university. Lyman and Alexander Agassiz both rejected the idea of joining the Harvard faculty themselves, but Lyman advised Eliot that giving an appointment to a person paid by the museum might be a good way to soften Louis Agassiz's hegemony. Hagen received the title "Professor of Entomology" in 1870. Lyman, as the go-between in a negotiation between Eliot and Agassiz in August 1870, was "somewhat surprised to find that Agassiz' ideas are not precisely what I had always supposed them,"[9] perhaps because Agassiz insisted that the museum "is not called on to furnish a teacher; but, on the contrary, its 'employés' are persons who are expected to pursue original scientific work."[10] If Eliot wanted zoology teaching in the college, he would have to pay for it. The transition was deceptively easy. Harvard would assume responsibility for Shaler's salary, hitherto paid by the museum. His title changed from "Assistant in Paleontology" to "Professor Shaler, Assistant in Charge of Instruction." He gave up his curatorial work on fossils to devote himself full-time to his courses, but he was still an "assistant" in the museum.

But Eliot gave Shaler more than a professorship. He seems to have given him the means to hire his own assistants. For the academic year 1870–71 Shaler reported,

> Instruction in my department in microscopy, under the superintendence of Mr. Tuttle, has also been provided for beginners and for advanced students. This teaching extends over one-half the year, and enables the student to become master of the practical detail of microscopic work, so far as is required in all ordinary investigations.
>
> A course of elementary instruction in the study of insects has been given by Mr Edward Burgess, who has acted a college instructor in this branch. [Tuttle and Burgess would continue this teaching the following year]. . . .
>
> I am also indebted to Dr. Hagen . . . for assistance given to particular students in the prosecution of especial parts of their work.[11]

Ironically, the initial sources of conflict grew out of the combined success of Louis Agassiz's pedagogy and his museum-building. As a student

Shaler had been so impressed by the effectiveness of Agassiz's unusual teaching methods that he consciously decided that every one of his own students should handle specimens.[12] Agassiz himself had used this method only with the handful of young men who came to him determined to become professional naturalists; for large classes he was quite content simply to lecture with a blackboard.[13] Shaler reported with pride that in 1871–72 he gave "practical teaching in the laboratory" to ninety-five students.

The student is compelled to come at once into the position of an investigator, receiving only such assistance as may be required to help him to help himself. The first year's lectures are designed only as an adjunct to the other work. When the students have been carried, in a practical way, through one group of animals, becoming acquainted with its outlines by the use of typical forms, the group is taken up in the course of lectures and reviewed. In the second year's course the same system of practical work is continued, but each student is now required to take up some limited subject and devote time enough to his work to attain a thorough knowlege of it.[14]

His determination to make practical work part of every course, even the introductory ones, created the need for a special student workroom adjacent to the lecture room, with tables, microscopes, and specimens. His quest for specimens, however, immediately reencountered opposition.

Every step of progress achieved by the curatorial staff over the museum's first decade, in sorting, labeling, and cataloguing specimens, had the effect of making less feasible the liberal policy that had figured in Agassiz's inspirational talks. Hermann Hagen gave a public explanation as to why teachers could not expect specimens to be provided by the museum.

The arrangement of a scientific collection is the work of a long and careful study, and represents in itself the result of many years' investigation. Even if the Museum, for the sake of more rapid progress, buys a scientific collection, or parts of it, the money spent represents not only the specimens but also the time bestowed upon their identification. Of course a collection representing such a mass of work, and beside this so easily damaged, could not be intrusted to any one not accustomed to handle such valuable objects.[15]

Hagen suggested that instead of using the main collection, an undergraduate should be provided with specially selected series of specimens "containing the most striking forms of all classes taken chiefly from the country about him, perhaps with the addition of some important forms not represented in the native fauna."[16] It would be "an admirable training" if the students had to help enlarge such collections, Hagen added. There is no evidence that Hagen's plan for exposing beginners to systematics was ever implemented. Instead, the specimens used by beginners were in most cases a few objects for dissection, not a series large enough to exemplify the problem of how to separate related species and decide whether they belong to one genus or to several.

The extra space laboratory teaching required also competed with curatorial priorities. Agassiz told Shaler in 1871,

And now tell Lyman . . . to press on the building of the Museum, since you can have no fitting accommodation for the students before our addition is completed.

And again,

If I were to surrender that room to students' work the Museum would suffer materially. . . . You will remember that when I told you I could only give up one room for students' work, it was because the Museum assistants should be provided for.[17]

The following year new money from the state, and from Agassiz's son and son-in-law, relieved the pressure on space. A fifth floor was added, and a new segment constructed which doubled the length of the building (see figs. 19 and 22, below).

Shaler's commitment to giving students direct experience went beyond the classroom, again imitating, in magnified form, Louis Agassiz's teaching practice. In the museum's first years, several of his chosen students had been sent into the field, for the benefit of their own experience as well as for the sake of the specimens. Shaler, Hyatt, and Verrill had spent a summer collecting on Anticosti Island in the Gulf of St. Lawrence. Caleb Cooke was sent to Zanzibar. Shaler translated this, as early as 1868, into a wish to put even elementary students out-of-doors.[18] Thus when Agassiz expressed to Shaler his concern that the museum was not fulfilling its promise to educate schoolteachers, Shaler devised a plan for a seaside summer school on Nantucket Island.[19] The outcome, transformed by Agassiz's fund-raising, was the Anderson School of Natural History, held on Penikese Island west of Nantucket, in 1873 and 1874. In Agassiz's mind it was connected with the M.C.Z., but it was yet another separate legal entity, and its students were not Harvard students.

Lyman's promise to Eliot that he and Alex would "stand by" the museum was faithfully kept. Louis Agassiz had suffered a cerebral hemorrhage in September 1869, at the age of sixty-two, and, though he gradually recovered, Lyman and Alex bent their attention over the next few years to how the museum, without the energy of its founder, would meet its responsibilities, including the new undergraduate teaching. Alexander had been watching Eliot's university-wide reforms closely and wrote Lyman in the summer of 1872,

The present arrangement is *bad*. Shaler who teaches the undergraduates is perfectly irresponsible and [so is] Eliot's attempt to try to introduce in the Museum the same tactics which have given so much dissatisfaction in some of the Depts. of an Assistant who is not responsible to the Professor in charge of the Dept. . . . There are none of father's students whom I should care to see back here and take any share in running the machine.[20]

Lyman replied,

The great rub is a *teacher,* as you say. There are plenty of men to teach, "school fashion"; but not to teach as Papa's Museum expects teaching to be done. I have no fear of Eliot's doing anything really troublesome; the thing is to find a *teacher.* . . . You know McCrady was once suggested. He was a man of ability, with all his queerness. If he has solidified, and has continued his studies, he would be a man who would keep things lively, and would turn out some good students. We don't want a sleepy-head, you know.[21]

Before Eliot's reform, teaching within the museum had been haphazard, sometimes nonexistent, or limited to special topics. Now that geology and botany and zoology were Natural History courses in the Harvard College calendar, they could not be omitted because an instructor was ill or out of town. Shaler departed at the end of 1872, temporarily burned out, and in 1873 he accepted the position of state geologist of Kentucky. (He kept on close terms with his mentor, though, and would return to the museum in 1874, again on Eliot's payroll, as instructor and then professor, in paleontology). So in 1873 Louis Agassiz wrote to his former student John McCrady of Charleston, inviting him to take charge of teaching the Harvard undergraduates. By this time Shaler's mode of teaching, with practical work supplementing the lectures, was well enough established that McCrady was given two teaching assistants to handle the laboratory portion of the courses.

Poor McCrady, "with all his queerness," was not up to the opportunity. Before the War between the States he had indeed showed promise, publishing several papers on hydroids and giving close attention to the particulars of Darwin's theoretical structure, but the war shattered his constitution, his finances, and his mental balance. He arrived in Cambridge, aged forty-two, determined to use his teaching to rescue Christianity from the evils of Darwinism (a plan Agassiz warned him against pursuing in the classroom), anxious to resume the scientific productivity he had lost ten years before, but hopelessly out of touch with his peers. When Louis Agassiz died in December 1873, McCrady insisted that he had been promised he would be placed in charge of the museum. Agassiz had made no such promise on paper, though it would have been very like him to have encouraged McCrady to believe that he might find himself in a position to succeed Agassiz as professor. The next four years did not go well, and Eliot informed McCrady that the 1877–78 academic year would be his last.* What was stated publicly was that "Prof. McCrady's lectures on the zoology of the invertebrates were unfortunately interrupted during December [1877], owing to a dangerous illness, which compelled him to leave Cambridge for the south."[22]

McCrady's laboratory assistant, however, stayed on. It was none other than Walter Faxon, belated discoverer of the alternation of form in male crayfish, successor to Hagen as classifier of crayfishes. Born and raised in the Boston suburb of Jamaica Plain, an avid birdwatcher, he had followed his Harvard College undergraduate degree with a further year of study (probably under Hagen's supervision) earning a bachelor of science degree in 1872. In 1873 Louis Agassiz placed him in charge of the museum's crustaceans. When McCrady left, Faxon was working on his doctorate, which was awarded in 1878. Serving in the multiple roles of instructor, researcher, and curatorial assistant, Faxon would seem to illustrate a revival of the successful integration Louis Agassiz had achieved before the mutiny.

In the first few years of the museum's new regime, from 1874 to perhaps 1880, it seemed as though the skirmishes between Shaler and Louis Agassiz that Hyatt had interpreted as a battle to the death between the "conservators" and "the College side" had been supplanted by a peaceful and mutually rewarding cooperation. Alexander Agassiz, in spite of his constitutional crankiness about Eliot's "tactics," believed wholly that the museum could fulfill both its original mission as a center of research and advanced training and also its new role as the science department for Harvard College. He envisioned an integrated operation, where people under his roof, whether paid from his budget or from Eliot's, would divide their time among research, teaching, and curatorial work. "The practical knowledge to be gained from the care of a special department is a necessary requisite for a successful teacher," he declared, and after assigning a new instructor, E. L. Mark, to arrange the museum's worm collection, Agassiz reported with satisfaction, "The bottles have all been thoroughly cleaned, and alcohol changed when necessary."[23] To comprise the other side of this

*My information on McCrady was limited to the bare record of the *Annual Reports* until I learned of the wonderful detective work of Lester D. Stephens, who has studied McCrady's letters and diaries. Stephens has kindly sent me copies of his "John McCrady: pioneering embryologist in the Old South" (lecture delivered at the History of Science Society, Raleigh, North Carolina, October 1987), and "John McCrady: pioneer student of North American Hydrozoa" by him and Dale R. Calder (delivered at the International Conference on Coelenterate Biology, London, July 1989).

cooperative understanding, Charles E. Hamlin, a paleontologist on the museum payroll,

> has been detailed this term [and would be again the following year] to take charge of the undergraduate instruction in structural geology and physical geography. . . . This has somewhat lessened the amount of his Museum work, but the Curator [Alexander Agassiz] has cheerfully consented to this diminution in view of his increasing usefulness in another direction.[24]

In 1876 Lyman and Alexander Agassiz convinced the trustees of the Commonwealth to simplify administration by conveying its interest to Harvard. Agassiz's description suggests that he expected the change to reinforce the authority of the head of the museum over the Harvard College instructors who worked within it, for he stated that

> the instruction in natural history at the Museum has gradually been assumed by the College in exchange for the facilities given by the Museum in the way of collections and laboratories, the Museum retaining, however, the general direction of the educational interests.
>
> The Corporation of the College now pays annually, in salaries connected with the Museum, nearly as much as the original income of the Museum itself. This makes a closer official relation between the Museum and the University doubly important, were it only for the fact that the specimens necessarily accumulated by the instructors in the several departments of natural history are, under the present conditions, merely deposited in the Museum without becoming an integral part of its collections.
>
> . . . The direction of the instruction belongs to the Curator. . . . this centralization [including the new Peabody Museum of Archaeology and Ethnology] will undoubtedly build up, within a comparatively short period, a comprehensive institution of natural history, with facilities enjoyed by few like establishments elsewhere.[25]

What this "general direction of the educational interests" meant legally, Eliot was too canny to press to a test. As president of the University, he was ex officio a member (as was Agassiz) of the five-man Faculty which set museum policy for the curator, and Eliot made sure that Agassiz joined the seven-man Corporation which ran Harvard, which he did in 1878. In practice, what happened was that Eliot consulted Agassiz about decisions, Eliot himself did the hiring and firing, while both men left the instruction to the instructor.

Professor Mark Builds a School

To replace John McCrady, Edward Laurens Mark was hired in 1877. Many years later he recalled his first meeting with Alexander Agassiz, who had been away when Mark first settled into the museum. "Mr. Agassiz . . . looked at me with a quizical, half-humorous sort of an expression about the mouth which I have occasionally seen since—as much as to say 'What sort of a bird is this new venture of Eliot's?' "[26] The story, if a wordless look can be a story, does reflect Mark's awareness of the delicacy of his position. He was in no doubt about his first priority, though, and threw himself into teaching wholeheartedly.

Mark was the same age as his new colleague Faxon (both men just turning thirty), and they divided their work as equals. But equals they were not. Mark, with his new Ph.D. from the University of Leipzig under his arm, was determined to bring back home the latest techniques in cytology and the high standards he had learned in Rudolf Leuckart's laboratory.

Mark much later recalled that the publication of Huxley and Martin's textbook, *Practical Biology,* and the founding of the Biology Department of the Johns Hopkins University in 1876, had created a demand for courses aimed at biological principles, that is, phenomena common to plants and animals.* A major revision of the Harvard College courses was put in place during Mark's second year of teaching. A new course, prerequisite to higher studies in zoology, required "a large amount of laboratory work, use of the microscope, and mechanical preparation of specimens." The zoology half of the course consisted of a close study of a hydra, amoeba, sea anemone, starfish, lobster, oyster, and clam. "Besides seeing, describing, and drawing every point in the structure of the Hydra, each student is required to stain and mount on glass slides microscopic sections for permanent preservation."[27]

Though often away from Cambridge, and always distant in his manner, Alexander Agassiz gave enormous material support to the teaching enterprise in the museum. If Mark wanted new microscopes, or the latest microtome (a machine for slicing ultrathin sections of a specimen), Agassiz would reach into his own deep pocket.[28] Already in 1875 he had sketched out ambitious plans for the growth of the building, intending it to house not only collections but all of Harvard's teaching in botany, geology, anatomy, and physiology as well as zoology.[29] The copper mines were making

*Mark, "Zoology, 1847–1921," pp. 386–87. Mark erroneously places in 1874 the appearance of the Huxley and Martin text, which was first published in England in 1875 and in the United States in 1876.

Figure 19. The museum building as it was from 1859 to 1872, taken from the Divinity Avenue side. That the entrance faced the Harvard Divinity School was cited as indicative of the close relation between science and religion. The building was rectangular, with the entrance on the end, although foreshortening makes it appear rather square in this picture. (By permission of the Harvard University Archives)

Agassiz steadily wealthier, so the sketches bit by bit became reality. (For the growth of the M.C.Z. building, see figs. 19–36).

A new segment of the building, constructed in the summer of 1878, had rooms designed for teaching, but most impressive was the space set aside in the very big northwest "corner piece," with laboratories, offices, and a large sunny lecture room under the roof peak on the fifth floor (fig. 34). Completed in 1882, this section of the building (fig. 29) had two doors, anticipating the outside and inside of the planned open quadrangle (finally completed in 1913). The architecture was only slightly less workman-like than Louis Agassiz had begun with, but his son did allow himself the small ornamental touch of carved brick letters over the two doors, N.H.L., for Natural History Laboratories, over one (fig. 26), and his own initials over the other. Agassiz did take pride in having made Harvard such a handsome gift.

E. L. Mark's first year of teaching was the sophomore year of Theodore Roosevelt, whom the twentieth century would thrust into the presidency of

Figure 20. The museum building as it was from 1859 to 1872, taken from the Oxford Street side, across the athletic field. This photograph confirms the recollection of J. H. Blake, whose map (fig. 6) showed an entrance on this end. When fund-raising, Louis Agassiz declared, "Whenever any structure is put up for the museum, it should not be built in a corner where it cannot grow, but be placed on such ground as will never be an impediment to its indefinite increase" (*American Journal of Science* 27 [1859]: 298). (By permission of the Harvard University Archives)

the United States, but who was in 1877 a teenager and keen amateur naturalist. In his autobiography Roosevelt would later assert,

> I did not [continue in science] for the simple reason that at that time Harvard . . . utterly ignored the possibilities of the faunal naturalist, the outdoor naturalist and observer of nature. They treated biology as purely a science of the laboratory and the microscope, a science whose adherents were to spend their time in the study of minute forms of marine life, or else in section-cutting and the study of the tissues of the higher organisms under the microscope. . . . I had no more desire or ability to become a microscopist or section-cutter than to be a mathematician. Accordingly I abandoned all thought of becoming a naturalist.[30]

Although we can well imagine that Roosevelt, like many whose first delight in science blossoms out-of-doors, found work in Faxon and Mark's zoological laboratory uncongenial, we should mistrust the implication that this was the whole story. For one thing, in his junior year, Roosevelt became interested in his political economy courses, and also began a se-

Figure 21. The M.C.Z. as it looked between 1872 and 1878 from the Oxford Street side. Comparing this with the previous figure, we can see that besides doubling in length, the building has been enlarged by the addition of a fifth storey. Once again, the rough end wall tells how firm was the expectation of further growth. (This photograph was taken at the first intercollegiate football game, Harvard vs. McGill, 15 May 1874.) (By permission of the Harvard University Archives)

rious romance, which may well have altered his image of an ideal occupation. More relevant than the microscopes may have been the information that for a successful career as a professional biologist he would have "to study three years abroad," a prospect that made him "perfectly blue."[31] This appears in his diary in December of his junior year, which was before he had to face Faxon's hydras and fixation techniques. Mark recalled that he had been teaching for several years before he was able to expose students to the use of the microtome,[32] so this, too, would have been after Roosevelt's career plans had already changed. After graduation Roosevelt married and began law school.

Of course, Roosevelt never did abandon natural history. Endowed with an exceptional zest for adventure, love of the outdoors, and family wealth, he became a rancher, hunter, and writer. He produced a number of books, with titles like *Hunting Trips of a Ranchman* and *African Game Trails*. So enthusiastic, informed, and sustained was his interest in the natural world that we would not hesitate to call him a naturalist, for to us the word includes, indeed tends to imply, a nature lover who is not a professional

Figure 22. The M.C.Z. as it looked between 1872 and 1878 from the Divinity Avenue side. We can read some of the building's history from the pattern of its windows: a cross-section in the 1875 *Annual Report* shows the basement, first floor, first gallery, second floor, second floor's gallery, and, sitting above the original roofline, the attic floor. By 1880 many of the galleries had been floored over to increase workspace, and the stories numbered one through five. (By permission of the Harvard University Archives)

biologist. To Roosevelt himself, however, to be a true naturalist was a full-time calling, one his experiences in Agassiz's museum deflected him from.

The emphasis on the latest German cytological techniques in Harvard teaching increased when Mark became the senior figure in the zoological laboratory over the next decade. In 1885 Eliot appointed Mark to the Hersey Chair of Anatomy, left vacant by the death of Oliver Wendell Holmes.* The following year Faxon, aged thirty-eight, only eight months younger than Mark, resigned. "I presume he would stay if made Professor," wrote Agassiz to Eliot.† It seems likely, though, that having to pay two professors was just what Eliot was trying to avoid, and he may even have

*Harvard had split the professorship between Holmes and Jeffries Wyman, who had died eleven years before (Dupree, *Asa Gray*, p. 149).

†A. Agassiz to Eliot, 25 May 1886, Eliot Papers, Harvard University Archives. A curious feature of Faxon's lectures was his ability to take catnaps in mid-sentence (Parker, *The World Expands*, p. 45).

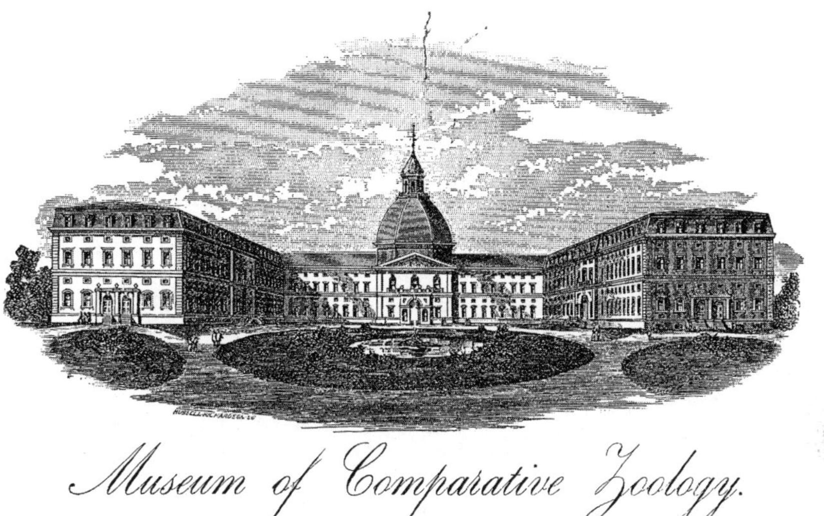

Figure 23. Vision of the M.C.Z. of the future, seen from Divinity Avenue. Although this drawing is undated, it must have been made between 1872 and 1876, for it shows the fifth storey (see previous two figures), but does not reflect the actual shape of the Peabody Museum of Archaeology and Ethnology, which in 1876 filled the site of the left end of this imaginary building. It probably predates 1875, when Alexander Agassiz included in his *Annual Report* a slightly different vision, with the same dome, but with the main entrance on Oxford Street, as it would later be constructed. The dome remained a fantasy. (By permission of the Museum of Comparative Zoology Archives, Harvard University)

impelled Faxon's departure. Agassiz discussed with Eliot the merits of Agassiz's assistants Jesse Walter Fewkes and Charles Otis Whitman. Both had doctorates, and, aged forty-three and thirty-six, either of these men would have been approximate replacements for Faxon. Another name, Howard Ayers, had also come up (presumably suggested to Eliot by Mark), but if he were appointed, Agassiz warned,

> we should be decidedly losing ground and trying an experiment of running the Dept with one head and assistants—to change hereafter constantly—as we cannot hope in this country to keep a man long as Assist. as in Germany. . . . I should feel safer in sounding Faxon and getting him to stay as Professor.[33]

Nevertheless, Eliot did choose the twenty-five-year-old Ayers, who had received his bachelor's degree from Harvard only three years before and was in the process of completing his German doctorate.

Agassiz's prediction proved perfectly accurate. Ayers stayed only two

Figure 24. The M.C.Z. between 1882 and 1886, from the Oxford Street side. Comparing the oldest segment, on the right of this photograph, with fig. 22, we can see one of the changes Alexander Agassiz made in 1878: the gallery windows were extended downward (*Annual Report* for 1877–78, pp. 3–4). In the foreground is the "corner piece" with bright granite stairs leading up to the entrance of the Natural History Laboratories. The dirt carriageway and board sidewalk lead back to the arched doorway to the public exhibits. The wall in the center of this photograph, its windows becoming doors, would sprout the next stage of the building in 1888. (By permission of the Museum of Comparative Zoology Archives, Harvard University)

years, to be replaced by an even younger man. George Howard Parker was hired at the age of twenty-three, his only degree a baccalaureate in science from Harvard. Parker reports how severely he was warned when he was hired not to expect to be kept on for more than a year. Hindsight tells us that Parker would finally "make good," become chairman of Harvard's biology department, and finally in 1921 replace Mark as director of the zoological laboratory, but none of that was in Eliot's plan. Parker had to agree with Agassiz's scornful remark, that the teaching of zoology at Harvard was entrusted to "a man and a boy."[34]

Parker's career actually illustrates in a positive sense the same pressures that had seemed discouraging to Roosevelt. It was the reputation of Louis Agassiz, though by this time he had been dead for ten years, that drew Parker to Cambridge in 1883. Having worked for several years in the butterfly

Figure 25. Same stage of the M.C.Z. as seen in fig. 24, from across Oxford Street and the athletic field, Holmes's Field on Jarvis Street (compare to figs. 20 and 21). (By permission of the Harvard University Archives)

collections of the Philadelphia Academy of Sciences, already skillful at identifying the bones of mammals, Parker had just the sort of enthusiasms the Swiss professor would have loved. Soon he was happily absorbed in studying the histology of sensory nerves. The necessity of including a European apprenticeship did not deter him. So pleased was he with his career in science that he titled his autobiography *The World Expands.*

Mark (fig. 37) was noted for pushing his students, even the undergraduates, to do original research and then publish their results. "From the time you took your research problem under him till the last punctuation point marking the completion of your printed thesis you were under Dr. Mark's eye."[35] He was fanatical about bibliographic thoroughness and precision and has been credited with introducing the form of citation now standard in scientific writing.[36] Mark's students wrote theses with titles like "The Development of the Pronephros and Segmental Duct in Amphibia," "The Segmentation of the Nervous System in Squalus Acanthia," or "The Maturation, Fertilization, and Early Cleavage of Bulla Solitaria." A number of articles in the museum's *Bulletin* from about 1890 are labeled "Contributions from the Zoölogical Laboratory," which meant that Mark had supervised the research and acted as editor, while Agassiz personally met the cost of publication. The label suggested the existence of an admin-

Figure 26. Recent photograph of the same entrance that appears in the center of fig. 24. Above the doorway can be discerned, entwined in leafy carving, the letters "N H L" for "Natural History Laboratories." This section of the building was the locus of Harvard's undergraduate "Nat. Hist."—courses in geology, botany, and zoology. (By permission of the Museum of Comparative Zoology Archives, Harvard University)

istrative unit, but it was one which had emerged in practice, not by official action of either the museum or the college (figs. 38 and 39).

Almost without exception, students of zoology at Harvard had nothing to do with the collections housed in the museum. Like Parker, those arriving with taxonomic inclinations had to find a more suitable topic. For example, in the fall of 1887 ichthyologist Carl H. Eigenmann arrived in Cambridge with his bride Rosa Smith Eigenmann, also an accomplished ichthyologist. He was a doctoral candidate at the University of Indiana, under the direction of David Starr Jordan. (Because Jordan had been at Penikese in the summer of 1873, Eigenmann considered himself an academic grandchild of the legendary Louis Agassiz, even making a pilgrimage to the island.)[37] His work in the museum consisted of a microscopic study of egg structure under E. L. Mark's direction.[38]

After Faxon's teaching position ended, no one paid by Harvard College did curatorial work in addition to teaching, and likewise it was rare for the M.C.Z. staff to do any teaching. The exception, Hermann Hagen, proves that the battle between college and conservators had to do with much more than the use of specimens. He had given lectures on entomology to small

Figure 27. Museum of Comparative Zoology from the Divinity Street side. In the center of the picture is a boxlike addition with an arched doorway, built in 1886–87 to house a grand double staircase for the public. Below the cornice of this addition, scarcely visible in this photograph but still sharp on the building today, are the words "Museum of Comparative Zoology." The only public access now is through the central doorway on the Oxford Street side, labeled "Agassiz Hall" and leading to the Botanical Museum's famous glass flowers. (By permission of the Museum of Comparative Zoology Archives, Harvard University)

classes of undergraduates since his arrival in 1868, but Eliot's famous reform of the undergraduate curriculum, which designated various science courses given outside Harvard College as optional credits for the B.A. degree, excluded entomology. Hagen, receiving an offer of a post in Berlin in 1876, refused it "on condition," so he thought, that his access to undergraduates would be restored, but in 1880 he wrote to the Harvard dean, Charles F. Dunbar, explaining that the system still had the effect of preventing him from teaching.[39] A few graduate students did seek out Hagen, but he supervised no doctoral degree.* It is hard now to trace whether this was "well-calculated" on the part of Eliot and his advisors, as Hagen told Dunbar, but under heavy pressures from his curatorial duties, Hagen had no energy to fight it. Mark had given up any pretense of assisting with cura-

*Arnold Mallis (*American Entomologists*) called Hagen a "notable" teacher of entomology and counts among his students Alber John Cook (1867–68), John Henry Comstock (summer of 1872), Henry Guernsey Hubbard (1873–74), Herbert Osborn (1867–68), and Charles William Woodworth (1886–88).

torial work on worms, so Alexander Agassiz saw no reason to encourage his staff to make gifts of their time to Eliot's enterprise.

When at last Louis Agassiz's rebellious student Alpheus Hyatt was invited back to Cambridge in 1885, he found the division between curating and teaching so deep that he was unable to bridge it. He was still active in theoretical investigations, his ideas exciting much interest among evolutionists up until his death in 1902.[40] Although he taught at both M.I.T. and Boston University, the closest he came to having a student at Harvard was Robert Tracy Jackson, who assisted him in the fossil collections and shaped his doctoral research and his later publications around Hyatt's theories. It was Shaler, though, who was Jackson's official instructor. Shaler hired Jackson to teach paleontology, which he did between 1892 and 1909. Hyatt

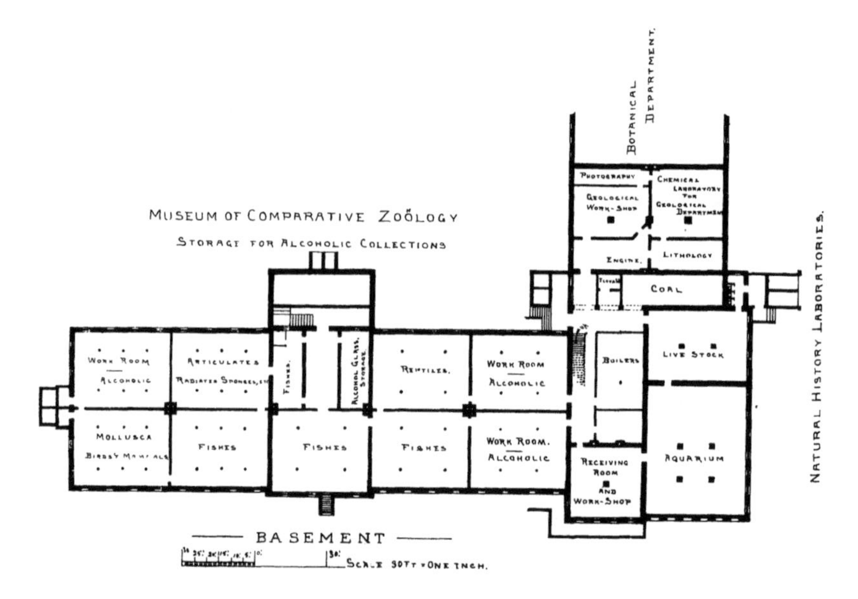

Figure 28. Plan of M.C.Z. from *Annual Report* of 1888–89, basement level. The first stage, built in 1859 with money raised by Louis Agassiz, consisted of the four rooms at the left of this plan. The next segment of about the same length was built in 1872, funded by the state and by gifts from his children, who were becoming wealthy from their investment in the Calumet and Hecla copper mines of northern Michigan. Alexander Agassiz's copper money paid for the next two stages: the pair of rooms labeled here "Work Room—Alcoholic," built in 1878; and the "corner piece," in construction from 1880 to 1883, delimited by two outside stairs. Although in this plan two rooms in the corner piece are assigned to "Aquarium" and "Live Stock," instructors complained that neither the museum nor the college provided funds for keeping living specimens (*Annual Report* for 1892–93, p. 34). Nevertheless, some work with live animals was carried on, including W. E. Castle's experiments in genetics.

continued to look after the fossils in the M.C.Z. until his death in 1902, often as a volunteer. An obituary noted that he

> enjoyed no opportunity to teach the young men who were pursuing the higher courses in zoology at Harvard. Through an unfortunate arrangement those who had charge of the various collections in the Museum of Comparative Zoology were not encouraged to give lectures to students, and they worked on throughout the years, their voices silenced, yet with active young minds eager to listen and to learn always near them.[41]

This memorialist, Hyatt's son-in-law Alfred Goldsborough Mayer, was in a position to know, having done his doctorate in Mark's laboratory and worked as Alexander Agassiz's assistant for several years.

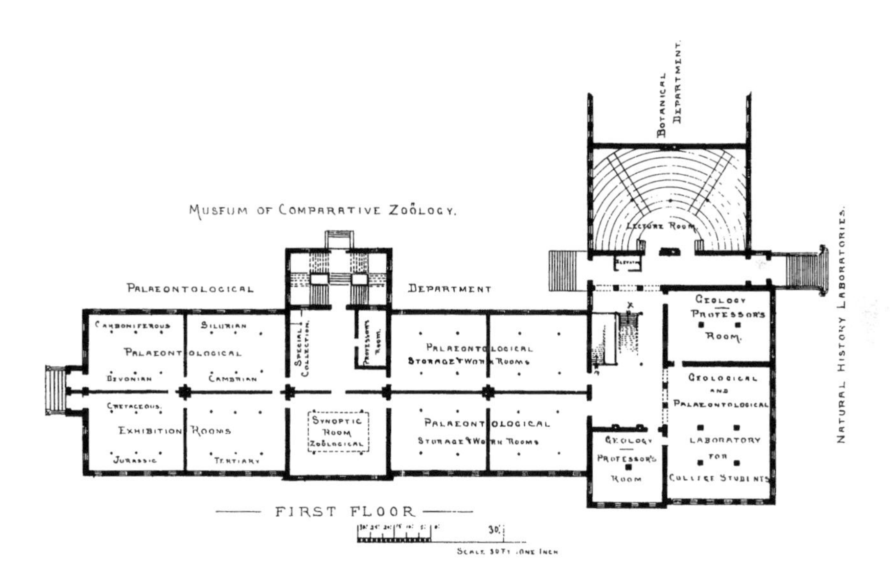

Figure 29. Plan of M.C.Z. from *Annual Report* of 1888–89, first floor. Directly below the words "Museum of Comparative Zoology" is the little wing, three stories high with skylights, which was constructed in 1887 to house a double staircase. In the museum's first expansion, stairways in what are here "Palaeontological Exhibition Rooms" were replaced by stairways and elevator, located where this plan shows "Professor's Room" and "Special Collections" (written vertically). Space never ceased to be at a premium, and in this century the little wing, raised to full height, has been filled with laboratories and offices, while the stairwell in the "corner piece" has been made more compact.

Thomas Barbour, so avid a herpetologist that he spent every spare moment of his undergraduate years in the collection, was determined to do his doctorate on a taxonomic topic, though he knew that "a spermatogenesis paper would probably please Dr. Mark and the other Zool. Lab. people better."[42] After completing his degree, he explained to a colleague,

> conditions of work have been pretty ticklish here since the Zoological Department of the University and the Museum are not in the closest relationship; and as Mr. Garman had actual charge of the Museum collections, and of course he was not an instructor, so that using the collection was not quite so simple as it would have been if I had had charge of it myself.[43]

Paleontologists, like Jackson and Jackson's student J. A. Cushman, did theses on systematic topics, as did botanists, but after Birge's 1878 degree, no other doctoral student in zoology worked on the systematics of living animals until Barbour, who received his degree in 1910.

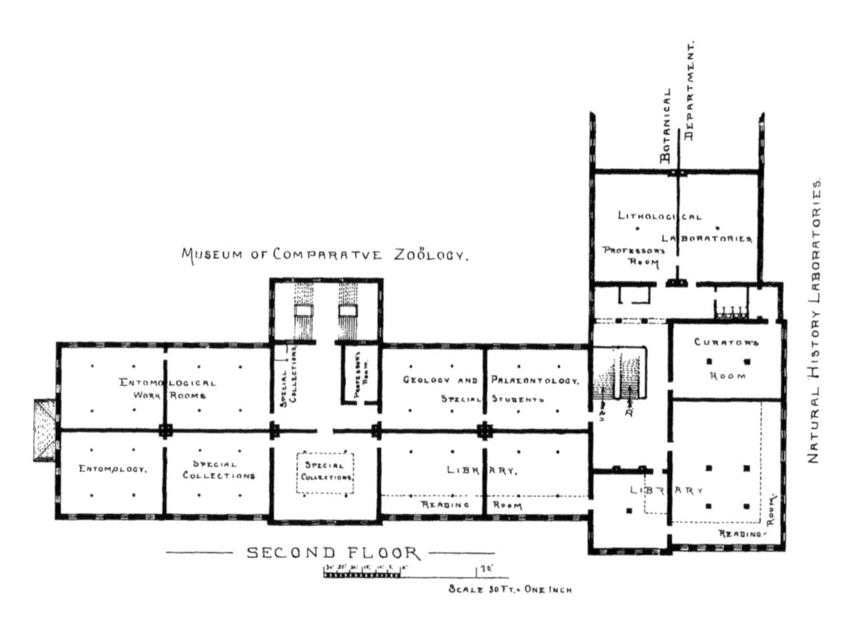

Figure 30. Plan of M.C.Z. from *Annual Report* of 1888–89, second floor. The galleries of all the main floor exhibition rooms save one have been floored over, creating the second floor rooms where in Louis Agassiz's day there had been galleries. The words "Special Collections" within a dashed box indicates a gallery above the first floor's Zoological Synoptic Room. The gallery design of Louis Agassiz's original building (see fig. 19) reflects the lack of separation of public exhibition space from research collections, common in museums in the mid-nineteenth century.

Naturalists Object to Harvard's Biology

In the spring of 1893, Theodore Roosevelt, by now commissioner of the Civil Service and author of two books, paid a visit to Harvard. Following its standard practice, the university was conducting a review of biology instruction and had formed a small committee, consisting of Walter Faxon, who was at that time "assistant in charge" of the M.C.Z.; Clarence J. Blake, a physician; and Roosevelt, in the chair. Professor Mark, who was assisted in his teaching by Parker, William E. Castle, and two others, hoped the Visiting Committee would convince the university to give them more equipment, particularly an aquarium, in order to keep up with the biology departments of rival universities like Johns Hopkins, Chicago, Columbia, and Clark. But as far as Roosevelt was concerned, what had been wrong with the biology he had been given as a student had only gotten worse. Besides listening to Mark, he questioned William Brewster, volunteer cura-

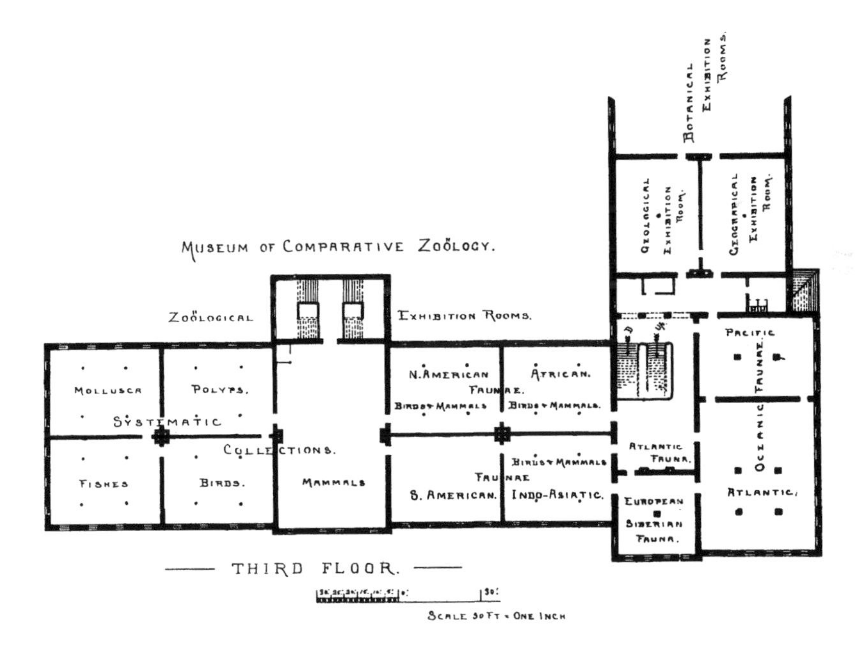

Figure 31. Plan of M.C.Z. from *Annual Report* of 1888–89, third floor. Alexander Agassiz was following the general outline of his father's idea, that a museum should have a number of parallel collections, by carefully distinguishing the systematic rooms, arranged by taxonomic group, from the faunal rooms, arranged geographically. A skeleton of a whale extended the whole length of the room labeled "Mammals," where it may be seen to this day.

tor of birds in the museum. Returning to Washington, Roosevelt evidently discussed the state of affairs he had found at Harvard with his friend C. Hart Merriam.

Head of the U.S. Department of Agriculture's Division of Economic Ornithology and Mammology, Merriam had become a naturalist by the old-time route, having trained and practiced as a physician while learning his natural history by apprenticeship and reading.[44] On 30 June 1893, Merriam published a polemic entitled "Biology in Our Colleges: A Plea for a Broader and More Liberal Biology." It contained the same argument, in wording too close for coincidence, as the confidential report Roosevelt was drafting for Harvard.[45] Both men hated what biology teaching had become.

In the pages of *Science,* Merriam decried the sweeping aside of natural

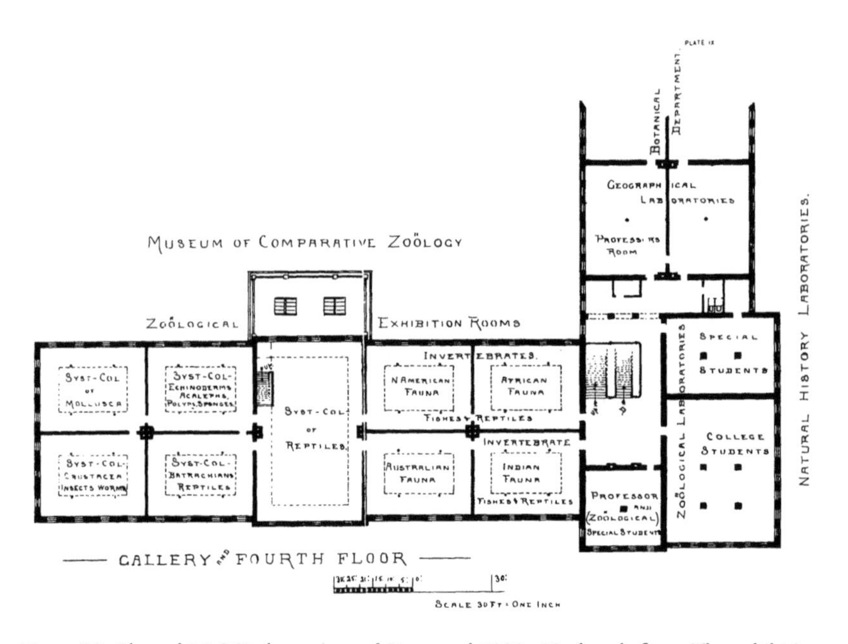

Figure 32. Plan of M.C.Z. from *Annual Report* of 1888–89, fourth floor. The exhibition rooms of the third floor all extend to a gallery level overhead, so that a fourth floor existed only in the "corner piece," occupied by laboratories for college students and special students. The teaching laboratories, under the supervision of E. L. Mark, grew into an independent fiefdom, finally formalized when Harvard gave him the title "Director of the Zoological Laboratories" in 1900. The large central room, here labeled "Syst[ematic] Col[lection] of Reptiles," is today the only one that retains its original two-storey height.

history by worshipers of the microscope. A naturalist will mature, said Merriam, only if he happens not to attend college, for there he must fall under the influence of "our modern teachers of biology. These teachers have deflected into other channels many a born naturalist and are responsible for the perversion of the science of biology." Roosevelt, whose spare-time activities included founding the Boone and Crockett Club to preserve the vanishing wilderness, was to Merriam a perfect example of a born naturalist.[46] Merriam complained that

> the term "naturalist" fell into disuse to be replaced by "biologist," and some would have us believe that even the meaning of the word biology is no longer what it was. . . . Is it not time to stop and inquire into the nature of the differences between the naturalist and the modern school of instructors who call themselves "biolo-

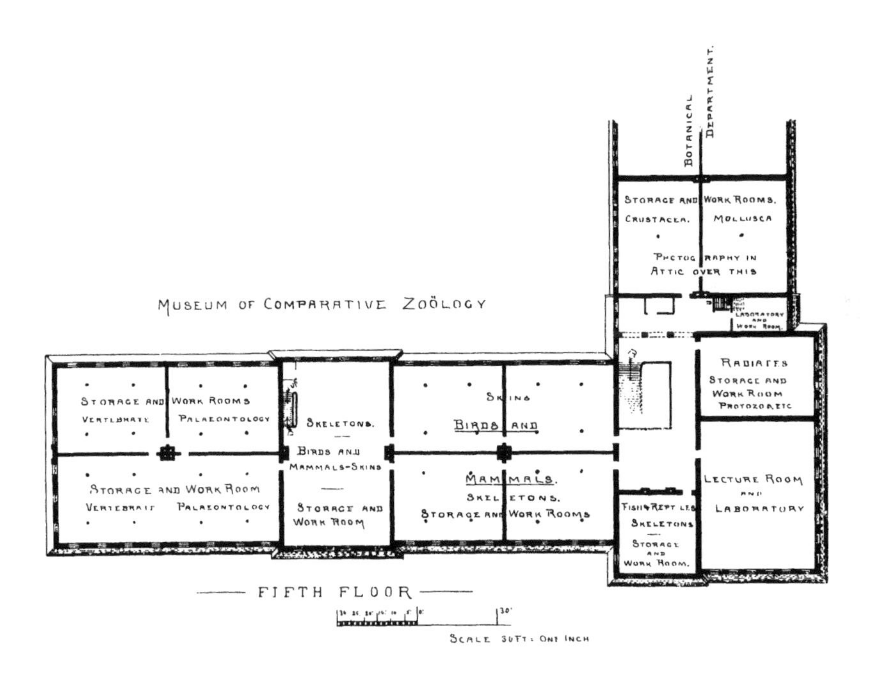

Figure 33. Plan of M.C.Z. from *Annual Report* 1888–89, fifth floor. Of the corner room at the far right of this plan, also shown in fig. 34, Professor Mark declared, "I don't know of a pleasanter lecture room in all Cambridge." It was home to all Harvard's Natural History courses from 1882 to 1889, when the second lecture hall, indicated in fig. 29, was built, and it served students of zoology until the construction of the Biological Laboratories in 1931. The wall below the words "Photography in attic over this" is the end wall seen in fig. 24.

Figure 34. The lecture room on the top floor of the M.C.Z.'s "corner piece." The diagrams illustrate the structure of echinoderms: holothurians on the far left, then echini, brittle stars, and starfish, and crinoids at the top right. Of the diagrams drawn in white on a very dark background, the two on the right show the jawlike action of pedicellariae. On the blackboard the instructor has sketched the homology between polyps and jellyfish. Louis Agassiz gazes from an oval portrait at the far left. This photograph, as well as fig. 40, were made for display at the 1892 Columbian Exposition in Chicago (*Annual Report* for 1892–93, pp. 4, 33). (By permission of the Museum of Comparative Zoology Archives, Harvard University)

gists"; into the causes that have brought about so radical a change, and into the relative merits, as branches of university training, of systematic biology compared with the things now commonly taught as biology. . . . the generally accepted meaning of the word biology has come to be restricted to physiology, histology, and embryology . . . [plus] morphology and the supposed relationships of the higher groups.[47]

His recommendation was extreme. There should be a new curriculum which would relegate the use of the microscope to an introductory "elementary course in general biology, including cell structure and the structure of the less complex tissues of animals and plants." The other courses would all deal with the comparison of whole organisms: taxonomy, geo-

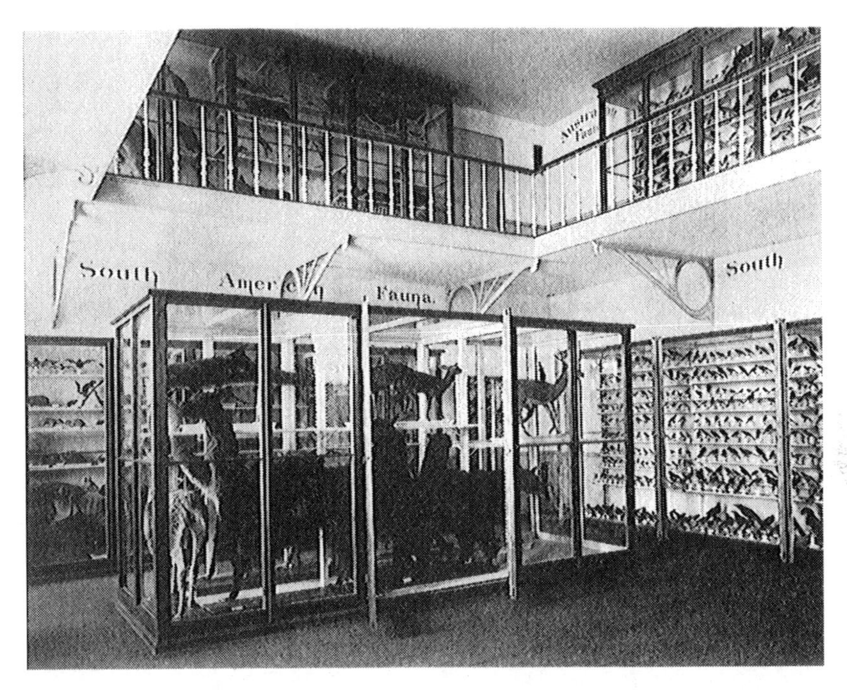

Figure 35. One of the M.C.Z.'s public rooms, 1892. Its location is shown in fig. 31, marked "S. American" and the gallery, marked "Australian Fauna," in fig. 32. After visiting the M.C.Z., Alfred Russel Wallace declared its series of small rooms a perfect example of "how a museum should be constructed and arranged" ("American museums," p. 349). Alexander Agassiz had put into effect J. E. Gray's recommendation that only selected specimens be exhibited, clearly labeled and intelligently arranged, while the majority of the collection was separately stored, available only to specialists. (By permission of the Museum of Comparative Zoology Archives, Harvard University)

graphic distribution, paleontology. Even the final year's "Lectures on the principles and philosophy of biology" were to make no mention of metabolism or microstructure but were to treat "evolution, heredity, migrations, special adaptations."

Roosevelt's Visiting Committee Report was more realistic, suggesting that systematics be added to the existing curriculum rather than displacing it:

The anatomical microscopist has a high and honorable function to fill in the scientific wor[l]d; but it is certainly no more, and is probably decidedly less, important than that of the systematist and the outdoor collector and observer of the stamp of Audubon or Bachman, Baird or Agassiz.

Figure 36. Aerial view of the University Museum from Oxford Street, about 1957, the M.C.Z.'s "corner piece" in the foreground. Mirroring the museum's open quadrangle are the Biological Laboratories, built in 1930–31. Between them on the left is the flat-roofed Gray Herbarium, crowding close to the oldest segment of Louis Agassiz's museum, the front entrance of which has been bricked over. A large laboratory wing has since been added to the M.C.Z. (By permission of the Museum of Comparative Zoology Archives, Harvard University)

His feelings about what people who study living things should be called matched Merriam's:

While recognizing fully the need of the laboratory worker, we must not forget the need also of the man who can collect, observe, and record his observations, in the open; who can work both in the laboratory and afield; who is a naturalist, in the fine old acceptation of the word, and not a latter-day "biologist"—a mere histologist and embryologist.

And his own preference was clear:

The highest type of zoölogist is the naturalist, the man who loves outdoor work as well as the work of the laboratory, and who studies and delights in animals and

Figure 37. Edward Laurens Mark, who taught zoology at Harvard from 1877 until 1921. Among his many students were Herbert S. Jennings, Robert M. Yerkes, and William E. Castle. He was director of the Bermuda Biological Station from 1902 to 1931, and a familiar figure there and around Harvard until his death in 1946, at ninety-nine years of age. (By permission of the Museum of Comparative Zoology Archives, Harvard University)

Figure 38. Faculty and graduate students of the Zoological Laboratory, about 1891. Standing, from left: unidentified, Herbert Haviland Field, Charles Benedict Davenport, Herbert Parlin Johnson, Edward Laurens Mark. Seated, from left, Emanuel Roth Boyer, William McMichael Woodworth, William Emerson Ritter, George Howard Parker. This is one of a series of such group portraits, reflecting Mark's pride in his research students, one of whom later recalled his "extreme method": "From the time you took your research problem under him till the last punctuation point marking the completion of your printed thesis you were under Dr. Mark's eye" (Parker, *World Expands*, p. 65). (By permission of the Museum of Comparative Zoology Archives, Harvard University)

plants, considered with reference to nature as a whole, and with regard to their own habits and interrelationship of structure.[48]

Harvard had at hand, Roosevelt pointed out, an invaluable tool with which to right the situation: the M.C.Z.'s collections. The other two members of Roosevelt's committee submitted a dissenting report defending the status quo, though admitting that the addition of a course in systematics would round out the college offerings.

Biology teaching at Harvard was no more imbalanced than it was elsewhere, but its location in Agassiz's museum might make the situation seem more ironic there. It was as though cytology had forced out the taxonomic portion of biology, like the cuckoo chick expelling the nest builder's own

Figure 39. The Zoological Laboratory, 1892. Front and center is a microtome, the tool for slicing tissue for the microscope, which symbolized the difference between scientific zoologists and the outdoor naturalists. (By permission of the Museum of Comparative Zoology Archives, Harvard University)

young. And why does that chick, scrawny and blind, behave so heartlessly? It acts unconsciously and without malice, but consistently and vigorously, because only with a monopoly on the food can it grow strong.

One of the replies to Merriam's "Plea for a Broader and More Liberal Biology" hints that the definition of an emerging profession was at stake. If Merriam had his way, the botanist J. Christian Bay contended, students would get a superficial knowledge of some pleasant natural history but would know nothing of the true science of biology:

> Natural history has become, in our century, so broad that no man possibly can become a "general naturalist" or a good "faunal naturalist" any more; he will, at least, not be able to treat all the questions that arise in any other way but in that of the amateur. . . .
>
> . . . it is possible to take a farmer's boy and make out of him "a general naturalist of the present day." . . . He will be no scientific man. . . .

> . . . So long as biological courses do not include a proper course in experimental physiology of animals and plants, they cannot be called properly scientific.[49]

Besides being a reply to Merriam, this was part of a larger debate filling the columns of *Science,* stimulated by the 1891 publication of a federal government report on biology teaching in American colleges.[50] Why, one might ask, was the way of the "amateur" to be shunned and only the "properly scientific" to be admitted in education?

A tug-of-war over the true meaning of terms signals a contest over territory; an attempt to define a word is a kind of claim of ownership of something worth having. What Merriam and his contemporaries were embroiled in was what Everett Mendelsohn has called the "second scientific revolution": the professionalization of science.[51]

One element of professionalization is, of course, the funding of jobs. Without these, a career in biology is open only to those of independent means, like Alexander Agassiz, Robert Tracy Jackson, and Thomas Barbour. The science such people do may be excellent, but the fields they work in will have few contributors. A second and intimately related element of professionalization is the restriction and self-definition of qualification for such jobs. If it were admitted that college students could get as much value from the teachings of an amateur naturalist, then colleges would not need to hire professionally trained biologists. It was in the interest of leaders in science to keep tight control over the definition of proper training.

Merriam and Roosevelt listed Louis Agassiz in their pantheon of great naturalists, contrasting him with narrow self-styled biologists. Being a great lover of nature was certainly one of the images Agassiz himself had cultivated, but behind the scenes he was an uncompromising elitist. Active in the American Association for the Advancement of Science and a founder of the National Academy of Sciences, he strove to create jobs for which scientific knowledge was the qualification and to raise the standards of entrance to the profession. His creation of the M.C.Z. was at the center of his endeavors to professionalize biology in America. The new institution was to train the next generation of zoologists, and to train them to the highest modern standards, so that their publications would be respected by European specialists. He made clear to his sponsors and to his students that he would not rest content with traditional natural history but aspired to a new science, more sophisticated in its aims than anything done by amateur describers of species. From the vantage point of 1893, Louis Agassiz may have looked like a naturalist of a comfortably antique kind, but to his contemporaries in the 1850s he was unmistakably an agent of change.

At the end of the century, Alexander Agassiz replayed the battle between his father and Shaler, bitterly proud that his side was undefeated:

No Professor who gives instruction will ever look upon a collection in any other way except as material to illustrate his teaching. I am sure that were the Museum Collection left to the tender mercies of Shaler [by now Dean of the Lawrence Scientific School], [Professor William M.] Davis, Mark and their assistants we would by this time have nothing left of value. They would take out of the exhibition-cases anything and everything. It might find its way back to the cases but probably not.[52]

The Museum and its departments had better lead a separate existence rather than go to the dogs by pulling together, as would surely be the case if the teaching staff had their way.[53]

No one knew better than Alexander Agassiz the several purposes the museum was intended to serve; it was to house an important collection, but it was also supposed to provide teaching at the cutting edge of modern zoology. German microscopy indisputably occupied that edge.

Once a year the instructors went through the formality of submitting to Agassiz a report, which he incorporated into his own summary of the year's activities and, beginning in 1880, printed in the *Annual Report* of the M.C.Z. alongside the reports of the assistants in charge of the various parts of the collections. For appearance's sake the *Annual Reports* changed little as the decades slipped by, but by the time Agassiz retired in 1898, he looked back on a divorce long since finalized:

The slender thread which connected the Museum with the teaching departments at the time of its incorporation with the University is practically severed, the administration of the Museum being no longer in any way concerned with instruction, as was originally included in the articles of agreement between the University and the Museum.[54]

Since instruction in geology and in zoology was under the auspices of Harvard College, these reports need no longer appear in the *Annual Report* of the museum, he suggested. But this was Agassiz's last year to write that *Annual Report,* and his successors kept up the pretense for another twenty-four years.

8

"Results Unattainable by Museum Study Alone"

Louis Agassiz's friends, organizing a fund drive in his memory in 1874, declared, "Every workman must have his tools: the tools of a zoölogist are collections of natural objects systematically arranged."[1] These loyal supporters were repeating the message Louis Agassiz had drummed into them for more than two decades, summed up in a powerful rhetorical image. Richard Owen had found a related image useful in his appeals on behalf of the British Museum: "The great instrument of zoological science, as Lord Bacon points out, is a Museum of Natural History."[2] The lover of nature delighting in an array of butterflies has been replaced by a resolute laborer, striving to build but needing ax and hammer. This is a picture that deserves closer examination.

The workman and his tool is a critical image for the history of museums and for the history of systematics. We have already heard others describe the M.C.Z. as a cathedral, a monument, a baby, even an elephant. Now we have the barest metaphor of all, seemingly just a description, which is part of its rhetorical effectiveness. Indeed, sometimes the collections were quite literally the tool for zoological work. In the hands of men like Hagen and Allen, dead specimens produced fresh information about the living world.

What we should notice about this fund-raising declaration, though, is not the tool but the workman. We are expected to call to mind a serious man with a challenging job to do (two lines later we are told that natural classification "taxes the powers of the greatest genius"). In the preceding three paragraphs the writers have insisted on the practical applications of pure research, the twitching of a frog's leg leading to the benefits of electricity. "'Museum,' a word that commonly suggests little more than a collection of curious objects, is scarcely an appropriate name" we are told.

They are distancing Agassiz's museum from the old natural history, the Sunday diversion of ladies and children or the pious recreation of the idle. Their zoologist is fully committed to his work. In keeping with Victorian categories, a difficult task that promises great material reward for civilization calls for a masculine effort. Louis Agassiz would have applauded this rhetoric, for it implies the reshaping of amateur natural history into professional zoology.

Seaside Stations Supplanting Collections

Zoology was indeed becoming a serious and demanding business, as Theodore Roosevelt was disappointed to discover, but rhetoric had not the power to fix in place its classic methods. Alexander Agassiz knew very well that the innocent claim of his father's friends was no longer true. By the mid-1870s the most influential of European zoologists were working with other tools. Naturalists and their collections continued to exist, of course, and museums of natural history held a higher place than ever in public esteem, but the novel word "biology" was being bandied about as signifying something more scientific than taxonomic botany and zoology. In 1876, in his first report as head of the museum, Alexander Agassiz announced that his policy would be different from his father's. Explaining why he would be trying to minimize the museum's stock of perishable specimens preserved in alcohol, he said,

> The constantly increasing facilities of travel, the comparative economy with which fresh specimens can be studied, the superiority of such work (with proper appliances) to that of the Museum, the daily increasing number of workers who are able, on the seashore or in the field, to produce results unattainable by museum study alone, shows that the time has come when large collections must naturally be supplemented by zoölogical stations. These . . . will enable museums to dispense with much that is now exceedingly costly. They will become, for certain departments at least, chiefly depositories where the record of work done at the stations—the archives of natural science, so to speak,—will be preserved, so that . . . they must hereafter be useful to the original investigator in a somewhat more limited field.[3]

An ambitious young man should have no trouble choosing between a field "somewhat more limited" and work of "superiority." Archives are a trust to be honored, but they are inclined to get dusty. The role of housekeeper belonged to women, not only symbolically but literally. A number of women were already working in the museum in Louis Agassiz's day, doing secretarial jobs and routine maintainence under supervision by the curatorial assistants.* Alexander Agassiz would later toy with the idea of dispensing with the assistants altogether, reasoning that the "work of preservation of collections could in future be perfectly well done . . . by cheaper assistance—Women—."[4] The arena for adventure and challenge was elsewhere.

Agassiz's allusion to workers at seaside stations accurately described the direction professional zoology was taking. Baird, in his role as commissioner of fish and fisheries, was already operating a summer research laboratory in Woods Hole. The important event, though, was the 1872 founding of the Stazione Zoologica at Naples by Anton Dohrn. This was a research facility, not a school, bringing together biologists employed or enrolled elsewhere for a period of intense self-directed work. Their investigations seemed modern and daring, for they subscribed to Haeckel's doctrine that the way to uncover the evolutionary links between groups was through the morphology of embryonic development.

The Newport Marine Laboratory

Coming across such a pronouncement as "the time has come when large collections must naturally be supplemented by zoölogical stations," we whiff again the scent of rhetoric, yet the consequent plea for financial support for the necessary marine station for Harvard does not come. Quite the contrary! The next year Agassiz announces that he has built a gem of a laboratory, beautifully equipped, on the perfect spot of seacoast, and what's more, he intends to share it. The description of the "Zoölogical Laboratory" at Newport, Rhode Island, complete with plans and photograph, in the M.C.Z.'s *Annual Report,* is as attractive as a sales brochure.† So gener-

*At the end of 1873, workers in the museum included Miss Slack (librarian), Mrs. Millson, and Misses Anthony, Atkinson, Harris, Olmsted, and Thayer (Lyman to Eliot, 31 December 1873, Harvard University Archives).

†"The tables for microscope work are three-legged stands of varying height, adapted to the different kinds of microscopes in use. The whole of the northern side of the floor upon which the work-tables and microscope-stands are placed is supported upon brick piers and arches independent of the main brick walls of the building. . . . This gives . . . the great advantage of complete isolation from all disturbance caused by walking over the floor. This will be duly appreciated by those who have worked in a building with a wooden floor, where every step . . . was sure to disturb any object just at the most interesting moment. . . . The centre of the large room is occupied by a sink, on each side of which extend two long tables, three feet by twelve. These are covered with different colored tiles, imitating mud, sand, gravel, seaweed, black and white tiles, as well as red, yellow, blue, green, violet, to get all possible variety of background. A space at each end is covered by a glass plate, allowing the light to come from underneath, thus enabling the observer to examine larger specimens from the under side. . . .

ous and enlightened was Agassiz that three of the six people invited to his laboratory in 1877 were "ladies."[5] So much for our preconceptions about Victorian bias.

No, we should trust our preconceptions. Women had been present at Penikese, too, because the purpose of that summer exercise had been to fulfill a longstanding promise that, among its many other activities, the museum would serve the Commonwealth by reaching out to schoolteachers. The role of women in teaching was well established, and it was "teachers of our common schools" whom Alexander Agassiz invited to Newport. This sentimental echo of Penikese quickly faded, however, for the younger Agassiz had no interest in organizing a school. He wanted it clearly understood that anyone who came would have to be "sufficiently advanced to study for themselves with profit."[6] He repeated this experiment one more year only. Alexander Agassiz was prepared to be liberal in regard to women in science (and welcomed another woman into his laboratory in the summers of 1882 and 1883), but the substantial percentage of females at Newport in 1877 was a sign that the facility did not yet have a clear role in the training of professional scientists.

Agassiz also announced that he would invite to his laboratory students of the museum, which he did, thereby also opening a new career pathway. Among the men who began as Harvard students, did research at Newport, and were employed as Agassiz's personal assistants were Walter Faxon, Jesse Walter Fewkes, and William McMichael Woodworth. It is hard now to reconstruct a complete list of those who visited the Newport laboratory, but it included men who later ran their own laboratories or otherwise influenced biology: W. K. Brooks, William Ritter, C. B. Davenport, C. A. Kofoid, C. O. Whitman, and E. B. Wilson.* When the history of marine

---

In the attic there is a large tank for salt water and another for fresh; the rest of the attic space will be eventually devoted to photographic rooms and room for an artist. [Agassiz pioneered the use of photography for scientific illustration.] The Laboratory is supplied with salt water by a small steam-pump driven by a vertical boiler of five horse-power: this is kept going the whole time day and night, the overflow of the tank being carried off by a large pipe. [It would be interesting to compare this to other laboratories, for Agassiz's experience with mines must have taught him a great deal about pumping systems.] The water is . . . drawn up at a horizontal distance of sixty feet from the shore in a depth of some four fathoms, the end of the suction pipe standing up vertically from the ground a height of five feet, and terminating in an elbow to prevent its becoming choked. The water is led through iron pipes coated inside with enamel. . . . From the tank, the salt water is distributed in pipes extending in a double row over the central table, over the long narrow tables for aquaria, and along the whole length of the glass shelves on the south wall" (*Annual Report* for 1877–78, pp. 13–14).

*Besides G. R. Agassiz's *Letters and Recollections of Alexander Agassiz* (pp. 151–56), my sources have been limited to the relevant paragraphs in the *Annual Reports,* and the acknowl-

laboratories is better chronicled than it yet is, the facilities in Newport, in seasonal operation continuously from the summer of 1877 until Agassiz's death in March 1910, will likely be found to have set a high standard.*

With respect to Harvard, to the M.C.Z., and to the status of systematics, however, the dominant fact about this laboratory was that it was part of Alexander Agassiz's summer estate. He had continued to share his father's Nahant laboratory after his marriage in 1860, to carry on the marine research he loved, but after his bereavement he found its associations too painful and sought a new location for a summer home. The site he chose, like Nahant, was on the tip of a long peninsula with a commanding view of the Atlantic. For those given space to work in his Newport laboratory, Agassiz made available his steam launch for collecting with net or dredge. He provided daily transportation back and forth to the town of Newport, where lodging was available.† (He did not underwrite living expenses, nor the cost of travel from Boston.) In those days, of course, a train ran between Boston and Newport.

Perhaps a nearer model for the Newport laboratory than Louis Agassiz's study in Nahant (the connection to Penikese being rather slim) was the facilities of the United States Commission of Fish and Fisheries which Baird developed at Woods Hole, Massachusetts, beginning in 1871.[7] Theodore

edgments by authors of articles in the museum's *Bulletin,* as well as a few letters cited individually. It seems very likely that a logbook and other records of the laboratory would have been kept, and these may still exist (although Mrs. Mabel Agassiz, the widow of Alexander Agassiz's son George, is said to have had a "famous fire"). Undoubtedly, much could be gleaned from letters of the people known to have worked at Newport.

*According to the *Annual Report* or the *Bulletin,* Brooks was in Newport in 1875 (though the laboratory building did not exist until 1877, Agassiz had workspace in his house and Brooks acknowledges his hospitality); Faxon was there in 1876 or 1877 and 1878; Fewkes was there in 1878 through 1886; Barnes in 1882; Tuttle in 1882; Whitman in 1882 and 1883; Nunn (Mrs. Whitman) in 1882 and 1883; Wilson in 1888; Ritter in 1889 and 1890; Woodworth in 1889 through 1892; W. Whitney in 1889 and 1890; Davenport in 1889 through 1892; Ward in 1891 and 1892; W. S. Nickerson in 1891; Kelly in 1892; Kofoid in 1892; Mayer in 1892; Weysee in 1892; Gerould in 1892; Castle in 1893 and 1894; Parker in 1890 and 1895; and Mark in 1890. This is certainly far from the complete tally. A letter of Mark to Davenport of 20 June 1891 in the American Philosophical Society, Philadelphia, names as users that summer, besides Davenport himself: Woodworth, Dr. Thos G. Lee, Lucas, [Frank?] Smith, and [Henry Baldwin] Ward. H. S. Jennings shared the experience in the summer of 1895 with W. E. Castle, A. G. Mayer, H. V. Neal, E. V. Wilcox, Seitaro Goto, and W. B. Cannon (T. M. Sonneborn, "Herbert Spencer Jennings," p. 158).

†E. L. Mark to C. B. Davenport, 9 September 1890, American Philosophical Society, Philadelphia; W. E. Castle, "Summer work in marine zoology at Newport," *Science* 22 (4 August 1893): 60.

Lyman had visited there in its first summer, and noted in his diary that Baird and his wife

> keep a sort of table d'hôte for stray naturalists. . . . Mrs. Hamlin who keeps the house is an excellent cook, and her chowder was most grateful to a hungry "Commish." Baird has all the apparatus for a complete investigation, a preparer, photographer, stenographer, together with boatmen, &c. He has the Gov. Light-House wharf, with alcohol, cans, &c.&c. He gets side help, too, from Verrill, Hyatt, [J. W. P.] Jenks, & Dr. [W. G.] Farlow, who are making collections by his aid. Furthermore there is a steam launch, and a sail boat or two, with dredges, seines & trawls. . . . Verrill was there with microscopes &c, studying dredgings, especially ascidians.—Baird & I had much interesting talk over our weir fisheries."[8]

People associated with the M.C.Z. went to Baird's laboratory as well as to Agassiz's, and when Baird developed the practice, used also by Naples, of inviting other institutions to "subscribe to a table" (rent workspace), Alexander Agassiz personally paid for a Harvard subscription.*

Professor Mark's Separate Territory

The split that was deepening in Cambridge, between the instruction over which Mark exercised such careful control and the rest of the museum, had its counterpart in the Newport laboratory. The first public statement of tension was Agassiz's statement in the 1893 *Annual Report* that "if it is necessary to carry on an elementary summer school of Zoölogy" the less advanced students would have to be accommodated elsewhere. This brought an angry protest from Mark, one of the few men who did not adopt a tone of subservience to Agassiz, that his students did not deserve such an insult, to which Agassiz replied,

> . . . the receipt of your note has been only one other reason for me to take account of my stakes here. For twenty years I have spent considerable time—at the sacrifice of many of my own scientific interests—in attempting to help the Nat. Hist Depts. of the College.
>
> It has become self evident that my plans do not work well. I am constantly placed on the defensive—while attempting to assist the University—to protect myself against this or that.[9]

*Acknowledgments in the *Bulletin* tell of work in the Fish Commission laboratory by Faxon (1876), Fewkes (1880 and 1881), Parker (1889 and ca. 1896), E. R. Boyer (1889), C. W. Prentiss, Davenport, Henry B. Ward (1891), and Castle (1893 or 1894)—certainly an incomplete list; use of the Woods Hole facility is mentioned in the *Annual Report* for 1887, 1888, 1889, and 1892.

Agassiz's efforts on behalf of advanced students were real and involved the expenditure of money as well as time. In 1891 he had enlarged the Newport laboratory building so that it could accommodate more investigators.[10]

Beginning in 1883, some of the items appearing in the M.C.Z.'s *Bulletin* were designated "Studies from the Newport Marine Zoölogical Laboratory [later Newport Marine Laboratory], Communicated by Alexander Agassiz." As if in reply, in 1887 other articles in the *Bulletin* began to be labeled "Contributions from the Embryological Laboratory [later Zoölogical Laboratory] of the Museum of Comparative Zoology, under the direction of E. L. Mark," or sometimes, after 1895, "E. L. Mark, Director." No less than the metaphor of a workman and his tool, these words were rhetoric, meant to affect the reader's beliefs. Neither laboratory was a legal entity. It was not until 1900 that Harvard officially designated Mark "Director of the Zoological Laboratory," overlooking the detail that no such unit had ever been instituted.

In spite of his many low moments, Agassiz had intended his time and effort to benefit zoology at Harvard, as they certainly had done, and he had seen the Newport laboratory as an essential part of that enterprise.[11] Certainly the physical facilities for research were outstanding at Newport (figs. 40–42), but the spirit of excitement and mutual encouragement so attractive at Penikese, Woods Hole, and Naples cannot have found a home on Agassiz's estate. When he strolled over to the laboratory after his butler had served him breakfast, the group who boarded in town must have been hard-pressed to get the sense of community that the beloved Baird generated automatically.* In spite of Agassiz's cold demeanor and ferocious temper, he was truly generous in inviting others to Newport, but there was never any question that they were guests on private property.

The work done at Harvard and Newport, as recorded in the M.C.Z.'s *Bulletin,* belonged to four separate categories: systematic studies on museum collections by museum staff, inventories of Alexander Agassiz's oceanographic expeditions, research at Newport by Agassiz or his employees, and doctoral dissertations by Mark's students. It is remarkable how little overlap there is between these four. Mark's students almost without exception described the arrangement of cells, usually during development, as revealed by techniques of fixation and microtome slicing, and

*Dall, *Baird,* pp. 441–42. With the late Mrs. Vincent P. Jones (née Marie Prince), a great–granddaughter of Alexander Agassiz, and Ann Blum, I had the privilege of visiting Agassiz's Newport home, now the attractive Castle Hill Inn. Innkeeper T. Paul McEnroe kindly let us into the laboratory building, which is now a private summer cottage. I am indebted to Ms. Blum, whose sensitivity enabled me to see how very different the research atmosphere must have been at Newport compared to Woods Hole.

Figure 40. The Newport Marine Laboratory, seen from the south side, on the grounds of Alexander Agassiz's summer home at Castle Hill, Newport, Rhode Island, as it looked from 1877 to 1890. A photograph from the other side, along with a detailed description of the building's construction and contents, appears in the *Annual Report* of the M.C.Z. for 1877–78. Among the researchers who worked here between 1877 and 1897 were C. O. Whitman, C. B. Davenport, C. A. Kofoid, W. E. Ritter, and E. B. Wilson. (By permission of the Museum of Comparative Zoology Archives, Harvard University)

the basis of their publications was work done in Mark's laboratory. His apparent enthusiasm for sending students to Newport (apparently he did not often accompany them) makes all the more striking how rarely they made use of marine material. Of the twenty-five Harvard doctorates in zoology before 1900, the first five predated Mark's doctoral teaching (W. K. Brooks, 1875; Jesse Walter Fewkes, 1877; E. A. Birge, Walter Faxon, and Charles S. Minot, 1878), and three of these (Brooks, Fewkes, and Faxon) used material from Newport in their dissertations. Of the next twenty, only three acknowledge Newport material in their dissertations, though the majority had spent time there. The exceptions, interestingly, were also those with the liveliest minds: George Howard Parker (1891), Charles Benedict Davenport (1892), and William E. Castle (1895).

However slight the link between Newport and Mark's teaching, the link between the museum collections and student or faculty research was even weaker. The single example of an instructor making use of the the museum's collection seems to be the analysis by Howard Ayers (whose doctorate was from Freiburg) of the morphology of blood vessels in a specimen Samuel Garman had identified as belonging to an otherwise extinct type of shark. Alfred Goldsborogh Mayer used the M.C.Z. collection in his dissertation on color patterns of butterfly wings (this was a study in biophysics, not taxonomy).

In 1892, Agassiz proposed that Harvard should build a second laboratory building at Newport, plus a dormitory. A large seagoing launch and a public aquarium would be desirable, and ideally a 200-foot steamer as well.[12] He promised to leave to Harvard in his will the laboratory he had built, for it was clear that Harvard would have to acquire a marine laboratory somehow if it were to keep up with other universities. (Agassiz was deliberately ignoring the existence of the Marine Biological Laboratory at Woods Hole, which had opened four years before with Whitman as director. This was not unreasonable, for at that date the M.B.L. was not a research facility for advanced students, as Newport was, but offered only introductory zoology courses. Its funding was inadequate and unpromising, and remained so until 1901.)* Because the offer he put to Eliot was so advantageous to the university, Agassiz was sure an agreement could be negotiated. This facility, its expenses entirely borne by him, had been for years a significant part of science at Harvard. He was furious at the total lack of interest which greeted his proposals. He felt sure, especially after seeing the inadequacies of the European marine laboratories he visited in 1895, that his offer of Newport had been a wonderful opportunity for science at Harvard, and he resented Harvard's indifference to it. President Eliot responded in the tactful and conciliatory tone with which he had over the years responded to other angry, exasperated, or discouraged letters from Agassiz, deploying sincere appreciation, flattery, and encouragement, but making no commitment. Possibly to emphasize his claim about the importance of the facility, Agassiz kept the Newport laboratory filled with students every summer in the 1890s, until 1898, when he excluded them. He continued to use the laboratory until his death in 1910. Because Newport was the site of advanced research in biology as far as Agassiz was

*Whitman did regard Newport as a threat to the development of the M.B.L., writing of "rival schemes" and complaining that "men of high scientific respectability and influence held aloof" ("A marine observatory, the prime need of American biology," p. 810). Trying to attract support for his Woods Hole enterprise, Whitman argued that "the advantages of a strong central station are so immeasurably superior to those of many weak ones, that we are bound to encourage the former and discourage the latter" ("A marine biological observatory," p. 462). Agassiz was, very understandably, incensed by this article (A. Agassiz to Scudder, 16 June 1893, Museum of Science, Boston).

concerned, he took Harvard's rejection very hard. He confided to his secretary, "I cannot quite make up my mind that Eliot is not two-faced after all my experiences with him."[13]

Alexander Agassiz's Assistants

A man with money and scientific projects and pressure on his time could hardly avoid hiring help, just as Agassiz did in his household and his mines. In 1878 he engaged a promising young zoologist named Jesse Walter Fewkes, a native of Massachusetts and graduate of Harvard College who had studied with Leuckart in Leipzig. Agassiz wanted help with the description of Atlantic jellyfishes, on which he had worked and published since the early 1860s. He sent Fewkes to the islands of Key West and Tortugas at the southern tip of Florida in the early spring of 1881, and to Eastport, Maine, in 1885, but usually Fewkes was based in Newport. In 1882 or 1883, he was given status in the museum as "Assistant in charge of Radiates." In Newport Agassiz put Fewkes in charge of the daily business of the laboratory and, in particular, asked him to supervise the visiting students.[14] Agassiz later believed that it was the presence of Fewkes that caused Mark to stay away. This seems plausible, since Mark did like to ex-

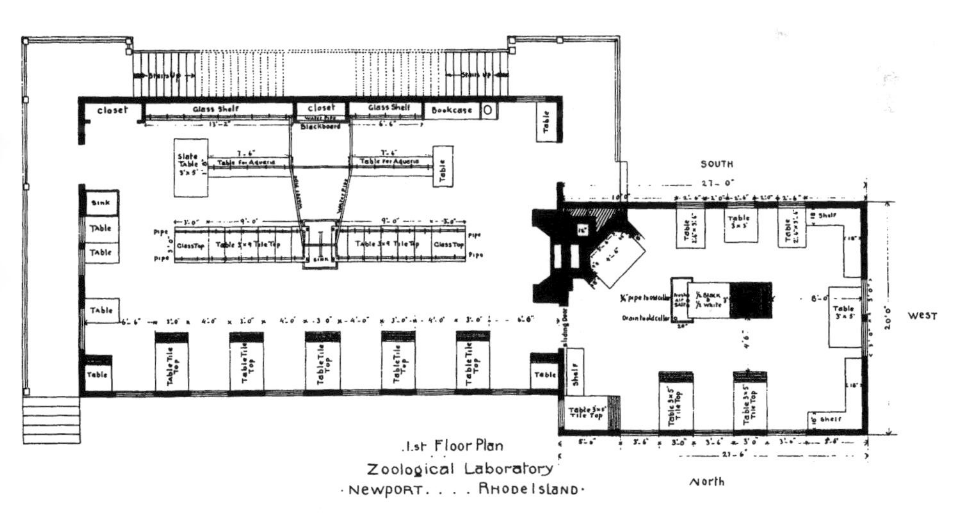

Figure 41. Plan of Newport laboratory, ground floor, after its 1891 enlargement. The fireplace and the room to the right were added so that Mr. Agassiz could work undisturbed by his guests, most of whom were Harvard graduate students.

Figure 42. Interior of Newport laboratory, main room, looking west toward Mr. Agassiz's room. Running along the center are pipes carrying sea water, which a windmill pumped up from nearby Narragansett Bay. The glass bowls allowed observation of live animals, to be viewed against the black background of the tile-covered tables. This photograph, and others reproduced in G. R. Agassiz's biography of his father, were probably made to help convince Harvard University to accept responsibility for the laboratory's future. (By permission of the Museum of Comparative Zoology Archives, Harvard University)

ercise close control over his students, and Fewkes, who held no faculty appointment at Harvard, was accountable only to Agassiz.*

Fewkes and Agassiz both initially saw Fewke's position as a mutually rewarding and solid beginning for his career as a zoologist. Beginning in 1880, he published a steady stream of articles, totaling sixty-five by 1890, mostly arising from his responsibilities in Newport and in the M.C.Z. Around 1888, some sort of problem developed between employer and em-

*This account is based on meager bits of evidence: the acknowledgments in Fewkes's publications and statements in the *Annual Reports*. The critical clue is a letter by Agassiz to his secretary, Elizabeth Hodges Clark, dated Newport, 19 June 1891, in which he writes, "The heading of Mark's paper amused me. If Harvard Univ. keeps him going o.k. I think some kind of notice of Museum might be appropriate. It's to keep even with Shaler's Dept. I presume. I may be very sensitive but I don't like it any more than his not taking any part here for nearly ten years because Fewkes was here—and the result is we have no end of Sea Labs which would perhaps not have sprung into existence." I am indebted to the late Mrs. V. P. Jones for sharing this letter with me.

ployee. It may be that the death of Fewkes's wife in 1888 resulted in his neglecting his duties,[15] or perhaps, since Agassiz's respected British colleague P. H. Carpenter had a low opinion of Fewkes's writings, Agassiz may have had second thoughts about his abilities.[16] It is also possible that Fewkes sensed that his work for Agassiz was leading him nowhere and began exploring other options. He did find an alternative source of patronage in Augustus and Mary Hemenway, who paid for his trip to California in the summer of 1888.[17] After hearing that he had been offered employment elsewhere (which may not have been so), Agassiz, in June 1889, informed Fewkes that his association with the museum was over.[18]

Two years later, Agassiz was astounded to see an article by Fewkes in the *Bulletin of the Essex Institute* titled "An Aid to the Collector of the Coelenterata and Echinodermata of New England," which included a diagram copied from Agassiz's *Seaside Studies,* along with data on dredging compiled at Newport while Fewkes had been in Agassiz's employ. Furious at what he considered a breach of trust, Agassiz demanded that the Essex Institute recall every number of this publication and delete the diagrams and the ten pages of data.[19] Fewkes immediately made a full apology, protesting that he had no idea that Agassiz had cared about this material. The offending volumes were recalled and reprinted, at Fewkes's expense.[20] The outraged Agassiz told Eliot,

> My friend Fewkes for whom I had done more than for any other student ever connected with Museum turns out to be a consummate sneak and hypocrite. . . . it is discouraging to be treated in that way by an assistant after being for so many years on an intimate footing. I am evidently not intended to get on with the average American student but I am fast learning.[21]

Agassiz grumbled that he was sorry to

> have been compelled to take such measures against one who once was so intimately connected with me and that I should have to protect myself in any way is not encouraging in regard to future dealing with Students.[22]

His choice of words is revealing. Agassiz refused both the distinction and responsibilities of a teaching appointment, but his assistants all began as students or instructors in his museum. Fewkes had done four years of graduate study in Europe before entering Agassiz's employment, and had earned his doctorate fourteen years before Agassiz wrote this letter.

Fewkes's apologies to Agassiz were emphatic, including the underlined declaration, "My zoological career is ended and I shall never put pen to paper on a zoological subject again."[23] This was true, for his transformation into an ethnologist, begun with Mary Hemenway's support in 1889, was already well underway. Agassiz was not a forgiving man, and he used his influence to prevent Fewkes from working at Harvard's Peabody Museum of Archaeology and Ethnology.[24]

Whatever lesson Agassiz was "fast learning" from disappointments like the Fewkes affair, he was fated to have others. Like his father's disappointments, these affairs, obviously linked to personal characteristics of the individuals involved, suggest something more significant, and once again, though the indications are not quite as sharp as with H. J. Clark and the rebels of 1863, the underlying theme seems to be the opposite interests of an established senior scientist versus those of a person striving to build a career in science.

Another assistant with whom Agassiz had a falling-out was Charles Otis Whitman, who was approaching forty years of age when he came to the museum in 1882. Having attended both sessions of Penikese and spent some time at the Naples zoological station, Whitman had earned his Ph.D. from the University of Leipzig. Because he had taught at the Imperial University of Tokyo, Agassiz referred to him as "Prof. Whitman," but his appointment at Harvard came from the museum Faculty, not the college. He did some collecting for Agassiz in Key West, and in the summers of 1883 and 1884 he worked in the Newport laboratory.* Agassiz expressed his pleasure in a letter:

> I have this summer gone back to my Fishes and with the help of an excellent assistant, C. O. Whitman, have some good things. We shall hope to publish early next year a good Memoir, made up of all my accumulations for twenty years and a careful revision by an outsider during two years.[25]

Did Agassiz's description of Whitman as an "outsider" portend the troubles to come? Possibly he meant this in a positive sense, expecting Whitman to contribute an independent viewpoint, but at the very least the term reveals a lack of sympathy. The project, on embryological development, was a direct continuation of Louis Agassiz's work, which Alexander had been nursing in the moments he could escape from his administrative burdens. Seemingly careful to avoid his father's mistakes, Alexander Agassiz would allow Whitman billing as coauthor, even though he was paying Whitman for his time and providing the materials.

Certainly Agassiz appreciated Whitman's technical skills. Their first

*In the summer of 1883 Whitman became engaged to E. A. Nunn, who was also doing research in the Newport laboratory. As Mrs. Whitman, she continued her research in Newport the following summer (Mark, recollections of Whitman, accompanying a letter to F. R. Lillie, Archives, University of Chicago Library). Her brother would later become a supporter of Whitman's efforts to control and finance the Marine Biological Laboratory.

joint article appeared in the *Memoirs* of the M.C.Z. in September 1885. When Edward P. Allis, Jr., asked Agassiz to recommend a good scientist, Agassiz mentioned Whitman. The opportunity to direct the new Allis Lake laboratory drew Whitman to Milwaukee in 1886, the fish work still incomplete. A second joint paper did appear in June 1889, based on the Newport work done six years before, but by then Whitman was deeply involved in plans for a summer school at Woods Hole. Whitman promised to finish the remaining articles on fish embryology that he and Agassiz had planned, but he failed to keep his promise. In July of 1892 Alexander Agassiz angrily swore off all communication with his former assistant.[26]

Another in the list of those who fell afoul of Agassiz's temper was Thomas A. Jaggar, Jr. (A.B. Harvard 1893, Ph.D. Harvard 1897), instructor in the geology department. Agassiz saw red when Jaggar ignored his instructions not to give a story to the popular press about their expedition to Hawaii. It was Eliot, as president of Harvard and chairman of the museum Faculty, who fired Jaggar, at Agassiz's insistence.[27]

No direct confrontation is recorded between Agassiz and one other museum student, but there are significant hints that Charles Benedict Davenport's promising career was derailed by Agassiz. Officially under Mark's direction, Davenport's dissertation, completed in 1892, was perhaps the most original the zoological laboratory produced, and one of the very few based mainly on work in Newport. An instructor beginning in 1890, Davenport was a frequent visitor to Newport, published impressively, and seemed destined to go far. In 1898, however, he failed to get the expected professorship at Harvard. So strong was his belief that it was Alexander Agassiz's personal dislike of him that had prevented the appointment, for he knew that he had Mark's enthusiastic support, that he confronted Agassiz directly. But the only official explanation he could elicit was that his research was "physiological" rather than zoological, surely a distorted characterization of the range of Davenport's investigations.[28] Mark shared Davenport's suspicions but saw there was no remedy and urged him to be confident of a career elsewhere.[29] After a year of anxiety, Davenport did obtain a position at the University of Chicago. Understanding what drives evolution was his passion (he would later head the Cold Spring Harbor Laboratory of Experimental Evolution), so Harvard's failure to find a place for him seems particularly unfortunate.

It may be that Agassiz's own path to a career in science—which included academic degrees, high standards, and significant commitment of time, yet depended upon an income earned outside of science—made him insensitive to the pressures on those who hoped to make their livelihoods from science. Not that he expected every young scientist to find himself a copper mine: Agassiz encouraged the idea of endowments for university chairs which would leave an individual free from obligations of teaching, and he willed such professorships to Harvard. He opposed, however, government support for science that could be done by private enterprise.[30] The value of an institutional buffer between a patron of science and the professional worker was something Agassiz did not understand, blaming instead the character of the "average American student."

9

"Collections Never of Use to Anyone"

The first decade after the death of Louis Agassiz had been a time of impressive growth in the size and activity of the Museum of Comparative Zoology. Alexander Agassiz, believing himself committed to fulfilling the promises his father had made, and devoted to science himself, freely poured cash from his copper mines into construction of the teaching laboratory and exhibitions. For an engineer who conquered unforgiving rock, a manager who kept his eye fixed on long-term profit, a businessman who led his fellow mine owners in a cartel to control the price of copper,[1] Agassiz's expectations for the museum were ill-defined and sentimental. He seemed to imagine that the beneficiaries, Harvard University and the general public, would be grateful. If they were, Alexander Agassiz had no idea how to harness their gratitude. Admission to the public exhibits was free, Louis Agassiz having promised this when begging the original state grant. Harvard did pay the salaries of instructors, but any further expenditures, such as for heat or equipment, could be extracted only by persistent complaint. The ideal of mutual benefit projected in 1876 had proved illusory, although Agassiz never ceased to believe it had been possible. During 1883 and 1884 he surveyed his Cambridge kingdom with deeply mixed feelings: satisfaction at the changes he had wrought and concern for its future. The weakest part of the whole structure from his perspective was the very one in which his father had placed his highest hopes, the research collections.

The doubling of the building, purchase of specimens for exhibit, and equipping of the teaching laboratory had all been paid for directly by Alexander Agassiz. The salaries of the curatorial and maintenance staff, upkeep of the physical plant, and supplies like alcohol to preserve the collections came from the museum's own funds, that is, the interest on its capital. Agassiz reported in the fall of 1884 that this amount had decreased from $35,000 a year to about $25,000 a year.[2] And this was not a time for financial optimism in general. Bank failures and a stock crash beginning in

August 1883 grew to panic and depression in 1884. In 1883, various factors began to erode the price of copper; the cartel finally collapsed into a price war, which seriously depressed the value of copper for several years. Perhaps the money question, or perhaps the completion of the building and exhibitions, but probably both factors, caused Agassiz, never keen on indoor science, to grow positively hostile to the collections. He wrote toward the end of 1884,

> There are stored in the cellars of the Museum immense collections of Fishes and Reptiles which have never been of use to any one except the assistants in charge of them. A very large part of this material, collected and maintained at great expense, ceases after a time to be of value for scientific purposes, and every year we are obliged to throw away as absolutely worthless a large number of specimens.[3]

Most of these specimens had entered the building in kegs and barrels, unpacked by Agassiz and Lyman while Louis Agassiz and his assistants steamed up the Amazon on the Thayer Expedition. They demanded manpower to sort, expensive glassware and alcohol to preserve, and space to store, but Louis Agassiz insisted they would revolutionize ichthyology and challenge Darwin, and he convinced Nathaniel Thayer to donate yet again so that extra workers could be hired: Richard Bliss, Jr., James Henry Blake, Samuel Lockwood, and Caleb Cooke, besides the visiting expert Steindachner. Upon Louis Agassiz's death, his successors had to see if the operation could be managed on its own income.

In 1874 the M.C.Z.'s collections were being looked after by at least twenty people, including volunteers like Theodore Lyman, who made a hobby of caring for the brittle stars, and the Baron von Osten-Sacken, who incorporated his own great collection of Diptera into the museum. The employees ranged from women paid by the hour, men of limited training and ability like Samuel Garman, better trained and thoughtful individuals like J. A. Allen, up to highly respected experts like Hagen.[4] Less than a month after Louis Agassiz's death, printed notices were sent to a number of workers, "to inform you that the Faculty will not be able to pay you for your services at the Museum after the first of April next."[5] Frederic Ward Putnam, once a student of Louis Agassiz, for old time's sake agreed to keep an eye on the fish, but it turned out that his position as director of the new Peabody Museum of Archaeology and Ethnology prevented him from actually giving much time to it. The death of J. G. Anthony in 1877, and that of the genial volunteer keeper, Count Pourtalès, in 1880, reduced the effective curatorial staff to four: Hermann Hagen (insects), Charles Hamlin (mollusks), Samuel Garman (reptiles and fish), and J. A. Allen (mammals and birds). A good part of their time was spent developing the public exhibits, which by 1884 were nearly complete and were expected to need little further attention.

Alexander Agassiz had already declared in 1875 that modern zoology called for field stations rather than museums; in 1884 he elaborated this judgment into a brutally pessimistic cost-benefit analysis. The time and money spent on maintaining collections seemed all out of proportion to the actual original research based upon them. The cost of keeping perishable material such as jarfuls of fish, which need additions of alcohol, or dried insects or bird skins, which have to be protected from decay, "may stagger the most enthusiastic collector. Do the results justify such large expenditures?" Agassiz asked. While acknowledging the archival responsibilities of the museum, he declared that the job of "supplying special investigators materials for their study . . . seems to me, nowadays, unreasonable to expect . . . of any museum. No naturalist who wishes to study fishes," he wrote,

> will expect to find in any establishment, no matter what its resources may be, the necessary materials. He will be compelled to travel, to collect in the various fish-markets of the world, and to study his materials on the spot. With the present facilities and the cost of travel, it would be far cheaper for an institution to supply the specialist with the necessary funds for such an investigation, if it be one of value and interest, than to go on for years spending in salaries of assistants, care of collections, interest on the cost of buildings, and so forth, sums of money which, if distributed to their ultimate object, would astonish the least prudent manager. Such accumulations of historical material are far too costly. The same sum spent in a different direction, in promoting original investigations in the field or in the laboratory, and in providing means for the publication of such original research, would do far more towards the promotion of natural history than our past methods of expending our resources.[6]

For all his sense of duty toward his father's plans, Alexander felt he must give first priority to making the museum the locus and instrument of contemporary biological research, and that meant laboratories, not collections. In times of economic belt tightening, material that interested only a few special investigators and imaginary future investigators could not compete with the needs of the scientific mainstream.

Agassiz's judgment was not based merely on his own research preferences but on his perception that the scientific community had shifted its priorities. He now felt, more strongly than ever, that the topics of current excitement, the research front of modern biology, needed laboratories, not collections. He declared that

> with the present tendency of science, original work cannot be based mainly upon the collections of a great museum. . . . With the requirements of to-day, collections

can only supply materials for investigations of limited scope; and while undoubtedly many most interesting problems require large collections for their solution, the more important biological problems of the day require materials prepared for special purposes in the laboratories of the Universities.[7]

Oppressed by the feeling of trying to sail a floundering ship weighed down with a cargo of useless specimens, Agassiz decided that it was unrealistic for a university museum to save up raw material for the doubtful possibility of future use. The only additions to the museum should be the private collections of respected specialists.[8] These would consist largely of specimens already preserved, identified, and labeled. Watching the growth of the Smithsonian with its government funding, the American Museum of Natural History, the Peabody Museum in Salem and the one at Yale, and even the Boston Society of Natural History, Alexander Agassiz felt that any efforts to enlarge his own museum's collections would be wasteful competition.

In 1886, when J. A. Allen was offered a job at the American Museum of Natural History, Agassiz urged him to go, on the grounds that there might soon be no money to pay curatorial staff in Cambridge.[9] In New York, Allen would prove that he had many years of leadership in mammalogy and ornithology ahead of him. Allen's responsibilities were taken over by William Brewster, quite a respectable ornithologist but an amateur, unable to commit much time in Cambridge. When Charles Hamlin died in 1886, Agassiz asked Alpheus Hyatt to care for fossils part-time, but in 1892 he announced, "Owing to our insufficient means, we have been unable to continue Professor Hyatt's salary at the Museum."[10] Hyatt, finding the collection and space useful for his continuing research, stayed on as a volunteer until his death in 1901.[11]

The pickled treasures of the Thayer Expedition, though lying largely undisturbed, were not forgotten by specialists. David Starr Jordan, a professor at Indiana University and leading ichthyologist, who was proud to count himself a student of the great Louis Agassiz after the Penikese Island idyll, spent "some weeks" in the summer of 1887 gleaning new marine species from Louis Agassiz's collection. The freshwater fishes were so plentiful he made no attempt to go near them.[12] Jordan had a doctoral student, however, who with his fiancée had been hatching a plan for nearly a year to see this fabled understudied collection.* The student was Carl Eigenmann,

*Rosa Smith to her mother, 10 October 1886, Eigenmann Papers, Lilly Library, Indiana University, Bloomington, Indiana. This collection was examined for me with care and intelligence by Ron Millen.

who married in August 1887 the accomplished ichthyologist Rosa Smith. Their letter of welcome from Alexander Agassiz reads, in its entirety:

In answer to yours of Sept 7 [1887] I would say that I have no objection to your working at this Museum with Mrs. Eigenmann providing it causes no expense to the Museum and that you comply with such regulations as are necessary for the welfare of the Museum.[13]

At the museum they met Samuel Garman, whom they thought "a fool in a general way and in every particular."[14] (Perhaps the fact that Garman had charge of the M.C.Z. fish explains why Jordan, who knew him from Penikese days, did not favor the idea of the Eigenmanns' visit.) Garman had arrived at the museum in the fall of 1872, having met Louis Agassiz on the wharf in San Francisco at the close of the *Hassler* voyage. He was later unable to recall much of what he had been doing before that.* In the early 1880s he did some fossil collecting in the West, resulting in a feud with Edward Drinker Cope.[15] Later generations can find little good to say about his work.[16]

Carl Eigenmann, candidate for a doctorate at Indiana University, used his time in Cambridge to take courses, as did Rosa Eigenmann. They also poked around in the cellar of the museum, looking for a promising topic based upon the Thayer collections.[17]

As Louis Agassiz had been almost a quarter-century before, the Eigenmanns were impressed by the great numbers of species in the Thayer collections. Even though they believed that Agassiz's observation of large numbers of extremely localized species "was in great part due to mistaking the variations of a species as distinct species," the jars certainly contained many hundreds of undescribed species.[18] They decided upon a complete revision of the Characinidae, the family as characteristic of

*Asked to supply biographical information, Garman wrote, "Of early life hardly a distinct recollection of childhood or a village school is left to me; all are lost or vague and dreamlike. The mists of forgetfulness obscure even the preparations for advanced education; trainboy; adzman; foreman on construction; fireman and running an engine. Vital reminiscences date little earlier than a course in a State Normal University with graduation followed by teaching. Principal of a State Normal School; Professor of Natural Sciences in a Seminary." Samuel Garman to Samuel A. Eliot, n.d., Harvard University Archives. The best account is that of Hubert Lyman Clark (*Dictionary of American Biography*), who identifies the Illinois State Normal School, the Mississippi State Normal School, and Ferry Hall Seminary of Lake Forest, Illinois. He earned a bachelor of science degree from Harvard in 1874 (not, as some sources say, 1875), and Harvard gave him an honorary M.A. in 1899 (Harvard University Archives, Overseers Records). One obituary claims he was with Major John Wesley Powell in Colorado (*Boston Transcript*, 1 October 1927), and this fits with the fact that Powell taught at Illinois State Normal University, leading parties of students to the Rocky Mountains of Colorado in the summers of 1867 and 1868.

South American waters as marsupials are characteristic of Australian land. "Unfortunately," they reported in May 1889, their work on this material "was interrupted shortly after it was begun," perhaps by the job offer which brought them that year to California.[19]

Much to the annoyance of Alexander Agassiz, the Eigenmanns were tactless enough to set down in print what ichthyologists worldwide already knew, namely, that a precious opportunity once housed in Agassiz's museum had been forever lost. Because comparison to type specimens is essential for accurate determination of new species, the convenience of having those specimens in one place is considerable:

> The collection of the Thayer expedition is by far the richest collection of fresh water fishes ever made in South America. It is scarcely necessary to enumerate here the advantages which would have accrued to American students and to this museum had this enormous collection been studied shortly after it was made. No systematic attempt has before been made to examine all the specimens of any of the families of fresh water fishes collected during this expedition. The result has been that most of the new forms having been rediscovered by other collectors the types of numerous species are now scattered through various museums in America and Europe which would be found in the Museum of Comparative Zoology if this collection had been studied at once. Even those specimens which have hitherto served as the types of new species are not so labeled, rendering very difficult the identification of the rest at this time.[20]

Alexander was incensed at their impudence. How easy for them to say what others should have done! Who was to pay the army of specialists such studies would require?*

Certainly there was no opportunity (though much need) for another professional ichthyologist in Cambridge. Care of the collections was dropping to its lowest point. Hagen, who was in his mid-seventies, had been increasingly absent from the museum due to ill-health and would soon become paralyzed.[21] Agassiz complained in his 1890–91 annual report that, without his extra gifts of cash, the museum's budget was inadequate for what was expected of it.

> We are unfortunately compelled for want of funds gradually to restrict the work of the assistants of the Museum to the mere care and maintenance of the collections,

*_Annual Report_ for 1890–91, p. 6. The various requests and proposals Carl Eigenmann made to Alexander Agassiz over the next decade were all rebuffed. He was informed on 31 March 1890, that a manuscript he had sent was not suitable for publication by the museum, and on 20 July 1892 that the museum could not afford the time to send material away from Cambridge for study. Also, Agassiz wrote to Nathaniel Thayer, the son of the sponsor of the Brazil Expedition, on 10 August 1906, advising against giving financial aid to Eigenmann (M.C.Z. Archives).

and to limit their use by specialists . . . [and] to refuse many of the constant demands made upon the administration of the Museum for an opportunity of working up or of examining parts of our collections, the value of which to science is thus greatly diminished. It is true we have a number of volunteer assistants [Walter Faxon, D. D. Slade, William Brewster, Alpheus Hyatt], but we cannot expect these gentlemen, who have kindly undertaken the general supervision of special departments, to spend their time in the drudgery necessary to meet the legitimate demands which every fairly organised Museum ought to be able to meet, and which are naturally made upon our collections and assistants. With the exception of Dr. Hagen and Mr. Garman, there is no assistant upon whom I feel at liberty to call for work of that kind.[22]

The next year was a bit better: Agassiz relented in his attempt to make the museum function on its own to the extent of hiring Faxon (called "Professor Faxon" in the custom of the day, because of his former status on the Harvard faculty) as maid-of-all-work. In 1893 he was given the title "Assistant in Charge."[23] Also, the university consented to grant a small pension to Hagen, releasing part of his salary to engage the part-time services of the entomologist of the Boston Society of Natural History, Samuel Henshaw (fig. 43), who started in November 1892.* Utterly dedicated to keeping order, Samuel Henshaw soon impressed Agassiz as a man who could be trusted, no matter that he was rigid, humorless, lacking in ideas, and antisocial.

The Eigenmanns were doing well in the 1890s, publishing on a variety of icthyological topics, including South American fishes. When Jordan accepted the presidency of the new Leland Stanford Jr. University, Carl Eigenmann replaced him in the professorship at Indiana. Year after year, the recollection of the unexploited wealth he and his partner had seen in the basement of the M.C.Z. kept nagging him. Early in 1895 he wrote to Charles B. Davenport, who had been the laboratory assistant when the Eigenmanns were at Harvard and was now an instructor in the zoological laboratory in the museum, that this would be the perfect material for the study of variation, a topic just then attracting evolutionary theorists.[24] He told Davenport that in the subfamily of Tetragonopterines there were variations in color; number of teeth; number of dorsal and anal rays; number of scales along, above, and below the lateral line; completeness of the lateral

**Annual Report* for 1891–92, p. 7; Henshaw is listed as the assistant in entomology in the *Annual Report* for 1891–92, which explains why obituaries give 1891 as the date of his appointment to the museum, but he had only just been hired when Agassiz wrote the report in October 1892, and he later tells a correspondent that he actually began work in November 1892 (Henshaw to E. T. Cresson, 6 September 1913, Academy of Natural Sciences, Philadelphia).

Figure 43. Samuel Henshaw, photographed unawares in a corner of the entomology department. Head of the museum from 1904 until 1927, Henshaw kept every key in his personal control. A woman works in silence at the extreme right, symbolic of the subordinate workers, many of them female, on whose labor the museum depended. (By permission of the Museum of Comparative Zoology Archives, Harvard University)

line; and overall proportions. Eigenmann's own attempts to return to the M.C.Z. had met with no encouragement; the museum would not pay his expenses, nor even, as was often done in return for the labor an expert spends sorting and labeling an unworked collection, would he be allowed to take away a series of duplicates.[25] He gave Alexander Agassiz the standard argument, that his work would increase the value of the M.C.Z.'s material by creating type specimens, besides adding to the museum's reputation, but to no avail.[26] We cannot know whether Alexander Agassiz still bore a grudge for the Eigenmanns' impudent criticism, but knowing Agassiz, we may be sure it had not been forgotten.

In 1898 Alexander Agassiz, at the age of sixty-three, decided he had finally had enough of the drudgery and annoyance of heading the museum. (He had thought about resigning at least as early as 1883, eliciting an eloquent plea in elegant German script from Hagen, who wanted to always have an Agassiz as leader.)[27] However, it was the trouble Agassiz wanted to be rid of, not the advantages. His plan was to continue the research in oceanography and marine zoology he so enjoyed. Carefully noting what aspects of the museum he would not like to lose, Agassiz offered Harvard a gift of his books, specimens, and all the objects he had already given the museum over the years. (Louis Agassiz had once been careful to stipulate that anything collected by an officer of the museum was automatically museum property, but no one was so tactless as to suggest that Alexander was offering to give the museum things that already belonged to it).* In exchange for donating his specimens and books, he wanted to be sure of keeping his office, secretary, outlet for publication, and rooms for the artists and assistants he employed. He wanted also the right to hire museum employees to assist him on his cruises, manage the material collected, or work in Newport or Cambridge on scientific projects in which he was interested.[28] Of the paid staff of the museum in 1898—Samuel Henshaw, Samuel Garman, Alfred G. Mayer, Walter Faxon, and William McMichael Woodworth—every one except Henshaw acted at one time or another as personal research assistant to Agassiz.

Agassiz wanted to give up the executive office but not the ultimate power. With the help of George Lincoln Goodale, a secret negotiation assured him lifetime control in exchange for a gift of several million dollars. His position as member of the museum Faculty gave faint public indication of the power he held until his death in 1910. His son George Russell Agassiz was respectfully consulted by Harvard presidents and M.C.Z. directors until his own death in 1951, when the role was assumed by George's widow, Mabel.

Agassiz (fig. 44) was one of the wealthiest individuals of his day, and he enjoyed unlimited control over the management of his own money. No one questioned his absolute right to spend it as he pleased, for the risks and grit and work associated with the first years of the Calumet and Hecla copper mines were real enough, and so was the labor of managing them thereafter. Every spring and fall he was in Michigan, actively supervising both the engineering and management of the mines. To classify Agassiz as a "robber baron" would be harsh; though firm in dealing with strikers, he took pride in ensuring the welfare of his employees.[29]

The feelings of his contemporaries, officials of Harvard and scientists everywhere, and the feelings of all those in later generations who benefited or are benefiting now from the existence of the M.C.Z., tend to sanctify Agassiz's generous support (gifts predating tax credits) of the museum. After all, there were other wealthy Bostonians, and neighbors of his in

* "No one connected with the Museum shall be allowed to own a private collection, or to traffic in specimens of Natural History, except for the benefit of the Museum. If an officer of the Museum, or a student working for the Museum, possesses a private collection with which he is unwilling to part, he must deposit the same in the Museum during his connection with the institution" (*Annual Report* for 1864, p. 48). Some such principle is now widely considered a healthy precaution against conflict of interest for the staff of museums of all kinds.

Newport, who devoted their money to clothes and parties and yachts. We may note, though, with no diminishing of gratitude, that unlike his sister Ida's husband Henry Lee Higginson, whose copper money went into a symphony orchestra he neither conducted nor played in, Agassiz's charity went to an operation in which he was deeply involved and which he controlled.

The playthings of the very wealthy have never been limited to private pleasure but often include institutions of various kinds. The man who loves horses may provide a polo field or horse show available to other horsemen of modest means, and enjoyed by the general public. If money is given for art, or for science, the donor enjoys enhanced prestige as a benefactor of culture. Much of what Agassiz did for the museum, in his expenditure of time and effort and money, was indeed a cultural gift, but frequently his gifts served his own interests as well, often quite directly. Eliot, not only with an eye to future legacies but out of genuine respect for Agassiz, was naturally glad to accommodate him.

In these circumstances, what was wanted for the M.C.Z. was not a new leader but a caretaker. "Billy" Woodworth, first a student and then instructor in Mark's laboratory, had worked at Newport beginning in 1888, prepared taxonomic reports of worms collected by Agassiz, and he had accompanied Agassiz on cruises in the Pacific in 1896 and 1897, serving as photographer, collector, and aide-de-camp. At Agassiz's retirement on 1 August 1898, Woodworth replaced Faxon as "Assistant in Charge," retitled the next year "Assistant in Charge of the Museum." The printed letter announcing Agassiz's retirement specified that policy would be "guided by a Committee of the Museum Faculty, consisting of Dr. H. P. Walcott and Professor George L. Goodale."[30] Woodworth's role was merely to follow Agassiz's wishes and protect the museum against innovation. He again accompanied Agassiz on a long cruise in 1899–1900 and a month or two in 1901–2, museum business being delegated to Samuel Henshaw. In 1900 he gently suggested that it might be a problem that his position was so informal: ". . . as it is, I have nothing to show, no paper of any kind, that I

▶

Figure 44. Alexander Agassiz. When fashionable Boston society was weighing his merits as a scientist against his father's, the "autocrat of the breakfast table," Oliver Wendell Holmes, recalled reading about a custom of the Fiji Islanders, whose parents expected and welcomed murder by their offspring (M. A. DeW. Howe, *Memories of a Hostess,* p. 48; cf. T. Williams, *Fiji and the Fijians,* pp. 144–60). Holmes was speaking not of the M.C.Z. but of evolution, which Louis Agassiz's students, Alexander Agassiz among them, accepted despite their teacher's most cherished beliefs. (By permission of the Museum of Comparative Zoology Archives, Harvard University)

have legal right to act for the Museum. Such a thing might be useful sometime."[31] So the Harvard Corporation named him "Keeper." The usefulness was most likely for his dealings with his former teacher, E. L. Mark, whom Harvard named "Director of the Zoological Laboratory" the same year.*

Woodworth was unable to command the respect of others in the museum; acquaintances would later call him "by nature a man of a sensitive and almost childlike disposition."[32] At times during their travels Alexander Agassiz found his agreeable nature exasperating: ". . . as for Woodworth he must be getting cracked. He has no pluck no endurance, no grit and gives right in at the least unpleasantness and becomes a wet rag. I don't see how he'll ever accomplish anything."[33] In 1904, two members of the museum Faculty visited Woodworth to help him "see plainly that a position in which he held strained relations with all the staff was one in which no man could do himself justice."[34] He tried to explain "that the task given to him was an impossible one—thought Henshaw responsible in large measure for the present catastrophe—did not explain how."[35] Henshaw had become a full-time employee in 1898, when the librarian, Frances M. Slack, retired, taking on her job in addition to the insects. Woodworth was demoted to "Assistant in charge of Worms" (his salary of $400 a year was paid by Agassiz, not the museum).[36] He did accompany Agassiz on two more oceanographic expeditions.

Samuel Henshaw would not accept a role as problematical as "assistant in charge" or "keeper"; he insisted on the title both Agassizs had held, "curator," which gave him ex officio membership on the museum Faculty. His tenure began on 1 September 1904. Giving such authority to a man "who, by training and temperment, was utterly unfitted" for such a post was correctly pronounced, many years later by the subsequent director, "one of Mr. Agassiz's extraordinarily stupid acts."[37]

Eigenmann decided to try his luck with the new curator and found Henshaw very sympathetic to good taxonomic work. He also wrote to the sons of Nathaniel Thayer, one of whom had gone to Brazil with Louis Agassiz forty years earlier, in hopes that they would pay the cost of colored illustrations, based upon Jacques Burkhardt's field sketches.[38] "They seemed ready to do it if Mr. Agassiz will recommend it," Eigenmann told Henshaw.[39] Agassiz may have listened to Eigenmann's proposal during the International Zoological Congress in Boston in 1907, of which Agassiz

*In the same letter to Agassiz, Woodworth rejected the idea of being given the additional appointment of assistant professor, as putting him "under Mark" (who was a full professor). Later correspondence with Harvard president A. Lawrence Lowell shows the great reluctance with which Mark finally retired in 1921, making way for G. H. Parker to take over as director of the zoological laboratory.

was president and where Eigenmann gave a talk on South American fishes in the section on zoogeography. Twice Eigenmann was encouraged to believe the Thayers would help.[40] However, they took a dislike to him; John E. Thayer scrawled on one of Eigenmann's letters, "I have no use for this cuss."[41] In his various publications, Eigenmann acknowledged financial aid from various other sources, but never from a Thayer nor from Alexander Agassiz.

One day when Eigenmann was letting off steam to Davenport about Garman's poor judgment in differentiating fish species, he found out that merely taxonomic issues seemed silly and insignificant to his old friend, now keen to investigate evolution through genetic experimentation. Davenport had recently founded the Laboratory for Experimental Evolution at Cold Spring Harbor on Long Island, New York (with Carnegie money), and was there testing Mendel's laws as applied to poultry and other animals. (His enthusiasm for eugenics was still in the future.) Eigenmann could not let Davenport's attitude pass:

We were talking as the train pulled in about Garman's judgment about species and you expressed your general disgust about such matters of no moment. I did not have time to reply but can't resist the temptation to attempt to show you how very wrong you are. I will give you two imaginary cases that I hope will put things in the right light.

You are attempting to get at the transmissibility of certain characters or the purity of the transmissibility of any characters. To do this you have to keep an endless number of uninteresting but exact records, describe birds, regulate your incubators and feed your chicks. Your end results will enable you to generalize and predict *what will happen next.* Now supposing [William] Bateson [leading English geneticist] or someone else who has not seen your birds, through slovenly reading or stupidity, would conclude that the different *y*-combed chicks [hybrids between birds with single and with *v*-shaped comb][42] in which you justly take pride were all alike and your conclusions wrong. I think you would be one of the first to recommend a lunacy commission or an entire disregard for such vaporing.

Now I am laboriously examining a lot of dead smelly fish and writing formulae that enable me and should enable others to distinguish one form from another. Don't imagine for a moment that this is any more edifying or pleasurable than turning eggs or regulating brooders or feeding chicks but it is my method of keeping records and it is *just as important* for me to distinguish between minute differences found in nature as it is for you to distinguish slight differences in the *combs you produce.* My records ought to enable me to trace the lines of evolution and lines of dispersal and I am ready against all comers to maintain that they do. Now comes some bull into my chinashop and smashes all of my artistic crockery—not only that but puts his mess on record as an advance over previous work. Unfortunately systematic zoologists have such a tender conscience that everything that has once appeared in print is sacred and however foolish it must be dealt with by all future work.

All of us make mistakes and I have possibly made even more than the share legitimately mine by the law of chance. But I think when a man habitually makes a mess we are entitled to register a kick and I think you will by now agree that lack of judgement in regard to species is not so trivial a matter as it may seem—especially in an individual who by nature and training ought to possess it. Any fool can describe one of your chicks or a specimen of a fish before him but it requires judgement to determine whether it is worth while to place it on record in such a way.[43]

The regulations the taxonomic community imposed on itself were not a matter of "a tender conscience"; they aimed at the stability of names by giving strict precedence to the earliest published description (using Linnaeus as the starting point). The side effect was burdensome, as Eigenmann said.

In 1904, Henshaw refused Eigenmann's request to borrow a section of the Thayer collection, but a few years later he changed his mind. The long-awaited trainload of boxes of characins arrived in Indiana in 1907, and Eigenmann crowed that it was "like opening a Christmas box to unwrap the various lots."[44] The loan was made with the usual conditions: that the results be published in the *Memoirs* of the M.C.Z. and the collection be returned well identified. Eigenmann found private benefactors, and the Carnegie Museum in Pittsburg paid for a collector to cover the areas in South America Eigenmann said were needed to complement the route of Louis Agassiz's party. Eigenmann's monograph appeared in parts between 1917 and 1927, finally proving wrong Alexander Agassiz's prediction that these fish would never be of any use—more than fifty years after being collected.

Eigenmann's boast in his letter to Davenport, that he would be able to "trace the lines of evolution and lines of dispersal," was put to a severe challenge by the diversity of the fish Henshaw sent him. In 1908 he confessed to Davenport,

I have been wrestling all winter with one subfamily of the Characins. They present unquestionably the finest material for a study of divergent evolution, Whether I shall be able finally to do them justice remains to be seen. The task of describing species and handling the material is bewildering.[45]

What the student of divergent evolution would find in these fishes was hundreds of species that shared so many distinctive anatomical details that their family relationship was unquestioned, yet they ran the gamut of size, shape, teeth, and feeding habits displayed by North American fishes. There were characins imitating "our garfish, our pickerel, our top minnows, our pompano, our trout, our minnows, our suckers, our darters, our fresh-water herrings and shad," even one that could fly. Eigenmann had an opportunity on New Year's Day of 1909, at the AAAS celebration marking the fiftieth anniversary of the *Origin of Species,* to present this wonderful diversity as ideal material to settle evolutionary debates.[46] Was change gradual or by mutations? in haphazard directions or orthogenetic? Were geographic variants incipient species? Did isolation cause variation? The characins could supply answers, he promised.

With the Thayer material he had been able to disentangle the maze of species and varieties called *Astyanax fasciatus, A. taeniatus, A. scabripinnis,* and *A. intermedius.* Almost every river had its own varieties, and while in any one locality the coexisting species were easy to tell apart, "each undergoes so many modifications in different rivers that it is not possible to give a clear definition that will distinguish the species when specimens from all the rivers are considered."[47] Yet isolation alone could not cause variation, because these highly variable fish could be found quite identical in separate regions, as long as the stream environment was the same. Nor could orthogenesis be universal, for these divergent adaptations were modifications of one type in unrelated directions. The process of mutation so dear to Hugo de Vries and his followers was perfectly illustrated, Eigenmann claimed, by *Hemigrammus (=Hyphessobrycon) inconstans.* In some individuals belonging to this species, the lateral line is complete, in others incomplete, in some complete on one side and not the other, in others stuttering.[48] But another species, *Astyanax aeneus,* was found in some localities to be "varying in the old-fashioned way."[49] Evolutionists inspired by de Vries, Galton, and Bateson were caught up in a fever of breeding experiments, but Eigenmann pleaded, citing this rich material, "Experiment is the watchword of the day; but while we are experimenting in our back yards we should not lose sight of the beauty and the importance of the experiments in landscape gardening and zoölogical gardening, that are and have been going on in our front yards that extend from here to Cape Horn."[50]

Though he held an appointment at the Carnegie Museum from 1909, and though his research was all taxonomic, Eigenmann was employed not as a museum curator but as an educator. More than most of his museum-based fellow taxonomists, Eigenmann projected an enthusiasm for the theoretical implications of supposedly arid taxonomic methods. The characins, for example, gave the best demonstration one could wish for of the evolutionary process of adaptation. But as he learned when he begged the Carnegie Institution for grant support, most professional biologists did not share his vision.

He had on New Year's Day an audience of nontaxonomists. His formal taxonomic monographs on that fish family began to appear eight years later in 1917, the last part published posthumously in 1929, constituting volume 13 of the M.C.Z.'s lavishly printed *Memoirs*. Yet he had covered only about a third of the family, and did not arrive at the positive insights into either the process of evolution or its course in these particular fishes his letter to Davenport and his AAAS lecture had promised. Instead, he presented concrete evidence for a frequency of parallel and convergent evolution that undermined the identification of taxonomic groups with phylogenetic lines, and he constructed a chart of relationships looking nothing like the linear branching tree of the evolutionists. The very plasticity that made the characins so interesting to the evolutionist, that allowed them to evolve in so many directions, radiating out to fill every imaginable niche, meant that they would remain intractable to the discipline of the standard taxonomic monograph. When it came to the point of analyzing the result of nature's vast experiment, the tools of even this skillful taxonomist proved inadequate.

Eigenmann ran into the same problem within one subfamily of characins as had baffled Alexander Agassiz in the order Echini, although unlike Agassiz he did not conclude that the whole enterprise was hopeless. Apparently trying to make his conclusions more precise and accessible to nonspecialists, he listed seventeen characters along with their opposites. By making the assumption that a character appearing in more genera was primitive, he could then postulate which genus was closest to the ancestral form. He emphasized, however, that the characters could occur in almost any combinations. Indeed, taking the 6 most useful character-pairs, he pointed out that those could theoretically generate 2^6 or 64 possible combinations. The technique may be reminiscent of Alexander Agassiz's "bombshell," but Eigenmann's calculation was only meant to ask whether some combinations were more likely than others. To begin with, he had chosen to calculate from only six instead of from all seventeen characters, so as not to "needlessly complicate" the discussion. Finding only half of the possible combinations embodied in existing genera, all he could conclude was "that either many of the possible combinations have never arisen, or, having appeared, they have not been preserved."[51]

Like Alexander Agassiz, and more clearly, Eigenmann found such patterns of characters as could only be explained if it often happened that the "same" character had evolved several times independently. These fishes have thought nothing of inventing the same character over and over. As long as characters can be evolved more than once, the genera defined by the possession of such characters, as many were, stood a good chance of being polyphyletic. Such genera would give the same name to species converging from different ancestors (as Cope's *Orconectes* had done) rather than to the descendants of a common ancestor. If one followed Darwin's definition of natural groups, polyphyletic genera needed reforming, but a worker whose task it was to make sure each species had some name affixed to it could scarcely afford to be a purist.

As an evolutionist, Eigenmann wanted to reconstruct the past lines of relationship, but these he knew were highly conjectural. As a taxonomist, he wanted to find constant characters for his species, and several constant characters to diagnose his genera. The groups thus described would often fail to reflect the course of evolution, and yet, recognizing that a given genus was polyphyletic did not necessarily point the way to fixing it.

We recognize two types of genera, one a group of closely related species, descended from a common ancestor and having certain distinguishing characters in common. . . . The other, a polyphyletic type, consists of species having a certain combination of definite characters in common which easily distinguish members of the genus, but which, instead of indicating a single ancestral line from which the species have diverged, are acquired possibly one at a time along distinct lines converging to a common definition. Sometimes the polyphyletic origin can be detected, sometimes not.[52]

Since it is difficult, or impossible, to say in any case which of the given characters has appeared first, it is extremely difficult to point out lines of evolution leading to different genera or species. We can only insist that certain innate possibilities may become actualities anywhere along the line, possibly wherever they may prove advantageous, though the advantage, to say the least, is not always obvious.[53]

He had come up against the very phenomenon Darwin had told Alexander Agassiz had "haunted me for half a century."

When Eigenmann came to summarizing his views in a diagram, he made no attempt to sketch a phylogenetic tree, inventing instead a peculiar squashed wheel (fig. 45). The genera estimated to be nearest the ancestral forms are in the center, with other genera radiating out from them in all directions. Some genera are connected to the inner oval by straight lines, indicating his confidence in their descent, but others are attached only by dotted lines, and many float about unattached. The outer oval is used to indicate the loss of the primitive character of complete lateral line, a stage of evolution many genera have crossed independently. The diagram is a vivid illustration of the struggles and compromises of an evolutionary taxonomist:

It must be quite evident from the foregoing that the subfamily [Tetragonopterinae] is a paradise for the student of divergent evolution. But the very conditions that make it of interest to the student of evolution make it the despair of the systematist whose object is to express relationship by grouping the species in an orderly array of genera and the individuals in an orderly array of species, always, if possible, in the form of the conventional phylogenetic tree. . . . The Tetragonopterinae seem to form an interlacing fabric rather than a branching tree."[54]

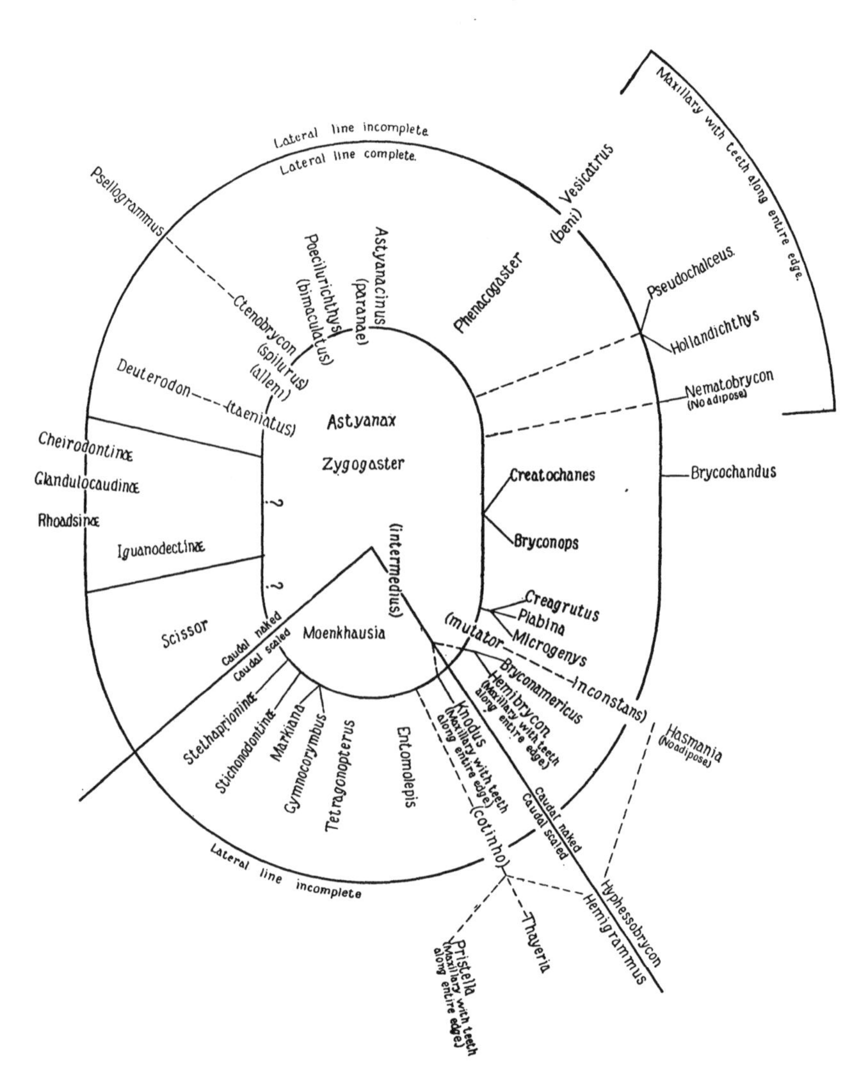

Figure 45. Carl Eigenmann's chart of characin relations ("The American Characidae," *Memoirs of the M.C.Z.* 43:49). Expecting his taxonomic analysis of South American fishes to provide a revealing laboratory of evolution, Eigenmann instead found himself stymied by parallelisms and convergences.

Thus although Eigenmann bragged of the characins, "there are few groups of fishes within which the lines of evolution are so clearly portrayed by existing forms,"[55] his exhaustive monograph could not after all make the same sort of contribution to undertanding the evolutionary process as the experimentalists could. His explanation about the complications created by parallel and convergent variations of characters was much clearer than Alexander Agassiz's had been, and he certainly did not share Alexander Agassiz's hostility toward cultivators of phylogeny. Nevertheless, Eigenmann ran up against the same difficulty Agassiz had tried to quantify: that comparing individual characters and tracing the descent of combinations of characters are two very different endeavors.

When Eigenmann died on 24 April 1927, Samuel Garman, more reclusive than ever at the age of eighty-four, still had charge of the M.C.Z.'s fish collection, and Samuel Henshaw was still director of the museum at the age of seventy-five. Before the year was out, Garman too had died, and Henshaw, much against his will, retired. The South American fish monograph was completed by Eigenmann's student, the late George S. Myers, who became in his turn one of this century's most eminent ichthyologists. Myers wrote appreciatively of the wisdom of Alexander Agassiz in protecting and encouraging the use of the great Thayer fish collection, and privately chided me when I ventured to doubt Agassiz's support of Eigenmann.[56] It is a virtue to honor one's ancestors, and it is certainly true that without Alexander Agassiz's loyalty to what his father had begun, all the M.C.Z.'s collections, indeed probably the museum itself, would not have survived.

10

"Dependent on the Personal Feelings of the Authors"

In 1905, the museum hired a curator who was not trained as a "museum man," a professor of zoology committed to his teaching, whose publications reflected his interest in wider issues of diversity and evolution as well as straight taxonomy. Hubert Lyman Clark was not hired, though, because of these wider interests but because of Alexander Agassiz's uncompleted projects in surveying the world's oceans. The young curator immediately found himself in the middle of a heated taxonomic disagreement. His own wider interests did continue for the first few years of his employment in Cambridge, but they gradually faded away, leaving the taxonomic core of his work unadorned by the spirit of inquiry.

Clark's first love had been ornithology, which remained his hobby even after his graduate work had led him into the marine invertebrates so popular in academic zoology of the day. He had had the best professional preparation available in America, earning his Ph.D. at Johns Hopkins University under W. K. Brooks. His dissertation and some of his research was embryological and morphological, including some experimentation, but after writing a few short field guides to the neighborhoods of Woods Hole and Jamaica, he soon found himself asked to identify the collections of various colleges' summer collecting excursions and then the collections of the U.S. Fish Commission's research vessel *Albatross*. This kind of work, even though more of a preliminary to theoretical science than deep scientific research in itself, he found challenging and interesting. He remained interested in evolutionary issues, especially the "ontogeny recapitulates phylogeny" theme of his dissertation, but soon he was giving most of his research time to the taxonomic monographs for which there was such evident demand.

As a thirty-four-year-old zoology teacher at Olivet College in Michigan,

Clark was surprised to receive a letter from Samuel Henshaw, curator of the Museum of Comparative Zoology at Harvard College, offering him the job of looking after its echinoderm collection.[1] Flattered though he was, he had ambivalent feelings about going. He enjoyed teaching and knew that his students liked him. He had a laboratory and time for research during the school term, as well as summers free for studying living marine animals in Woods Hole, Bermuda, or Jamaica. He had recently been made dean of the college, and when the president of Olivet learned of the Harvard offer he was ready to raise Clark's salary.[2] Part of his temptation was that his family roots were back in New England.

Recognizing that Henshaw's letter represented a major choice in his career, Clark wrote for advice to Alexander Agassiz. He had every reason to look upon Agassiz as a judicious counselor. They had corresponded since 1901 about material of mutual interest collected by the *Albatross*. Agassiz had confided to Clark that the *Challenger* report Agassiz had published in 1881 had been marred, due to the time pressure he had been under, forcing him to concentrate on characters of the animal's "test" (skeleton); he explained to Clark his intention to publish improved descriptions of species he had already named, by adding illustrations of extra features such as spines, pedicellariae, and even internal anatomy and development, where possible.[3] An artist had already made numerous anatomical plates under his direction, but Agassiz was beginning to realize that he would not find the time to study the details of the pedicellariae himself.

Pedicellariae can be seen with a good simple lens on the surface of a living sea urchin: tiny waving pincers, like the beaks of blind and snake-necked birds, plentifully scattered over the test between the spines. These organs look so much like independent creatures that they had been described as parasites in the eighteenth century, and Louis Agassiz had thought, in the early 1840s, that he was watching the budding off of a peculiar kind of larva. Several workers, beginning with Delle Chiaje in 1825, realized that pedicellariae are appendages.* Alexander Agassiz gave a nice description of their action in his 1874 "Revision of the Echini."

> If we watch a sea-urchin after he has been feeding, we shall learn at least one of the offices which this singular organ performs in the general economy of the animal. That part of the food which he ejects passes out of the anus, an opening on the summit of the body in the small area where the zones of which the shell is composed converge. The rejected particles, thrown out in the shape of pellets, are received on

*A. Agassiz, "Revision of the Echini," p. 659. Many of Alexander Agassiz's notes, drawings, and engravings for the anatomy of echini were destroyed at the printer's in the great Boston fire of 1872.

> these little forks, which close upon them like forceps, and they are passed from one to the other, down the side of the body, till they are dropped off into the water. Nothing is more curious and entertaining than to watch the neatness and accuracy with which this process is performed. . . .
>
> On watching the movements of the pedicellariae we find that they are extremely active, opening and shutting their forks unceasingly, reaching forward in every possible direction, the flexibility of the sheath enabling them to sweep in all the corners and recesses between the spines, and occasionally they are rewarded by catching hold of some unfortunate little crustacean, worm, or mollusk which has become entangled among the spines. They do not seem to pass their prey to the mouth . . . but merely throw it off from the surface like any other refuse matter.[4]

He did not mention—perhaps it did not occur to him—how important this action must be in preventing the larvae of encrusting organisms like sponges and barnacles from making their home on the urchin's test.

In his "Revision" of 1874, Agassiz had made pedicellariae the subject of an original little study in comparative zoology, claiming that their growth and homologies proved them to be "nothing but highly specialized spines."[5] He noted with satisfaction that this undermined a recent argument that pedicellariae were an instance of "the sudden appearance of organs for which no utilitarian motive could be given."[6] Thus, in this case, he fended off an attack on Darwinism by applying the classic method of embryology and morphology practiced by his father. Alexander had also supplied information about the constancy and variation of these organs, giving illustrations of the pedicellariae of dozens of species, but he had not used them as taxonomic characters.

Agassiz's letters to Clark nearly thirty years later show that it was only lack of time, not an intellectual decision, that had made him neglect the possible taxonomic value of pedicellariae. In June of 1903 Agassiz wrote to Clark,

> But of course the study of the Pedicellariae ought to be made and I should very much like to have my work on the '91 Expedition [of the *Albatross*] supplemented by a study of the Pedicellariae and also make some arrangement by which the pedicellariae of the Hawaiian Echini could be worked up as well as the anatomy."[7]

Such a thorough study, though, would require more time than Agassiz could spare, and he was clearly relieved to find this potential coauthor.

Collaboration with Agassiz would be an honor and an advantage to Clark, whose work would appear in the most sumptuous style, with handsome plates. He was pleased to be invited, in July 1903, to luncheon at Agassiz's Newport estate, and while there he must have admired Agassiz's seaside laboratory. Afterward, books Clark had been unable to find for sale at any price appeared on his doorstep. That fall Clark agreed to become Agassiz's collaborator and immediately set to work. Agassiz would send specimens—whole animals or unattached pedicellariae—to Michigan along with instructions on how best to prepare his material and make his drawings. There is a striking difference between the ill-defined drawings of Clark's earlier papers and the beautiful illustrations he did for Agassiz.

Thus it was perfectly natural for Clark to turn to Agassiz for career advice upon receiving the offer from Henshaw. Clark confided that he would greatly regret giving up teaching but that he was attracted by the salary, the extra time for research, and the opportunity to continue his collaboration with Agassiz. He worried whether routine curatorial duties would crowd out research and begged Agassiz's opinion.[8] He appears not to have realized how much control Agassiz still exercised in the museum.

Shortly after Clark arrived in Cambridge as the new "assistant in charge of lower invertebrates," it became his embarrassing duty to write to Theodor Mortensen, a zoologist from the university museum of Copenhagen, refusing him permission to examine the specimens in the M.C.Z. Such a fair and kind man as Clark cannot but have been pained to receive Mortensen's reply:

> I am . . . surprised that you seem to think it right and natural to forbid me to make studies at the collection of Echini at the Museum. Of course, I find it a natural thing that the collections of the "Albatross" etc which are just being worked out by Prof. Agassiz and yourself cannot be placed at my disposal—and I have never thought of that, except possibly the favor of just seeing a few types of them. But that the other collections of Echini of the Museum are also to be "taboo" for use, that really surprises me. I think, indeed, no other museum would treat a fellow worker in that way—a fellow worker, who spends time and money to go even from Europe to America to see specimens and assure facts which it is impossible to gather from literature or in any other way. I suppose, you cannot have realized the fact, and hope that you will, however, find means to assist me and to let me see and examine a few of those types, which have caused so much trouble to Echinologists.—that the collection is under rearrangement could, I suppose, scarcely prevent me from seeing such types. Otherwise I must think that my unhappy personal controversy with Prof. Agassiz is the real cause, why I am forbidden to make such researches.[9]

Mortensen was perfectly right that his exclusion was a breach of fundamental scientific ethics.

By the end of the nineteenth century, the standards of professional taxonomy demanded that an author, when naming a new species, carefully note exactly which specimens his description was based upon. No description, no matter how detailed, however well illustrated, can be perfectly complete. Generations of zoologists have had to learn the humbling lesson that a new species may turn up, matching in every detail their published description, yet differing by some other character they had not thought to mention. The care with which curators preserve those particular "type specimens" only make sense if other researchers are able to reexamine them. Just as the ethics of experimental science require that one's methods be laid out as well as one's results, so that other experimenters may, in theory at least, check for accuracy by repeating one's procedures, so too in taxonomy, considered by its advocates to be essentially scientific, publication of results entails by implication free access to type specimens for one's peers.

The evident duty of a museum to provide access, however, must share a tense coexistence with other reasons to refuse it. Obviously it has always been the first concern of every curator or keeper to protect unique material from damage. Stories were told about people, some of them fanatical amateurs but a few quite reputable professionals, who would handle material carelessly, never get around to returning specimens they had borrowed, or pocket the odd item for their personal collection. Other considerations were commonly invoked as well. Mortensen referred to one in his letter: it was proper for the owner of a set of specimens not yet "worked up"—carefully examined and the species unknown to science noted—to decide who should publish the descriptions of new species and other novelties. But though Mortensen was correct that the older material, and especially type specimens, should have been available to him, there was no higher court to enforce the scientific ethic. Nor indeed was its violation so rare; for example, the ichthyologist Francis Day had a very similar frustration in the British Museum.[10] Mortensen, too, was denied loans there, after criticizing in print the work of curator Jeffrey Bell.[11]

Mortensen was doubtless correct that Clark was refusing him access to specimens in the M.C.Z. on Alexander Agassiz's instructions. Agassiz was by nature quick to anger and powerful in his hatreds. Even his devoted son George could not forbear including in his biography some indication of this aspect of his father's character.

> One day at Newport he was looking over some charts with one of his sons. . . . Suddenly Agassiz looked up and beheld in the distance a buggy tied to a tree, and a couple of men fishing on the rocks.
>
> "There," said Agassiz, "they are at it again! The way everybody drives all over the place and ties his horses everywhere is perfectly outrageous. The worst of it is that when they are spoken to, they are so insolent!"
>
> With that he stalked off toward them, and was well across the lawn before the thought that he might need protection from the insolence of these intruders oc-

curred to his son. When he finally come to his father's aid, he found him upbraiding the trespassers for their iniquities in plain Anglo-Saxon. These unfortunates, whenever they could get a word in edgewise, endeavoured to make the most abject apologies, while their attitude suggested a caterpillar, who when poked with a stick curls up and tries to disappear. As quickly as possible they got into their buggy and drove off, two deeply humiliated men. Then Agassiz turned to his son and observed in perfect good faith, "You see how insolent they are!"[12]

But even a man of more placid temperament would have had his virtues sorely tested if he saw in print, as Agassiz did in Mortensen's monograph of 1903, that his statements were not to be relied upon. On page after page of his *Ingolf* report, Mortensen, at times sarcastically, condemned Agassiz. A few of his phrases may have been harsher than he intended, the fault of the translator of his Danish prose, but the general tone arises from the meaning, not the choice of words. For Agassiz to give details of a specimen that is admittedly perhaps not distinguishable from a known species is "a rather superfluous work!" Or with respect to the inclusion of doubtful species in a geographical list, he writes that "this way of proceeding is very objectionable." Again and again, "Agassiz says [so and so]. This is incorrect," "The following very remarkable statement is found in Agassiz . . . this is very improbable and Agassiz has not proved it," "This notion of Agassiz's is quite illusory." Mortensen's treatise begins with the declaration that all the existing literature just creates despair and confusion for anyone trying to identify echinoids, but that he himself after a "profound and careful attempt at penetrating into the mysteries of the relationship of the Echinoids" had succeeded where his predecessors had failed. Even Clark had felt this "I've settled it all" manner was offensive when he had first read Mortensen's work.[13]

Alexander Agassiz's copy of Mortensen's monograph is filled with penciled underlinings and cries of "Oh!" "D—— impudent," "He had better shut up," and "What in hell shall I do to please your majesty." Agassiz used his next major publication, the "Panamic Deep Sea Echini" of 1904, to reply to such insolence, declaring "his assumption of omniscience is offensive to the utmost."[14] Calling Mortensen "childish," Agassiz labeled some of his statements "gratuitous misrepresentations of facts." Mortensen fought back in the second volume of his monograph, filling twenty-five quarto pages of print with point-by-point expansions of his criticisms.

Clark did his best to reconcile the two contenders, writing long letters explaining to Mortensen why any mortal in Alexander Agassiz's position should have found his language personally insulting. Mortensen replied that he demanded as a gentleman that Agassiz stop calling him a liar, which is how Mortensen interpreted the words "gratuitous misrepresentation." Since neither man would apologize first, the battle was ended only by Agassiz's death in 1910.

Mortensen protested, to Clark and in print, that he was simply doing his scientific duty by correcting errors as he found them, and he claimed that Agassiz should have been able to separate personal feelings from scientific regard for truth. Years later, Mortensen expressed satisfaction that the Smithsonian echinologist Austin Hobart Clark was friendly even though Mortensen had torn apart his pet theory, telling H. L. Clark that "it is such a fine thing that he does not take scientific criticism personally—I remember a very famous Echinologist, who could not rise so high."[15] But that ideal separation of scientific and personal matters Mortensen had not achieved himself. The real clue to Agassiz's anger is to be found, indeed, not in the substance of Mortensen's criticisms (Agassiz knew there were shortcomings in his published work) but in the temper of them. It was Mortensen who had started the fight when he corrected Agassiz not in a tone of civility but with passion. Mortensen explained his mood in a letter to Clark. He had begun his studies, he said, with nothing but respect for the famous author of the "Revision" and the *Challenger* report.

> By and by, however, as I realized how unpardonably carelessly the latter work especially is done, my respect turned into anger at seeing what damage is done to science by such a bad work being generally taken as a standard work, whose statements nobody ventured to doubt.*

The battle of these two men is representative of a type of contention, drawn out over the years, sometimes between men who had never met, involving deep pride and sense of righteousness on both sides, by no means as rare in the international taxonomic community as the image of objective science would have us imagine.[16] The International Commission for Zoological Nomenclature found it necessary in 1913 to draw up a document,

*Mortensen to H. L. Clark, 14 November 1908, M.C.Z. Archives. It is not clear whether the *Challenger* report was really quite so sloppy, by the standards of its time; it was neither a revision nor a monograph, only a notice of the species collected on the voyage. Some British scientists argued that all the collections should be described in Britain, so there could have been bias in the criticisms of Agassiz's report by F. J. Bell and P. M. Duncan. Bell, echinoderm specialist at the British Museum, was distinctly incompetent. Criticism of Agassiz's *Challenger* work has been passed on from generation to generation of echinologists, including the accusation that Agassiz had written his descripions from inaccurate drawings rather from the specimens, which is not so (David L. Pawson, personal communication); the *Challenger* echini did go to Harvard for study (*Annual Report* for 1877–78, p. 12; and for 1879–80, p. 6).

later adopted in its formal regulations, "On the need for avoiding intemperate language in discussions on zoological nomenclature."[17]

Mortensen's lack of tact was not only personally inconsiderate, it was scientifically unfortunate. It had the effect of making it impossible for Agassiz to compromise. His injured pride now became inseparable from the question of the taxonomic usefulness of pedicellariae. At one time he had acknowledged that the test by itself gave inadequate characters for a natural classification, and he had felt himself remiss when he had limited himself to those characters. He had even commissioned Clark to draw the pedicellariae of all his species. But the novelty of Mortensen's work was its heavy dependence on the pedicellariae, and now Agassiz began to insist that these organs are of no taxonomic value at all. So vehement was his scorn that Agassiz, in effect, served notice that any zoologist who dared to agree with Mortensen was risking his wrath.

Mortensen claimed, without much pretense of modesty, that after finding nothing helpful in the existing literature he had wiped the slate clean, basing his classification on his own direct observations with no reference to previous arrangements. There was nothing novel in his underlying methodology, just that he claimed to find after trial and error that some useful characters had been overlooked, while dependence on other misleading characters had led to artificial groupings. The new genera Mortensen created within the group of regular sea urchins (now the order Echinoida) scattered species that had been judged similar and gathered together some that had been classed in different families.

Agassiz, in his choice of taxonomic characters, had followed the lead of his father's old associate Edouard Desor, who had created categories based upon how many pores pierce each plate of the test. Conveniently permanent characters, these paired holes mark the site of another kind of appendage also unique to echinoderms, the "tube feet" that, acting like suction cups, enable the animal to feel, taste, pick up food, and cling to a rock. The pores serve as the entry and exit of the fluid supplying pressure to the tube foot. Echinoids were divided into the "polyporous," with more than three pairs of pores per plate, and "oligoporous" with three or less. According to Mortensen, however, details of the pedicellariae yielded much more natural groupings. His suggestions were immediately adopted by the long-established German echinologist Ludwig Döderlein as well as by several other Europeans.[18]

Agassiz had specifically defended the division of oligoporous versus polyporous as being not an artificial convenience but a fortunate indicator of a more fundamental distinction. "This division, although it appears a

Figure 46. Two great echinologists, Hubert Lyman Clark and Theodor Mortensen, in the Caribbean. Clark was caught in the cross fire of a war between Mortensen and Alexander Agassiz. Clark and Mortensen did fieldwork together in Tobago in 1916, when this snapshot may have been taken. (By permission of the Museum of Comparative Zoology Archives, Harvard University)

numerical one, is yet one of great physiological importance, as the mode of growth of the poriferous zone in these two families is totally unlike."[19] In several joint studies, Agassiz and Clark argued that the pedicellariae are too variable to be useful, and that certain species, because they carry several types of pedicellariae, could be classed equally well in one genus or another. "What are we to do in the face of such disorderly pedicellariae?" they scoffed.[20]

At the same time, Mortensen and Clark (fig. 46) were corresponding as independent colleagues, Mortensen assuming that Clark would betray no confidence to Agassiz and trying to convert Clark to his views. Clark was caught in the middle. Mortensen wrote to him, "I perfectly realize that you have reasons to keep to Mr. Agassiz's classification as long as possible."[21] On one occasion Agassiz did assure Clark that authors in M.C.Z. publications were solely responsible for their own papers, but on another occasion we find Clark writing to Agassiz,

I thank you for your note received this morning. I see that my statement of my position is open to misunderstanding and I have accordingly modified it in this way: "I do not think that species with 'essentially different pedicellariae' are likely to be congeneric, any more than those with essentially different spines, ambulacra, or other structural features. . . . Further it is quite possible that very closely related species may occasionally reveal essential differences in any single morphological character."

Of course, I shall not even hint to Dr. Mortensen, or to any one else, that you and I either agree or disagree in this opinion.[22]

In spite of Clark's efforts at circumspection, European scientists did assume that his resistance to Mortensen's innovations was based on deference to Agassiz rather than on objective judgment. In the late 1920s, a Russian wrote,

Clark on the whole finds fault with Mortensen's classification, and asserts that the pedicellariae are of no great taxonomic significance. . . . Very typically, however, Clark's conclusions are quite close to the scheme proposed by Döderlein. With a few exceptions Clark recognizes the same genera as Döderlein and Mortensen, establishing only a few new ones."[23]

This criticism seemed exaggerated. Taking the example of the fat-spined urchins (the cidarids), we find that Clark had accepted only two of the new genera in Mortensen's 1904 monograph and had disagreed with 36 percent (15 out of 42) of his species assignments. The reasonableness of Clark's dissent from Mortensen is indicated by the fact that Mortensen himself later changed his mind for every one of the species Clark had disagreed with, and in four cases he even adopted Clark's genus.*

In general, it is true that over the years Clark moved step-by-step into closer agreement with Mortensen, but the details are quite complicated, since Mortensen himself kept making changes in his classifications. Clark was an intelligent, honorable, hardworking scientist, and sincerely believed himself capable of making objective judgments of fact. Still, just as Agassiz's and Mortensen's judgments were evidently distorted by emotion,

*Beginning with the eighteen species of Cidaridae found in Agassiz's *Challenger* report, I compared Mortensen's 1903 "Echinoidea," H. L. Clark and A. Agassiz's 1907 "Hawaiian and other Pacific Echini: the Cidaridae," Clark's 1907 "Cidaridae," and Mortensen's 1928 *Monograph*.

it is very possible that, in a more subtle but nonetheless significant way, Clark's own judgment of what he believed to be matters of zoological fact could have been distorted by feelings of gratitude, sympathy, or loyalty to Agassiz.

The death of Alexander Agassiz did not noticeably alter Clark's attitude. In fact, his opposition to the pedicellariae increased, because of the appearance in 1912 of Robert Tracy Jackson's monumental study of fossil echini. Jackson carried out this work convinced of the truth and value of the theories of Alpheus Hyatt, contending that forces of acceleration and retardation in embryonic development were the key to evolution. Clark was profuse in his admiration of Jackson's work.[24] The characters Mortensen was promoting were not even an option for Jackson, since pedicellariae are too small to leave much trace in the fossil record. But Clark insisted that his opposition to Mortensen's favorite characters was not based on this merely practical consideration but on a matter of great theoretical importance, namely, development.

No doubt a continuous tradition of belief could be traced, from Jackson to Hyatt back to Louis Agassiz, of belief in the deep significance of parallels between taxonomic rank, embryological stage, and paleontological development, just as Ernst Haeckel's "ontogeny recapitulates phylogeny" can be traced to pre-Darwinian sources.[25] But the attractions of this idea early in the twentieth century are not explained by its antiquity. To all those who found random variation and natural selection unsatisfying as the sole source of evolutionary change, the unknown forces at work in individual development seemed to promise an explanation that would have direction and intelligibility. Clark insisted that ontogeny was the fundamental clue to phylogeny, and thus the fact that young urchins have pedicellariae no different from adults proved, in Clark's view, that this character does not develop and so can tell us nothing of evolution.[26]

Mortensen was no less respectful than Clark of the value of ontogeny to the taxonomist. Indeed, he exerted himself, as Clark did not, to gather information about the embryological development of scores of species. Before long, these two men were admitting in their correspondence that the differences between them were not matters of principle but questions of perception as to what characters actually seemed to work in each different genus and family. They were both agreed about the double meaning of their activities as taxonomists. They had the practical duty of classifying to get on with, and at the same time they were cherishing the noble goal of investigating evolution. Both of them were seeking nice taxonomic characters, ones that would bring together species whose relatedness was indicated by an ill-defined series of subtle similarities. Mortensen wrote that he regretted Clark's giving primary importance to the oliogoporous versus polyporous character, and he objected to Clark's accusation that Mortensen cared only for the pedicellariae. "I do not regard the polyporous condition of so great value, because I think it has arisen separately in many different groups. Of course also the pedicellariae have developed the same type within different groups."[27] No single character can work all the time, he reminded Clark:

It is living organisms we have to do with, and they have to be regarded in the light of life. Any character may be liable to development and variation. We must regard each case for itself and see what the living nature will teach us . . . I am afraid I have been rather too hot! Please forgive me.[28]

A few years later they had made little progress on one another's interpretations, but their personal friendship was growing.

That you will not recognize the polyporous condition to have arisen in different ways surprises me very much indeed. . . . Of course, these matters of classification are in a high degree dependent on the personal feelings of the authors, and it is often hard to give definite reasons why we regard one character as of more prominent importance than others.[29]

How I agree with you that we must have *all* the data of *all* the sea-urchins! Then I hope, we shall agree—then, at least, we can discuss things fully. That the larvae will prove of great importance, I can see definitely. Pity, it takes so long time to get them all.—and that life is so short—and that we are so few workers![30]

It is very likely that, without the initial emotional confrontation between Agassiz and Mortensen, the different systems of echinoid classification would not have been so sharply drawn. Probably also Clark and Mortensen would have been able to find a compromise between each other's views sooner if Clark had not been employed by Agassiz. Instead, there was a deep difference of opinion between two of the world experts on sea urchins for several decades. There was thus exposed to view what is normally invisible, namely, that taxonomy consisted of the opinion of the experts, and beyond this there was no higher court of appeal to prove relationships objectively. Evolutionary theory offered no clear route of escape from this situation, for on the nature of classification and its relation to evolution, Mortensen and Clark were fully in agreement.

Clark suffered from a hearing problem; by the last decade or two of his life (he died in 1947) his deafness made ordinary social intercourse difficult. It was generally thought that this handicap was the reason he had given up teaching,[31] but there is substantial evidence to contradict this explanation. In 1929 he obtained leave to teach for a year at Williams

College, and he encouraged an effort by Vernon Kellogg to create a permanent post for him there. It seems to have been no decision of Clark's, but a financial decision by the trustees of Williams, which caused this hope to come to nothing. Again, in 1936, he left the museum for a year of teaching at Stanford University.

In 1911 Clark had published an article in *Popular Science Monthly*, "The Purpose and Some Principles of Systematic Zoology," implicitly acknowledging the low regard other zoologists had for taxonomy. He admits that some authors deal only with issues of nomenclature, or give descriptions without comparisons or discussion of interrelationships, and he recommends instead that mutual relationships, including geography and geological history, should be part of a proper taxonomic study. Still, he says, this field deals with form and not the cause of form. "It deals with the travelers, the routes traveled and the destinations reached in the animal kingdom, but it leaves for other zoologists to determine the means of transportation and the causes of traveling." In 1913, 1929, and 1932 Clark visited Australia, funded by the Carnegie Institution, delighted by the opportunity to observe colorful starfish, lively basket stars, and graceful crinoids in the wild. The resulting monograph, published posthumously in 1947, was a contribution to the Linnaean program, not to evolutionary biology. In choice of characters, Clark had moved closer to Mortensen, as Mortensen had to Clark, without claiming thereby to have understood phylogeny any better.

11

"I Made Up My Mind That Very Day to Be Director"

When Tom Barbour first visited the Museum of Comparative Zoology, in June 1899, he was already quite an experienced observer of natural history and of natural history museums, having spent time in the fascinating outdoors of Florida, the Bahamas, and the Adirondack Mountains, and in several of the great museums of Europe and America.[1] In the Agassiz museum, as he tells the story,

> I spotted some specimens which I thought were wrongly labeled—and as a matter of fact they were. I wrote with all the dignity of my thirteen years to Dr. Woodworth, then Acting Custodian of the Museum, who was rather infuriated by my temerity. As I look back on it, I don't blame him. I suspect that my letter was as fresh as green paint. I made up my mind that very day that if I lived I would be Director of the Museum."*

This marvelous story has the ring of truth. Recalling the position Woodworth was in, we can understand that he might have reacted defensively to criticism from a self-important boy. Certainly the point of the story is not simply that Barbour had fallen in love at first sight with the M.C.Z. and consequently resolved to work there when he grew up. Rather, the critical part of the experience may have been his exchange with Woodworth, for it can be a heady experience for a teenager to discover that he knows more than someone in authority. He may even have perceived that, had he himself been in Woodworth's position, he would have responded more sympathetically to a letter from a young person. Thus I suspect that it was

*Barbour, *Naturalist at Large,* p. 18. He was actually nearly fifteen. His recollection of the year as 1899 is corroborated by his recollection that the occasion was the Harvard commencement that was the twenty-fifth class reunion of his father's friend Theodore W. Moses, who did indeed graduate in 1874 (*Harvard University Directory,* 1913, p. 575).

the whole incident, and not just the afternoon wandering through the exhibition rooms, that gave Barbour the notion that he would someday rule the M.C.Z.

Another milestone in young Barbour's development was the advice he received from a schoolmate's father, Fairfield Osborn, head of the Bronx zoo, who told the budding naturalist to read Alfred Russel Wallace's classic book *The Malay Archipelago.** Barbour later recalled, "I read it over and over again until I knew it almost by heart. And my desire to see the Dutch East Indies became so all-consuming that I must have seemed a veritable monomaniac to my parents."[2] Family wealth from the Irish Linen Thread Company would enable him to realize his boyhood dream before he reached the age of twenty-three.

It was his love of the M.C.Z., Barbour tells us, that drew him from New York to Harvard College instead of Princeton, the school his parents preferred.† Arriving in Cambridge in 1902, he made straight for the museum and settled in like a barnacle. He managed to work his way into the good graces of the misanthropic Samuel Garman, then in his early sixties, who cut "quite an impressive figure" in his old-fashioned black coat.‡ Majoring in zoology, Barbour could take no courses from this herpetologist, who, like all the curatorial workers, held no faculty appointment at Harvard, but he spent every spare moment as a volunteer helper in the collections, especially with the reptiles.

By the time Barbour met him, Garman had cloaked himself in habits of seclusion, hiding his work lest anyone spy on his specimens and scoop his descriptions. Many years later, when Garman died, Barbour found in his room

> a jar full of little stickers bearing his name and address which he had cut from each copy of the Nation [mailed to him]. Another giant glass container was filled with his

*Fairfield Osborn's recommendations included also Henry Walter Bates, *The Naturalist on the River Amazon;* Thomas Belt, *The Naturalist in Nicaragua;* and the works of William Henry Hudson. In telling the story, Barbour names Professor Henry Fairfield Osborn, but as Ernst Mayr reminds me, that Osborn was a paleontologist, whereas Barbour says his advisor was head of the zoo and spoke with him as they rode back from the Bronx together.

†Professor Barbour, *Naturalist at Large,* pp. 18–21. He later declared, "I came here to Harvard College as an undergraduate in 1902, and from then until now I believe never a day has passed when I was in Cambridge and not ill, that the day or a part of it has not been spent here in the Museum" (Barbour, *Naturalist's Scrapbook,* p. 19).

‡Barbour was introduced to the suspicious and reclusive Garman with the help of Austin Clark, then a graduate student. He was "finally" allowed to work in Garman's room in the museum, but it is not clear whether this was in his undergraduate or graduate years (Barbour, *Naturalist's Scrapbook,* pp. 19–24). Barbour to E. W. Nelson, 1 November 1932, Barbour Papers, Harvard University Archives.

old rubbers. . . . another jar, at least three feet high . . . contained bits of bread, the uneaten corners of the sandwiches which Garman had brought for his luncheons for years and years.*

Garman's character showed itself in his publications; "his major works are little consulted as they were badly arranged without keys for identification and full of vain repetition."[3] His descriptions of reptiles and amphibians from each different island in the West Indies had been roundly criticized as a meaningless exercise of "splitting" species by naming local forms which were only trivially distinct.

Barbour also soon made a point of developing a cordial relationship with Alexander Agassiz, in spite of the fact that "he was held in awe, indeed was considered a terrifying and almost legendary figure by all the graduate students."[4] In the summer after his sophomore year, when Barbour with his friend Glover Allen chartered a sailboat to dredge off the Bahamas, they had the use of some of Agassiz's equipment.† And at the end of his senior year, when Barbour's father sent him a substantial check as a gift to the museum, Agassiz wrote to the young man: "His letter to you can not but touch deeply any father who has sons," and invited them both to Newport.‡

Among Henshaw's woes as head of the M.C.Z. was an inadequate operating budget. Barbour's cash contributions to relieve this situation, which continued all his life, began during his undergraduate years.

In October of 1906, following his graduation, Barbour married Rosamond Pierce, second cousin by marriage of Ida Higginson, Alexander Agassiz's sister. The newlyweds set sail on a series of steam passenger liners, first to the Irish branch of the Barbour family, then to the far side of the globe. In spite of advice from several quarters, including Alexander Agassiz, that collecting in the East Indies would not be feasible, they spent

*Barbour, *Naturalist at Large,* p. 139; one can imagine, as Barbour did (*Naturalist's Scrapbook,* p. 25), that Garman planned to feed the crusts to birds but never got around to it. Doubtless he meant to use the stickers as return-address labels and thought his old overshoes could be mended or worn doubled up.

†Barbour, along with Glover M. Allen, who would later be curator of mammals in the M.C.Z., and Owen Bryant, chartered a fifty-nine-foot schooner. "Through the kindness of Mr. Alexander Agassiz we were permitted to make us of a hand-winch and 350 fathoms of steel cable" (Barbour and Allen, *Narrative of a Trip to the Bahamas,* p. 7). Barbour offered to look out for specimens of interest to Agassiz (A. Agassiz to Barbour, 18 June 1904, M.C.Z. Archives).

‡Agassiz to Barbour, 30 May 1906, M.C.Z. Archives. William Barbour's gift, $5,000, was used to buy specimens, including bones and an egg of the extinct great auk and storage cases for the bird department (*Annual Report* for 1905–6, p. 4, and for 1906–7, p. 5).

months in the islands described by Wallace, accumulating quantities of birds, insects, mammals, and Barbour's beloved amphibians and reptiles. Barbour related their adventures in the *National Geographic Magazine*, highlighting the humor and danger of visiting places still exotic.[5] The significance of that trip for Barbour's career has little to do with adventure, however, and everything to do with Alfred Russel Wallace.

The Geographical Distribution of Animals

Wallace wrote three books devoted to biogeography: the popular *Malay Archipelago* (1869, second edition 1890), *The Geographical Distribution of Animals* (1876, two fat volumes), and *Island Life* (1880). Though packed with a mass of information, each work strongly conveys the message that the pattern of distribution of animal life is neither a matter of chance nor an insoluble mystery, but the historical outcome of a set of comprehensible processes, namely, geological change and evolution. What once had been mere natural history, a record of the wonders of the living world, can now become a science, Wallace promises, a reasoned investigation into causes.

On the assumption that species are fixed, so that their resemblances to other species of the same genus or family cannot be historical or blood relationships, we may still choose whether to regard the present natural range as the product of historical events. Much discussion in the mid-nineteenth century postulated an original center at which a given species first appeared and from which it subsequently spread to its full range. As long as the unit being investigated was the species, such a debate had no evolutionary implications. But once a process of historical change in the range of species was accepted, the biogeography of genera and higher categories became heavy with evolutionary implications.

Wallace was in a different position than Darwin had been when he had composed the *Origin*. In 1859 the challenge had been to assemble evidence that species had evolved, so the burden of Darwin's two chapters on geographic distribution was to show that the distribution of species could be explained by natural processes, and that the general pattern of the distribution of genera follows the same rules as the distribution of species. Naturalists had long assumed that ordinary wandering of animals and dispersal of seeds had produced existing patterns of distribution outward from the point of creation of each species, and many, following Buffon, were willing to imagine a similar process for the related species in a genus. Disjunct distributions, however, such as the mountain flowers of Labrador and Massachusetts, or forms on islands far at sea, seemed to demand miraculous explanation. Louis Agassiz had claimed that the creation of a single species occurred over its entire range, so that disjunct patterns would not be a problem, and the continuous range of a genus could not suggest the common ancestry of its member species. To make Agassiz's view seem less necessary, Darwin suggested several ways of explaining the anomaly of disjunct ranges. Mountain and arctic regions now far separated may be the refuges of species that once occupied the intervening land during the cold of an Ice Age, he suggests; species found far from their presumed center of creation, whether on islands or distant continents, could be descended from ancestors carried there by a sort of accident which although rare, Darwin argued, was not impossible. Not only seeds, insects, and birds, but other small animals are carried about by high winds; he collected instances of seeds and eggs clinging to the bodies of ducks, or remaining alive in their crops.

Darwin was careful not to claim that he could prove conclusively that a particular species had so traveled in any particular case. In this respect, his logical strategy was like his treatment of the origin of highly adapted forms such as bats and whales, where the actual sequence of intermediate forms had left no fossil trace. All he tried to accomplish was to show that the difficulty the evolutionary history of such creatures presents to our imagination can be removed. We should notice the degrees of modification of the various squirrels, from those with flattened tails to the so-called flying squirrels, whose wide expanse of skin lets them soar, and we should consider land animals which are partly aquatic, from an individual bear observed catching floating insects while swimming, to the otter. Darwin did not pretend that such examples show the actual steps gone through by the ancestors of bats or whales; what the examples were meant to do was to undermine the creationist argument, which says that each form is perfectly adapted, implying that an intermediate, imperfect one could not survive. Likewise in biogeography, Darwin's aim was merely to aid the imagination in seeing how disjunct distributions might have arisen, so that Agassiz's multiple centers of creation would no longer seem the only possible explanation.

Once we agree on a single center of creation for each species, then those facts concerning the distribution of the several similar species forming a genus may be added to Darwin's list of phenomena which would make sense if evolution were true but remain utterly mysterious and capricious on the creationist view. The strength of Darwin's approach was that he did not need to establish convincingly the exact historical events in any one particular instance, nor could he be refuted by a few puzzling cases. His aim was only to convince us that these difficulties, though "extremely grave," are "not insuperable."[6] The attractiveness of evolution was that it offered a transparently simple explanation for the general rule that species in the same genus are found in the same region.

Wallace's own conversion to evolution dated back to the mid-1840s, but now, in the late 1860s, he could write for an audience of fellow believers. In his *Malay Archipelago,* which he dedicated to Darwin, he expended no effort arguing for evolution but treated it as an accepted fact. The logical structure of his investigation was thus very different from Darwin's.

Given that not only populations of the same species but species of the same genus, and even, tracing further back into the remote past, species belonging to different genera but the same family, have diverged from some common ancestor, what can we infer from the present fauna of the Malay Archipelago about the geological history of the islands? Of course, if every creature can occasionally make its way from one island to another, the mixing back and forth would have long ago destroyed any pattern, but Wallace claimed that such migration is rare:

> Birds offer us one of the best means of determining the law of distribution, for though at first sight it would appear that the watery boundaries which keep out the land quadrupeds could be easily passed over by birds, yet practically it is not so . . . if we leave out the aquatic tribes. . . . As an instance, among the islands of which I am now speaking, it is a remarkable fact that Java possesses numerous birds which never pass over to Sumatra, though they are separated by a strait only fifteen miles wide, and with islands in mid-channel.[7]

To bolster his claim that migration from one island to another has been insignificant, Wallace also uses the fact, less direct but more weighty, that the distribution of different animals forms definite patterns. There are sixteen species of bats in Borneo, he tells us,

> and of these ten are found in Java and five in Sumatra, a proportion about the same as that of the Rodents, which have no direct means of migration. We learn from this fact, that the seas which separate the islands from each other are wide enough to prevent the passage even of flying animals, and that we must look to the same causes as having led to the present distribution of both groups.[8]

This cause was the physical changes the land and sea have undergone, a former sea of lower level having surrounded a great landmass of which the modern islands are the remnants.

The direction of Wallace's argument is bold and significant. What he claims is that the distribution of animals is on the whole so regular and orderly that the data of biogeography can supply information in the absence of geological data as to the past configuration of land and sea. In

Malaysia, the western islands are characterized by forms related to the fauna of the adjacent East Asian mainland, while those on the eastern islands are highly characteristic of Australia. Yet there is not a gradual shifting of the faunal proportions as one moves eastward across the island chain. Instead, Wallace declared, a radical shift from mostly Malay to mostly Australian occurs between the small islands of Bali and Lombok, separated by a channel only fifteen miles wide. Biogeographers still call this "Wallace's Line." The faunal division became a powerful confirmation of his picture of a former continent linking the western islands of the archipelago together, when he noted that an area of deep water cut across the otherwise shallow sounding of the archipelago, just coinciding with the boundaries between his two faunal regions. If the sea level had once been lower, as his supposition required, the land east and west of that line would have been still separated by a band of water. That the soundings should so agree with his faunal regions not only encouraged him to believe in the significance of such regions but suggested that the physical scientist might have to look to the biologist for help, a reversal of their usual levels of authority. Now, he said,

> it will be evident how important an adjunct Natural History is to Geology; not only in interpreting the fragments of extinct animals found in the earth's crust, but in determining past changes in the surface which have left no geological record. It is certainly a wonderful and unexpected fact, that an accurate knowledge of the distribution of birds and insects should enable us to map out lands and continents which disappeared beneath the ocean long before the earliest traditions of the human race. Wherever the geologist can explore the earth's surface, he can read much of its past history, and can determine approximately its latest movements above and below the sea-level; but wherever oceans and seas now extend, he can do nothing but speculate on the very limited data afforded by the depth of the waters. Here the naturalist steps in, and enables him to fill up this great gap in the past history of the earth.*

Wallace's 1876 *Geographical Distribution of Animals* repeated this argument, even featuring it in its subtitle: "Geographical Distribution of Animals with a Study of the Relations of Living and Extinct Fauna as Elucidating the Past Changes of the Earth's Surface." In spite of its subtitle, however, this and his later works on biogeography concentrated on the less contentious task of defining the main regions of similar fauna, Wallace ap-

*Wallace, *The Malay Archipelago* 1:14. He repeats this claim on p. 372: "[G]eology alone can tell us nothing of lands which have entirely disappeared beneath the ocean. Here physical geography and the distribution of animals and plants are of the greatest service."

parently deferring to Darwin's continuing arguments in favor of the possibility of living things surviving remarkable journeys. Darwin was reacting against biogeographers who were quick to imagine former land bridges without any geological evidence, "cooking them up as easily as a cook makes pancakes."

Barbour on his honeymoon relished the association with Wallace's early travels. He deliberately chose a route that would let them "see the greatest number of localities mentioned by Wallace."[9] In Ternate he met "a wizened old Malay man" named Ali who had been Wallace's faithful collector and companion fifty years before. The Barbours, keen photographers, sent a picture of Ali to the aged but still active Wallace. Barbour too employed natives to collect for him, as naturalists generally do, making up for the short time he spent in any one locality by hiring a large number of helpers.

Quite aside from his private enthusiasm for Wallace, Barbour had plenty of stimulus within the M.C.Z. for an interest in the biogeography of archipelagos. Alexander Agassiz was much interested in such matters.* He explicitly addressed one of Wallace's speculations in his 1888 *Three Cruises of the "Blake."*[10] Wallace had proposed in his *Geographical Distribution of Animals* that the West Indies may be the remnants of a great landmass, "Antillea," and thus that some of its flora and fauna were descended from a once continuous population, rather than having arrived upon uninhabited islands across water. Agassiz's soundings showed that by considering the forty-fathom, hundred-fathom, or five-hundred-fathom banks, we can imagine today's numerous small islands united into a few great islands, but he regarded Wallace's Antillea as highly speculative and leaned strongly to the belief that the modern flora and fauna were satisfactorily explained without invoking any significant change in sea level. During part of the *Blake* cruises, Agassiz was accompanied by Samuel Garman, collecting mammals, birds, fish, but most particularly reptiles for the museum.[11] Agassiz continued to be interested in the geology and biogeography of the West Indies. Beginning in 1894 he commissioned a series of field trips and monographs by the geologist Robert T. Hill, who studied Cuba, Jamaica, Costa Rica, and Puerto Rico at Agassiz's expense, specifically searching for evidence on the question of former continental connections of the Antilles.[12] (Hill's research would later earn him the title "Father of Antillean Geology.")[13]

*He had written in 1864, while his father was in Brazil, "If there is *anything* in geographical distribution, there is *nothing* in Darwin, and vice versa. The one flatly contradicts the other" (G. R. Agassiz, *Letters and Recollections,* p. 50). I suspect that here Alexander was echoing what his father was reporting from the Amazon.

Barbour's Career Begins

Barbour was determined to become a professional biologist, that is, a respected member of the discipline. He wanted to pursue the subject Merriam and Roosevelt would call natural history, but he wanted to do it seriously, full time, with the acknowledgment of his peers. For example, telling his father in 1907 why he would have to cut short a family visit to attend the International Zoological Congress, he had said, "I am sure you can understand what this means when I tell you frankly that it is as important for my career as affairs of a similar sort are for any business man."[14] He knew that collecting specimens around the world and making himself the expert on some group of animals was not enough for the role he wanted to play. After the turn of the century qualifications for scientists included a doctoral degree, so Barbour was planning on graduate school.

Barbour returned from his wonderful honeymoon just in time for the Seventh International Zoölogical Congress, held in Boston in August of 1907. Alexander Agassiz had grumbled about the burden of preparing the presidential address and tried at least twice to resign. Recalling the time invested in the "Ignorabimus" address of 1880, and the unpleasant controversy his talk had caused, he finally gave up his attempt to present a significant paper in favor of a brief review of oceanographic voyages.[15] Still, even without much of a contribution from Agassiz, a substantial portion of the 1907 meeting concerned biogeographical theory. Robert Francis Scharff of the National Museum of Dublin delivered a review "On the Evolution of Continents as Illustrated by the Geographical Distribution of Existing Animals," in which he cited the work of dozens of men, including Arnold Ortmann, Leonhard Stejneger, and George Baur.[16] Scharff believed that "all terrestrial animals progress slowly and step-by-step on a land surface, and that islands are not populated to any appreciable extent by accidental means of dispersal," identifying this as one of the issues currently in dispute among biogeographers.[17]

Shortly after the zoological congress Barbour enrolled as a doctoral candidate at Harvard. Barbour's carefully nurtured personal contacts enabled him to cross the no-man's-land between a Harvard degree and the M.C.Z. collections. Even though Garman and Henshaw were technically in charge of the museum, Barbour noted that it was "the kindness of Mr. Agassiz" that gave him "free access to all the Museum collections."[18] Another help, perhaps indeed the critical factor, making it possible for Barbour to do his thesis outside the instructors' areas of interest, was his personal fortune. He was not obliged to please Professors Mark or Parker and make himself employable. Then, after his degree, he could afford to wait in the wings while Garman and Henshaw aged. In all his official positions—associate curator (1911), curator (1925), director (1927)—he was never paid a salary.*

It was important for Barbour's aspirations to scientific respectability that he address current questions of theoretical interest. The historical biogeography of archipelagoes was a contentious and lively field, as Scharff's review had shown, and this was the topic upon which the young Barbour concentrated. Besides his trip to the East Indies, Barbour had several times visited the West Indies, including a childhood trip to Nassau, his undergraduate dredging excursion with Glover Allen, and a stopover of a few weeks in Jamaica and Cuba, but during his graduate student years most of his time was spent in the library and workrooms of the museum, evaluating the data contained in the old museum specimens, cataloguing and studying his new material, comparing the published descriptions with the specimens in hand, and reviewing the geological and biogeographical literature. He corresponded with Alexander Agassiz about former land connections in the Antilles, and with Scharff.[19] Around 1910 he was working on three major manuscripts, two dealing with the Caribbean, and one, his doctoral dissertation, on the islands of his honeymoon, the Malay Archipelago.[20]

These three early studies of Barbour's are full of the vigor and vision appropriate to a young scientist. The material he was working with was challenging, for museum labels and the publications based on them can be riddled with error. Also, the ranges of many kinds of animals was rapidly changing. For example, the mongoose recently introduced into the West Indies was exterminating the local lizards, while in the East Indies, trading in bird skins was endangering the birds of paradise. Barbour's interest, however, went beyond mere inventory. Rather, the same theoretical belief about the meaningful regularities of animal distribution found in Wallace's *Malay Archipelago* permeates these three works of Barbour's.

Like Scharff's review of 1907, Barbour's dissertation supposes that

*In the *Annual Reports* Barbour had been thanked for contributions of money and specimens since 1904, but his first appointment was announced in the *Annual Report* of 1911–12. Bigelow ("Thomas Barbour," p. 15) gives 1910 as the date of this appointment, but no mention is made in the *Annual Report* for 1910–11, and a letter of 8 December 1911 says that it is a pity he has no official position in the museum (A. L. Lowell to Henshaw, Harvard University Archives). After the death of Alexander Agassiz in 1910, Henshaw decided that all the curatorial staff should have titles more appropriate to their status. Henshaw himself changed from curator to director, the former assistants became curators, and Barbour became "Associate Curator of Reptiles and Amphibians." Since Garman was "Curator of Reptiles, Fishes, and Amphibians," Barbour belonged to his department, but in practice he functioned and was treated as a full curator, submitting a separate report. In 1925, when Barbour became "Curator of Reptiles and Amphibians," Garman's title became "Curator of Fishes."

what was until recently the orthodox view is in the process of being replaced.[21] The view which had become standard was that of Darwin: that except for the few cases where a definite continental shelf links islands to a mainland—such as the British isles to Europe, and Sumatra, Java, and Borneo to Malaysia—we may not imagine islands to have been formerly connected, but must assume them to have been populated by some means of dispersal. But now, Barbour says, just in the past few years, some geologists have again questioned the permanence of the deep oceans, and zoologists are doubting the possibility of various creatures enduring the experiences that transport by "flotsam and jetsam" requires. Even the historic examplars of islands that had arisen as volcanoes hundreds of miles from the nearest land, the Galapagos, were now said to present evidence of being continental remnants.[22] The reptiles and amphibians he had seen in the East and West Indies left no doubt in Barbour's mind that this new trend of thought was correct, and that the inhabitants of these vast archipelagos were silent testimony of an earlier continuous land mass secondarily broken up into islands. He was harking back to Wallace.

As an ambitious young scientist, Barbour was looking for a field of lively controversy in which his contributions would be important. He was sure we could tell which were the species of reptile or amphibian that could travel about from one place to another, as pets or stowaways in native boats, or blown on natural rafts of drifting vegetation; such animals were easily identifiable, their habit of clinging tightly to branches and their resistance to salt water correlating nicely with their irregular and far-flung distribution. Such "waif" species, such as the skinks, should simply be set out of consideration before serious zoogeography begins.

For the East Indies, former land connections have always been admitted by everyone, so the question becomes one of sorting out the sequences of connection and separation. Wallace's line had been challenged by several writers, as not running exactly between Bali and Lombok, where indeed the seafloor is not so deep as Wallace at first thought, and Barbour's herpetological evidence led him to agree that the division between the Asian and Australian type of fauna does not occur so sharply.[23] The general picture remains, however, but it is less dramatic, the result of several connections at different times. For the West Indies, on the contrary, former land connections have not been the orthodox belief, but Barbour argues this idea exactly as Wallace did in his *Malay Archipelago:*

> A peculiarity of the fauna of Jamaica is the fact that while its proximity to Cuba is practically the same as its distance from Haiti, the evident relationship of the island's fauna with that of Haiti is well marked, while with Cuba it has in common only species which range widely through the West Indian region. Now a possible

explanation of this offers itself when we examine a contour map of the Caribbean Sea . . . the Bartlett Deep, of over 3000 fathoms, extends between Cuba and Jamaica—doubtless a cleft of very ancient origin. The depth of water, however, between the great southern arm of Haiti and Jamaica is only from 500 to 800 fathoms.[24]

Barbour's conviction that the fauna of Jamaica was not the product of incidental transport over water was reinforced when he traveled to the interior of the island to collect the onychophoran *Peripatus.* This marvelous creature, which looks like an earthworm with fat legs, is thought to be related to the ancestor of the insects. Gently, Barbour put his living specimens in containers with the moss and earth in which they had been found, but by the end of the day, in spite of his care, they had all died; this agreed with the experience of a local naturalist, whose attempts to keep *Peripatus* alive had invariably failed. It was inconceivable, Barbour declared, that such a delicate creature could survive a trip floating in the sea, and yet the islands of Trinidad, Dominica, St. Thomas, Antigua, and Puerto Rico each has its distinct species of *Peripatus.* Another point in favor of an early land connection and against chance migration is that to find the closest relatives of the Jamaican *Peripatus,* and of Jamaican reptiles and amphibians, we must look not to the nearest mainland—Florida to the north or Venezuala to the south—but to Central America, even though the prevailing wind and ocean currents do not blow from Central America to Jamaica but the reverse.

Surveying the fauna of all the islands, Barbour finds the same sort of homogeneity that had led Wallace and others to assume continental connections in the East Indies. Barbour points to "the regularity with which the important genera occur on practically every island that has been scientifically explored, with a species peculiar to each" as showing that the populations on each have been long isolated. If immigration is common lately, the immigrants would tend to blur the distinctions between species, yet if such immigration had ever gone on, it should be happening still.

The view that Barbour was espousing is now called "vicariance," from the Latin for "substitute," referring to situations where a genus is represented by different species in different localities. Biogeographers who minimize accidental dispersal and see present distributions as evidence of an ancient continuous population now form a self-conscious school.

W. D. Matthew and the Calculation of Dispersal

Barbour's sense around 1910 that the trend of biological thought was against the "flotsam and jetsam" dispersal advocated by Darwin turned out to be wrong. In February 1911, a paleontologist at the American Museum of Natural History presented to the New York Academy of Science the abstract of an argument on the evolutionary history of the mammals. William Diller Matthew's treatise, called "Climate and Evolution," appeared in print in February 1915. It had to do mainly with the materials of Matthew's own research, mammalian fossils, but his overarching thesis, that mammals had invaded the southern continents repeatedly from centers in the north, driven by alternating periods of harsh and warm climates, was founded on the premise of the permanence of the pattern of land and water as now existing. He needed, therefore, to insist that the presence on islands of species that could not swim or fly was not evidence of former land connections.

The structure of Matthew's argument for the dispersal of animals across barriers, especially across the ocean, resembled Darwin's, in that he felt he could at least throw the burden of proof back onto the other side. Opponents of "flotsam and jetsam" dispersal were claiming that such imagined scenarios were highly unlikely. But since an unchanging geography would be a more conservative hypothesis than land bridges which have left no trace, all he needed to do was to show that natural rafts successfully carrying immigrant animals "is not an explanation to be set aside as too unlikely for consideration."[25] It was known that floating clumps of vegetation, torn out by floods and carried down to sea by great rivers, are sometimes formed and have occasionally been seen far out at sea. Assume, Matthew says, just one in a hundred such rafts to carry living mammals, assume a one-in-ten likelihood of landfall, assume probability of a passenger surviving the trip as one in three, landing at a favorable environment one in ten, chance of the passenger being a gravid female one in five, and allow about three rafts a year, then in the thirty million years of the Cenozoic, three hundred species may have thus obtained a foothold.

Barbour's immediate response to Matthew's paper was incorporated in a paper in the October 1915 *Bulletin* of the Museum of Comparative Zoology on the lizard genus *Ameiva*. Coauthored by his undergraduate assistant G. Kingsley Noble, the paper was aimed explicitly at the controversy over the geological history of the West Indies. The conjectured rafts, they said, would have to originate from the mainland, where great rivers running through forests could carve off pieces of their banks; most of the islands had no such conditions for generating rafts. If rafting happened frequently, the influx of new blood should have prevented formation or maintenance of distinct species on each island. If rafting is rare and fortuitous, then the relationships of various species of *Ameiva* found on the islands should be distributed haphazardly. On the contrary, Barbour and Noble argued, they display an orderly pattern.

They expressed their claim that the distribution of the various species of *Ameiva* is coherent, and thus cannot be the product of chance, in an unusual diagram, in which the names of the species are arranged spatially to correspond to the appropriate locality, as "if the whole diagram were superposed on a map of the Caribbean" (fig. 47).[26] Unbroken or dotted or wavy lines connect species judged to be related closely, or distantly, or "related but with evidently unknown species intervening"; the length and direction of these lines varies so as to position the species geographically. Barbour and Noble use the term "relationship" to mean closeness of evolutionary kinship and not mere similarity; they point out that convergent characters can "make two species, probably but distantly related, appear closely similar."[27] The exercise the reader was being asked to perform mentally, of laying their diagram on a map, does work pretty well, with only a bit of stretching and shoving (fig. 48). The authors insist that there was no reasonable way these lizards could have made their way by the haphazard process of crossing the salt water from one island to another, or to each

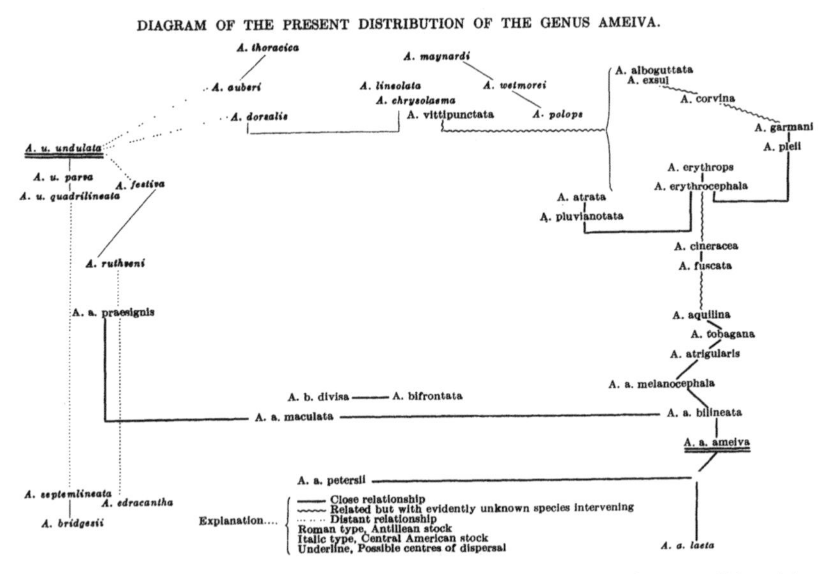

Figure 47. Barbour and Noble's chart of the taxonomic relationships of *Ameiva* ("A revision of the lizards of the genus Ameiva," p. 422). The authors asked their readers to imagine the species "superposed on a map of the Caribbean."

island from the mainland, yet distribute themselves in such an orderly arrangement.

Matthew had calculated that even if rafting were a rare event it would be adequate, but Barbour and Noble suggest that, even if this event were common, it would be useless. They propose sarcastically that

> we could lie-to in the mid-Caribbean and watch the rafts go by, speculating as a pleasant game as to which bore Onychophora [*Peripatus*] and earthworms and which cyprinodonts [freshwater fishes] or Amphibia, wondering how the little ponds on the rafts in which the fresh water fishes, mollusks, and crustaceans would have to be carried, are kept from becoming a bit, only a bit to be fatally, brackish.[28]

The *Ameiva* article was not very clear about exactly how these lizards had evolved into their present condition. Barbour and Noble declare that the "whole question of explaining the origin of this genus and its dispersal is difficult and unsatisfactory. . . . It is even far from easy to surmise which

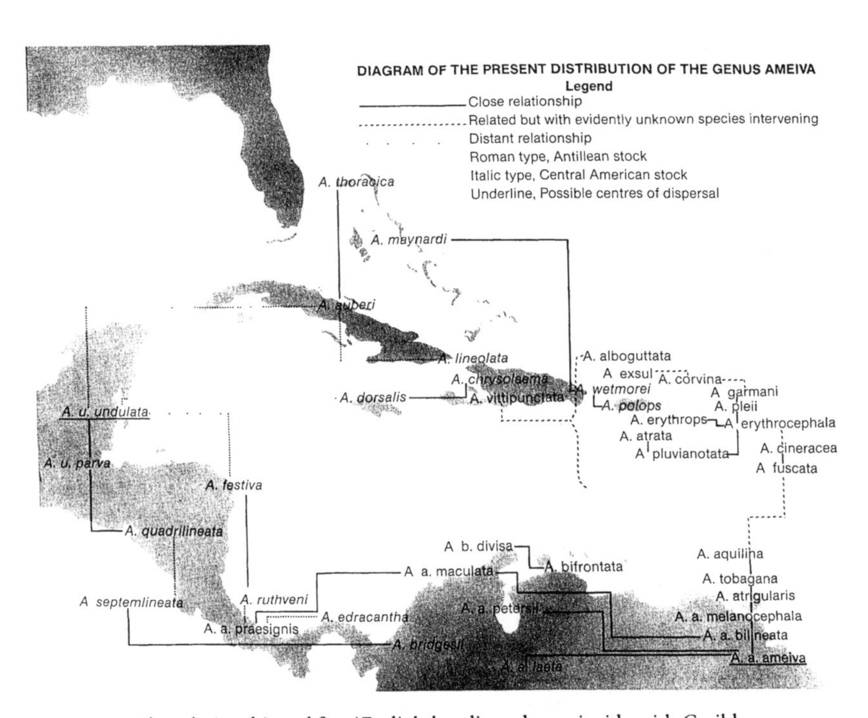

Figure 48. The relationships of fig. 47, slightly adjusted to coincide with Caribbean geography. Thomas Barbour, with his student G. K. Noble, argued that the orderliness of these relationships was evidence of evolution across a continuous land mass, whose sinking left the island populations as remnants.

are the more primitive types."[29] They do not seem to envision one ancestral species, broken up into several populations as the rising ocean created islands where a continent had been. They talk rather of "an orderly progressive migration, such as would only be possible over a continuous area of land,"[30] as if they were imagining the speciation process to have happened when the ancestral lizards first invaded Antillea, and the species to have been later frozen in time as the water invaded. If, instead, one ancestral species of *Ameiva* had covered a large Antillean landmass, and speciation had only come about after Antillea had been broken up into islands, there would be no reason to expect the species on adjacent islands to be any more similar than those on distant islands. The evolutionary model they are implicitly using is that of Joel Asaph Allen, David Starr Jordan, and Carl Eigenmann: that across a continent the local races of a widespread species grow to become distinct species, each of which is most nearly related to its closest neighbors.

In addition to the coauthored *Ameiva* paper, Barbour also responded directly, contradicting Matthews's reasoning on the question of rafting, in a note sent to the New York Academy for its *Annals* that fall. The note, which appeared in January 1916, did not indulge in sarcasm but argued vigorously that for delicate creatures strictly limited to their environment, like *Peripatus,* freshwater fishes, and burrowing amphibia, the calculation of probabilities becomes meaningless; even the enormities of geologic time could not convert something that cannot succeed into something that can. In the face of geologists' certainty, of course the biologist must yield, but in cases where there was a wide variety of geological opinion, the zoogeographer was entitled to make his inferences from faunal patterns. The groups in the West Indies are too evenly distributed, too homogeneous, to have been distributed fortuitously, Barbour insisted. Even mainland forms which would seem most capable of withstanding that kind of transport, such as the boa constrictor, which had actually been observed washed up alive on a beach after a storm, are not native to the islands.[31] For the most concrete evidence, Barbour referred his readers to the recent *Ameiva* article.

Matthew's response, full of courtesy, merely pointed out that they both agreed on "the need for securing more complete distributional data, towards obtaining which he has devoted so much time and energy."[32] To calculate the likelihood of successful transport of various invertebrates, he says, we surely need information on the survival of eggs, but for now we must admit that living eggs or young may be carried by high winds. To Barbour's claim that the regularity of the distribution of *Ameiva* was incompatible with a haphazard means of dispersal, Matthew made no public reply. He did admit privately to Barbour that he had not had time to transfer their chart of affinities to a map, as he would need to do to study their argument closely.[33]

In their printed exchange, and in a long and friendly correspondence, Matthew and Barbour each repeated his ideas in hopes of making the other see things differently, without effect. Each seemed sincere in acknowledging that hard evidence was scant. The outcome of the debate would be decided not by these men but by the scientific community. Exactly how that process occurred, though, is difficult to trace. Clearly, Matthew's treatise had the status of a major synthesis of a large amount of paleontological data worldwide, while Barbour's evidence had to do with one genus in one archipelago. Also, biogeographical indications could carry weight only in situations of geological uncertainty, and the doctrine of permanence of ocean basins took on new strength in reaction to the challenge of Wegener's theory of continental drift starting about 1920. There was a slowness and irresolution about the process, though, which may tell us that biogeography was not a discipline at all. Each contributor owed his first loyalty to some other field, like paleontology, entomology, herpetology, or geology.* There were no journals, departments, or societies devoted to biogeography. The process of reaction and testing of the questions raised by Barbour was, therefore, rather diffuse.

Barbour threw himself into the task of documenting the reptiles and amphibians of the West Indies, publishing a dozen or so articles on the subject in 1916 and 1917. When the United States finally gave up its neutrality in 1917 and joined the Great War, Barbour managed to do his military service on the island of Cuba.

A Useful Chore Man

Returning to the museum at the end of the war, Barbour kept up a stream of taxonomic contributions, but more and more of his time was taken up with administration. Samuel Henshaw depended heavily upon his advice and moral support, and soon everyone with dealings with the museum turned to Barbour when their letters to Henshaw went unanswered. Barbour was also deeply involved in many areas of the administration of biology at Harvard. He continued to contribute money to the museum—for the purchase of collections, for repairs, for hiring assistants. More important, he also began to attract gifts of money and specimens, inspiring other people, as

*"Scholars have drifted into the realm of biogeography from their various fields, often quite unconsciously or inadvertently, so that the left hand of the subject has hardly known what the right is doing" (Stott, "History of biogeography," p. 14).

neither Henshaw nor Alexander Agassiz could do, with his conviction that the M.C.Z. was a marvelous and delightful place and that the kind of biology it supported was eminently worth doing. He had wealthy friends, and he made friends easily. He agreed with Henshaw that the museum was hopelessly cramped for space, and could hardly accept new material when bulging at the seams already. So he campaigned, behind the scenes and in the *Harvard Alumni Bulletin,* for a separate building for the teaching laboratories.

It was no wonder that several people began to see in Barbour the next director of the museum.[34] But when and how the transition would be made was not a straightforward question, in an era when neither age of retirement nor pensions were formalized. When Henshaw turned sixty-five in 1917, Barbour, at the age of thirty-three, was clearly qualified to take over the leadership of the museum. Already the president of Harvard, A. Lawrence Lowell, was consulting with Barbour about the new biology building, about fund-raising, and about the museum, and when Henshaw's crusty habits became more embarrassing, Lowell consulted Barbour about how to get the old man to retire. Barbour's intimate letters to Henshaw and to others, however, show him quite content to forward the affairs of the museum from the sidelines. Sympathetic to the little entomologist whose life was so bound up in the museum that he had always refused to take even a single day's vacation, Barbour convinced Lowell to let the question rest a few more years.[35] In 1927, it was George Russell Agassiz, son of Alexander and member of the museum Faculty, who finally informed the seventy-five-year-old Henshaw that he would have to resign:[36]

> There was a last outburst of rage. Henshaw left the building and (to the regret of no one) never darkened its doors again, although he lived another 14 years. When he left he took with him masses of documents which, I am told, burned for two full evenings in his home fireplace.[37]

Certainly there is a vast gap in the M.C.Z.'s archives for the years of Henshaw's directorship.

The transformation of the M.C.Z. after Barbour's appointment in the fall of 1927 was like Dorothy in *The Wizard of Oz* waking up from dreary Kansas into the technicolor whirl of Munchkinland. Construction crews moved into the museum at once, creating new workrooms by flooring over the galleries of the exhibition halls, installing electric lights, and modernizing the heating. The exhibits were renovated and quantities of old specimens junked. Barbour loudly assured the curators that he wanted to help them do whatever they thought would promote the interests of their departments. They were given keys to locked doors, and the petty regula-

Figure 49. On the steps of the M.C.Z., July 1935. Left to right, William Morton Wheeler, Thomas Barbour (director of the M.C.Z. from 1927 to his death in 1946), and Henry Bryant Bigelow, all of whom proudly considered themselves naturalists of the old school. (By permission of the Museum of Comparative Zoology Archives, Harvard University)

tions that had oppressed their morale were swept away. He welcomed the museum refugees from the crumbling Bussey Institution—Charles Thomas Brues, Frank Morton Carpenter, and William Morton Wheeler.[38] Aggressive and creative in using the museum's funds, Thomas Barbour was also quick to use thousands of dollars of his own whenever he saw the need. When the Rockefeller Foundation promised partial funding for a new laboratory building, Barbour convinced Harvard to complete the amount; with the opening of the "Bio Lab" in 1931 the zoological laboratory vacated the museum.

Barbour's leadership—benign and hearty if autocratic—carried the museum through the years of the Great Depression and into the blackouts of the Second World War (fig. 49). Barbour saw clearly the tension, both historical and natural, Hyatt had described as "the battle between the conservators of the Museum on the one hand and the teachers in the College," and while he cultivated cordial relations with Harvard's Faculty of Arts and Science and the Harvard Graduate School, he held close to his early love of a museum as a sanctuary for natural history collections.

During that time he published herpetological descriptions and checklists, and notes and essays on a variety of topics, but nothing on the question he had debated with Matthew. By 1939, paleontologists regarded Matthew's "Climate and Evolution" as a classic, and it was reprinted by the New York Academy of Science. In 1943, collecting his *Atlantic Monthly* essays and adding some further reminiscences in a book called *Naturalist at Large,* Barbour repeated his opposition to "flotsam and jetsam" dispersal, even though he knew the subject was much more technical than the rest of the book. So he created an appendix, "For zoogeographers only," about the likelihood that the islands of the Greater Antilles were once connected to one another and to Central America, noting irreverently, "This is a matter of great theoretical interest with no practical application of any sort whatsoever."[39] He included sections from his 1910 article on Jamaica and cited with satisfaction a geologist of 1935 who supported his model. Opinion in biogeography was, he suggested, like a swinging pendulum.

Knowing how out of favor his beliefs had become, Barbour did not repeat them in professional journals, but he did give voice to his convictions over the years on the back steps of the museum, where the smokers gathered, rain or shine (smoking was forbidden in the museum), to talk shop and joke and challenge each other with obscure geographical names. (Several young scientists took up tobacco so they could join the group.) There in the late 1930s Barbour explained his reasoning to the entomologist Philip Darlington.[40] Since Barbour's viewpoint assumed the impossibility of extraordinary transport, and not only rafting but high winds had been suggested, Darlington proposed to challenge Barbour by dropping frogs from the top floor of the museum. The story was told and retold how, as each frog landed and lay inert on the grass at Barbour's feet, he called up to Darlington at the open window, "That one's dead!" "But almost immediately the stunned frogs began to recover and in a few minutes they began to hop about in all directions."[41] Darlington's courses and textbook explained that Cuba, Haiti, Jamaica, and the rest of the Greater Antilles belonged to the same category as the Galapagos, oceanic islands that had received all their flora and fauna by dispersal.[42] Ernst Mayr in 1944 called the "bridge-builders" the "naive" school, and declared that there was "ample evidence" that animals like frogs and freshwater fishes can occasionally cross over hundreds or even thousands of miles of ocean.[43] The views of Mayr and Darlington, ably supported by G. G. Simpson, triumphed completely throughout the '50s and '60s.[44]

When in 1958 Léon Croizat railed against the tyranny of the followers of Matthew, his undisciplined volumes were treated as the ravings of a madman.* According to Croizat, Barbour's objections to dispersal were perfectly correct, but he was ignored in spite of being right. Croizat complained that Matthew's "Climate and Evolution" had "managed to colour the whole of the learned world's perception of dispersal," not because of its intrinsic merits but by the happenstance of timing: North American paleontology was in 1915 at a pinnacle of reputation, and Europe was just beginning a debilitating war. Also, the coherence of a body of doctrine was attractive and Matthew's ideas "are presented with scintillating sales-appeal."[45] Only an equally comprehensive doctrine could make a dent in the dispersalist school, and this Barbour certainly did not provide. As Croizat said,

> An accomplished naturalist in the museum and the field, Barbour did know that the conclusions of Matthew and his like . . . were objectionable, and he forthwith stood up against them. However, Barbour did not have himself a critically elaborated plan of integration of what he otherwise knew, and so while he could tellingly criticize the Matthewian "zoogeography" *in detail,* he had nothing with which to replace it *in general.*[46]

The pendulum Barbour mentioned might be said to have swung back, because with the rise of plate tectonics, geologists can now provide land bridges as broad as Wegener's drifting continents. Among biogeographers there are now once again two schools of opinion, some prone to doubt dispersal by all but "waif" species (which everyone agrees to discount), and others willing to accept the presence of organisms on islands as evidence for their ability to disperse.[47] Once more there are those who find distributional and geological evidence to suggest overland connections in the West Indies, and even in the sacred Galapagos.[48] The debate has inspired some revealing polemic, each side imagining that the other is doing mere natural history and not science.

There can be no science of biogeography, however eloquent the theorizing, without healthy museums that house not only jars of specimens but taxonomists familiar with their contents. Charles Darwin knew this, for it was in discussions in the workrooms of the British Museum that the evolutionary implications of the Galapagos animals became clear to him, not out on the barren lavas of the islands themselves.[49] "One result was that he would never allow a depreciatory remark to pass unchallenged on the

*Croizat was born in Italy of French parents, was a U.S. citizen from 1929 to 1953, and died in Caracas, Venezuela, in 1987. He worked for a while in the Gray Herbarium, which is adjacent to the M.C.Z.; perhaps his interest in Barbour's criticism of Matthew's view dated from this period.

poorest class of scientific workers, provided their work was honest and good of its kind."[50] In 1881, Darwin provided the funds for a great register of botanical names, the *Index Kewensis*.[51] So privileged by his circumstances and talents that he could keep the deepest questions of theory always before him, Darwin had the experience, insight, and character to appreciate the toilers who made that grand kind of science possible.

When Barbour died in 1946, he was eulogized for his personal qualities and administrative achievements and as a naturalist, but not as a biogeographer and not as a scientist. He had indeed had a great impact: on the Barro Colorado Laboratory in Panama, the Atkins Institute (a branch of the Arnold Arboretum located in Soledad, Cuba), the Boston Museum of Science (in its metamorphosis from the museum of the Boston Society of Natural History), the Peabody Museum of Salem, Radcliffe College, the Carnegie Institution of Washingon, the Guggenheim Foundation, the Fairchild Tropical Garden in Florida, the Bishop Rhinelander Foundation, and the Woods Hole Oceanographic Institution. He had revitalized the M.C.Z., not only by his own financial contributions and considerable talent as an administrator but also by attracting to it other lovers of natural history, like John Phillips and Henry Bryant Bigelow, independently wealthy and serious about their favorite areas of biology. His vigor and tact and frank enthusiasm in forwarding these several scientific institutions made him much loved, but his achievements were not those of a creative scientist.

Barbour himself knew as well as anyone what was happening to him. As he assessed himself in 1934:

> My great weakness is, and always has been, the fact that I have been interested in pretty much everything, hence I have never been as profound a scholar as would have been highly desirable. On the other hand I have had more fun out of life and have perhaps been a more useful chore man around various departments of the University.[52]

The theoretical issues he had addressed in his youth, centering on biogeography, continued to interest him, but he pursued them only as a hobby.

Concluding Remarks

Buffon and Louis Agassiz made use of thought experiments to develop their ideas about the meaning of natural groups. Let us likewise experiment a bit with the relationship of museums to the science of biology. Can we separate the two in our minds? Very easily, for a museum is just the mature version of the well-known human instinct for collecting. Darwin collected beetles as a child, with no curiosity about the animal's life or function or cause of being; he was motivated only by the love of acquisition as well as a competitive urge to find things rare or undescribed. This sort of diversion (especially when it grows to an obsession) naturally gives rise to techniques for preserving and storing specimens, a network of fellow hobbyists, and the publication of aids to identifying specimens. Making a natural history collection is an innocent enough pastime, at least when the specimens taken are few in proportion to the wild population (which has, in fact, not always been so: collectors have actually been responsible for several extinctions). All it takes to transform one's private recreation into a full-fledged institution, complete with curators, publications, and exhibits is determination and money, as Lord Walter Rothschild proved by creating the Tring museum, which began as a childhood passion and ended superior to many long-established public institutions.[1]

My thought experiment supposes that museums have no necessary connection to science, which is rather an intriguing notion. Perhaps this may explain how they apparently managed to be so unaffected by major conceptual changes like evolution. It also accords with the important role that amateur naturalists have always played in the life of natural history museums. Certainly the collections and volunteer labor of people with no scientific credentials—their donations acknowledged in the *Annual Reports*—contributed significantly to the well-being of various departments of the M.C.Z. at every stage in its history. The mammal collection, for example, which had been neglected after J. A. Allen's departure (save for bits of caretaking by ornithologist William Brewster), grew substantially and gained a curator when Outram Bangs donated his specimens (he had stipulated that his services were part of the deal).* Every museum has benefited from the enthusiasm of amateurs, although the great Smithsonian administrator George Brown Goode warned in 1889 that it was not wise for institutions to become dependent upon volunteer curators.[2]

Having constructed this imaginary museum, emerging naturally from the instinct of collecting and the love of nature, it is time to consider the other partner in the marriage, the science of biology. We must pay careful attention to what we mean by "science." People use this word in dozens of different ways, many overlapping but some barely connected. One meaning is merely "knowledge about the natural world," another something like "cautious reasoning and systematic observation," and still another "the search for natural laws, causes, or testable models that resemble explanation." A simple and authoritative definition of science will certainly elude us, because in practice various scientists have held and now hold different notions, even while assuming that they and their fellows share a common understanding.

Certainly we know that particular beliefs about the world cannot serve us as the hallmarks of science. Thus, for example, phlogiston, the stuff of combustion which imparts negative weight, seems utterly silly today, but it quite properly belongs in the history of chemistry because in the late eighteenth century the most respected chemists of the day took it seriously. The miraculous creation of unchangeable species is an unscientific belief today, making no more sense to a biologist than phlogiston does to a modern chemist, but when Louis Agassiz conferred with Cuvier over fish fossils in Paris in 1832 that belief was perfectly scientific. Closer to the essence of science is something about its methodology.

Turning to my favorite scientist, Charles Darwin, we can find illustrated a wide gamut of methods. At university he developed a fascination with science, his reading of Humboldt and Herschel having fired him with a desire to contribute to this glorious enterprise.[3] He outgrew his beetle mania and learned that facts were meaningful only in relation to some theory. Yet he never lost his respect for the humble fact collector. His most respected colleagues were taxonomists, including Joseph Hooker and Asa Gray, and he devoted eight years to his taxonomic monograph of barnacles. Still, his

*The collection was purchased by friends of mammalogy for $5,000 and given to the museum. The transfer agreement included stipulations that Bangs be given a formal appointment to serve without pay (A. Agassiz to Eliot, 29 August 1900, Harvard University Archives). Bangs, who later switched to ornithology, was on the M.C.Z. staff from 1900 (not 1899) until his death in 1932 (Peters, "Outram Bangs," pp. 264–74).

own work was all heavily infused by the search for causal explanation, unifying laws, and general principles. His synthetic theory of evolution by natural selection, scorned by Louis Agassiz as too speculative to be good science, was admired by many others as belonging to the highest class of scientific thinking.

Surely what we learn from thinking about Darwin is that science is a mixture, perhaps a rather lumpy mixture, of activities that range from the fact-collecting end of the scale to the theorizing end. Individual scientists may work at different points of the scale at different times. Far from being destroyed, the scale is made richer by the recognition that there can be no bare fact not infected by theory, and no pure theory independent of facts. Arguably, the healthy condition of science depends upon a balance of the various approaches.

Balance is achieved, as always with human affairs, imperfectly, through struggle and contradiction, through opposition and compromise. Scientists' rhetoric at various times has sometimes extolled the superiority of exact, "positive" observation of particulars, sometimes celebrated the revolutionary or synthetic insights of great theorists. Empiricists love to disparage speculation, and rationalists are fond of scorning mindless data crunching. Yet, generally speaking, it is theories, causes, or laws—as long as they are convincing—that command the highest prestige in science. In unexplored territory the collection of raw data will be valued, and marshaling facts within the framework of an accepted theory is always respectable, but those scientists who expand our understanding rather than merely adding new information are the most admired.

Now we are ready to put together our carefully purified museum with our clarified image of science. Our imaginary collector steps into the realm of science, in one meaning of the term, by the mere act of choosing shells and butterflies instead of stamps, since he or she cannot help but gain some "knowledge about the natural world," even if that knowledge is limited to the inventory of diversity. A "scientific" collector keeps in view the goal of knowledge of nature, by attempting to associate with the objects assembled at least some information about them, such as the locale in which the animal was found, perhaps with notes about the habitat. In practice, however, much more is required. A collection is judged more or less "scientific" by how well it conforms to the standard practice of the surrounding scientific community. This adds to the elusiveness of simple definitions, for such standards change over time. For example, as Louis Agassiz told his assistants, fishes labeled "Brazil" used to be welcomed into reputable museums, but his requirements were now stricter. In this century curators demand still more precise data than Agassiz did. Systematists think of taxonomic rules, such as priority in nomenclature and designation of type specimens, as logical requirements of the project of inventory, but the rules actually arose bit by bit over time. And at every stage the rules served to sharpen the distinction between a scientific collection and one made for private amusement.

In relation to the scale from fact-collecting up to theorizing, it would appear that even the most carefully documented museum must be stuck somewhere near the bottom of the ladder of scientific prestige. Anyone who needs a chart treats the mapmaker's art with grateful respect, and all students of living things are glad to have a taxonomic key, but the gratitude we accord librarians is the best that the modest museum worker can hope for. There is nothing in taxonomy to stir the soul. Or is there? Louis Agassiz said there was. The scale of scientific prestige was something he understood very well indeed. He taught his students that science must be closely grounded on the elemental facts contained in natural specimens, yet must also reach beyond these facts to their ultimate meaning. They were urged to search for the homologies that define families, orders, or classes rather than to rest content with describing new species. What higher ideas could there be than the thoughts of God himself? When that notion wore thin, he promised that his collections would be the testing ground for Darwin's theory.

To those of us who see in evolution a far more elegant, rich, and interesting theory than anything Agassiz could offer, it would seem logical for 1859 to have marked a great step upward in the prestige of the kind of science that museum specimens make possible. This is what Ernst Mayr meant when he said, "One might have expected that the acceptance of evolution would result in a great flowering of taxonomy and enhancement of its prestige during the last third of the nineteenth century." Perhaps what is wrong with this expectation is that it is based upon a modern view of evolution. If we could put ourselves in the shoes of the mid-Victorians, we might see Darwin's theory as the majority of naturalists did then see it, even those who could not deny its logical force: cold comfort, higgledy-piggledy, lacking purpose, too heartless to be true. Among those who understood the theory best, there were many, like Asa Gray and Alfred Russel Wallace, who still felt the need to supply the transcendental meaning for which Darwin himself steadfastly refused to make room.

Most taxonomists did not recognize in Darwin's theory the breakthrough they had been expecting. Yes, it supported them in their feelings about affinity and analogy, embryological development, geography, and fossils. It let them get on with their work in peace, as Darwin predicted, no longer troubled by cases where nature seemed to taunt them by posing exceptions to her own rules. At the same time, Darwinism seemed to destroy any hope of finding general laws such as physicists could boast of. No longer could taxonomic relationships glitter and beckon with transcendental meaning. Anton Dohrn, for one, declared that "we can no longer believe that our system is an expression of the plan of creation, because Darwin has removed the last residue of 'thinkability' from this concept."[4]

As a reply to Mayr's expectation, this does not suffice, for in those same years there did live several naturalists who could appreciate the exciting riches the *Origin* was offering systematists. Henry W. Bates and Fritz Müller, who developed our explanation of mimicry in butterflies, were Darwin's personal favorites. Benjamin Dann Walsh, on a more modest level, is of the company. Such thinkers can be submitted in evidence to show that the conceptual potential inherent in Darwin's words, much of it unexploited until well into the next century, was in some sense available to Victorian taxonomists.

Mayr went on to make the provocative suggestion that the reason for taxonomy's slumber was its location in museums. When we recall Walsh, miles from anything like a scientific institution (as were Bates and Müller too), we should take this idea seriously. The contrast he was drawing, however, was between two kinds of institutions, museums versus universities.

> This [flowering and enhancement of prestige] was not the case—in part for almost purely administrative reasons. The most exciting consequences of the findings of systematics were studied in university departments, while the very necessary but less exciting descriptive taxonomy, based on collections, was assigned to the museums.[5]

This brings us back to Cambridge. If systematics in general was at a disadvantage from its "administrative" isolation in museums, then surely Harvard should have stood as the exception to prove Mayr's rule. There a museum was founded in association with a university by a charismatic teacher, so systematics should have flourished at Harvard if it could have flourished anywhere. Historian Philip J. Pauly has already pointed to the situation at Harvard in a general way, in support of the novel and thought-provoking thesis of his article "The Appearance of Academic Biology in Late Nineteenth-Century America."[6] He notes the difference between older, established universities, like Harvard and the University of Pennsylvania, where the presence of entrenched interests, such as medical schools, discouraged change, versus new institutions like the Johns Hopkins University, the University of Chicago, and Clark University, where the self-conscious new discipline calling itself "biology" did well. It is good to bring such structural questions to bear, if only to counterbalance our natural inclination to concentrate on the actions of a few important individuals, for no one can lead without followers. If only the second director of the M.C.Z. had been a person of more vision, we are tempted to think, someone who could really see masses of specimens as the testing ground for a great theory as Louis Agassiz had, rather than a man who deplored speculation as Alexander did, might not the generation after Darwin have witnessed after all a flowering of systematics, with the M.C.Z. showing the way? The Agassizs themselves were prone to blame the ill will of others for the troubles that beset them. They had begun their tenures committed to the vision that this museum would be more than a collection of specimens, that it would be a center of teaching and research. The ideal eluded them both.

Some larger forces were apparently at work. In Paris, where Cuvier had made the Muséum d'Histoire Naturelle the mecca for zoologists in the first third of the century, at mid-century the Faculté des Sciences was taking control of zoology teaching; according to a recent study the "Muséum was in ever greater danger of being relegated to a marginal existence," and by the end of the century it "was teaching an almost non-existent student body."[7] Biology's center of gravity had somehow shifted, leaving traditional approaches on the outskirts, or on the defensive.

The great advantage of university departments, of course, is the presence of students, which is invigorating in two distinct ways. First, young minds can challenge and probe, or merely by their presence spur one to reexamine one's assumptions, keep up with other fields besides one's specialty, and think about connections. Second, and probably even more important in the period when university degrees were becoming the required credentials of a professional scientist, subjects belonging to university departments had an overwhelming advantage over other subjects in a kind of Darwinian selection. Laboratory cytologists located in colleges produced not only publishable results but academic offspring. They could replicate their own interests in the next generation. As long as taxonomy stayed with the old-fashioned apprenticeship system, the juggernaut of professionalization would shove it to the margins of scientific respectability, irrespective of the intellectual potential of its subject matter.

Many individuals concerned with the status of systematics have identified its role in the degree-granting system as the key to improvement. Mayr negotiated hard to gain for M.C.Z. curators the status of membership in the Harvard faculty (not without cost to the museum, some would say). Elsewhere too—in the British Museum, the U.S. National Museum, the American Museum of Natural History—in the middle of this century curators began to be cross-appointed to academic departments. But who was it who had "assigned" to museums "the very necessary but less exciting descriptive taxonomy" in the first place?

It is time to return to my modest thought experiment, which is not really my own, I believe, but only what most of us half-consciously understand about natural history museums. Like Louis Agassiz with his lobster, all I have done is to make the generally held belief more explicit. My thought experiment began with collecting as a hobby, but approaching the matter from the other end gives a more fruitful idea of the relation of museums to science. We can just as well begin in our minds with the abstract quest for an understanding of life, the scientific curiosity about the nature of living things, a quest that must confront the great phenomenon of diversity just as it must confront the phenomenon of metabolism or heredity. Faced with the problem of varied form, the scientist would have to invent a means of systematic comparison. Science would have to invent museums. Their storage cases, record keeping, and network of exchanges would be just as truly an instrument for scientific investigation as the cytologist's microscope.

I find the notion of scientific instrument a very useful one, but I cannot leave our imaginary museum without pointing out that our exercise could be just as misleading as Agassiz's lobster was. We can conjure up a lobster in our mind, free of parents and free of cousins, but mother nature never makes a real lobster this way. Likewise, nothing we could call a museum would arise from the collecting instinct alone, independent of science. Though Tring was Rothschild's hobby, in forming it he drew upon the goals and standards of science. Agassiz began the M.C.Z. in close relation to scientific exemplars already existing—imitating them in most respects, improving on them in other respects; his supporters responded to his appeal because they understood that his plan, however bold, was not idiosyncratic but closely related to other scientific collections; his success would be measured against a yardstick wielded by the international scientific community. A museum like Agassiz's outside of the context of a scientific community is as impossible as a lobster without relatives. Equally so, the study of diversity could not have grown to be a proper science without the acquisitive instinct supplying its material.

Once we acknowledge that the museum is an eminently respectable piece of scientific equipment, a tool literally rather than metaphorically, we are struck by the dominant and highly awkward feature of this instrument. A museum is not a convenient item. We are happiest applying the term "tool" to things a person can pick up and handle, and with good reason. Larger things are not so easily within the control of a single individual. This piece of equipment is more than big: to do the job, it has to consist of more than bricks and shelves and bottles and catalogues; it must have people bound together in a structured workplace. It must be an institution.

Institutions are by their nature conservative. They resist change. They exert themselves not only for the purpose for which they were founded but for a new purpose as well—their own survival. From the standpoint of longevity—certainly a desirable quality for the preservation of taxonomic type specimens—conservatism can be a good thing. From the standpoint of intellectual revolution, it is an obstacle. People attached to an institution tend to be committed to the entity itself as well as to the abstract purposes it was created to serve.

The institutional nature of museums took over almost immediately after Louis Agassiz's fond dream for a museum of comparative zoology was achieved. Inviting Philip Uhler to Cambridge, when he had decided to replace the self-interested student-assistants with steadier workers, he wrote (see fig. 10, above), "I expect that those connected with the Museum *shall work for it* and ***not for themselves.*** The object in view being to erect a great Monument to science and not to foster the private objects of those connected with it."[8]

By the middle of the nineteenth century the study of diversity was a mature science. Its flowering produced, besides Agassiz's abortive "Essay on Classification," its choicest fruit in Darwin's great book. It had succeeded also in developing its great instrument, but because that tool happened to be an institutional one, systematics was not well placed to react to the challenge of change, nor to hold its own in the struggle for prestige among the sciences.

Notes

Preface

1. Genesis 7:14.
2. C. R. Darwin, *Origin of Species,* p. 411.
3. H. Daudin, *Cuvier et Lamarck;* E. S. Russell, *Form and Function;* E. Stresemann, *Ornithology from Aristotle to the Present;* E. Mayr, *The Growth of Biological Thought.*
4. T. A. Appel, *The Cuvier-Geoffroy Debate;* F. Sulloway, "Darwin's conversion."
5. M. P. Winsor, "The development of Linnaean insect classification"; idem, *Starfish, Jellyfish, and the Order of Life;* idem, "Louis Agassiz and the species question."
6. Winsor, "The impact of Darwinism"; idem, "A historical investigation of the siphonophores."
7. W. Bateson, "Heredity and variation in modern lights"; R. Pearl, "Trends of modern biology"; C. H. Merriam, "Roosevelt the naturalist."
8. Mayr and R. Goodwin, *Biological Material: Part I: Preserved Material and Museum Collection.*
9. Mayr, "The role of systematics in biology," p. 417.
10. D. Hull, *Science as a Process.*
11. A. Agassiz to C. W. Eliot, 8 August 1884, Harvard University Archives.

Chapter One

1. A review of definitions and issues is provided by W. Bechtel, "The nature of scientific integration"; I found food for thought in W. Coleman, "The cognitive basis of the discipline: Claude Bernard."
2. Appel, *The Cuvier-Geoffroy Debate.*
3. J. LeConte, *Autobiography,* p. 128.
4. E. C. Agassiz, *Louis Agassiz,* 2:548–49.
5. Ibid.
6. G. R. Agassiz, *Letters and Recollections,* p. 45.

7. T. Lyman to G. H. Shaw, 4 August 1856, Lyman Family Papers, Massachusetts Historical Society, Boston.

8. L. Agassiz, *Essay on Classification,* pp. 21–22.

9. T. H. Huxley, *Scientific Memoirs,* 1:538; Winsor, "The impact of Darwinism."

10. Appel, "Jeffries Wyman."

11. L. Agassiz, *Essay on Classification,* p. 12n.

12. D. G. Wayman, *Morse,* p. 120.

13. L. Agassiz, *Essay on Classification,* p. 12.

14. Ibid.

15. T. T. Bouvé, "Historical sketch," pp. 48–59.

16. The story of the M.C.Z.'s beginnings is well covered in E. Lurie's biography, with a few extra details in an extract he produced for its centenary, *The Founding of the Museum.*

17. L. Agassiz, *Essay on Classification,* p. 11.

18. Lurie, *The Founding,* p. 17; idem, *Louis Agassiz,* p. 219.

19. Lyman, "Recollections of Agassiz."

20. Lurie, *Louis Agassiz,* p. 226.

21. N. S. Shaler, *Autobiography,* pp. 93–99.

22. S. Scudder, "In the laboratory with Agassiz."

23. L. Agassiz, *Essay on Classification,* p. 6.

24. Ibid., p. 8.

25. Scudder, "In the laboratory with Agassiz," p. 3.

26. R. Ward, *Henry Augustus Ward,* p. 55; Joseph Leconte, *Autobiography,* pp. 128–29.

27. J. Lyon, "The 'initial discourse' to Buffon's *Histoire naturelle,*" pp. 145–46.

28. T. H. Huxley, "A lobster: or, The study of zoology."

29. Ibid., pp. 206–7.

30. Scudder, "In the laboratory with Agassiz," p. 4.

31. Mayr, *The Growth of Biological Thought,* p. 238.

32. L. Agassiz, *Essay on Classification,* p. 173.

33. Winsor, *Starfish, Jellyfish;* idem, "Linnaean insect classification."

34. Mayr, *The Growth of Biological Thought,* p. 207.

35. Winsor, *Starfish, Jellyfish,* p. 131.

36. L. Agassiz, *Essay on Classification,* p. 146.

37. [J. D. Dana], "Agassiz's contributions," pp. 332–33.

38. E. C. Herber, *Correspondence between Baird and Agassiz,* p. 94.

39. L. Agassiz, *Essay on Classification,* p. 179.

40. H. Strickland, "On the true method of discovering the natural system," p. 185.

41. Darwin, *Origin of Species,* p. 413.

42. Ibid., p. 52.

43. Ibid., p. 411.

44. Ibid., p. 478.

45. Ibid., p. 413.

46. L. Agassiz, *Essay on Classification,* p. 10.
47. Ibid., p. 9.
48. Ibid., p. 145.
49. [Dana], "Agassiz's contributions," p. 321.
50. Ibid., p. 332.
51. Ibid., pp. 340–41.
52. L. Agassiz, "On the Origin of Species," p. 151.
53. L. Agassiz, "Evolution and the permanence of type," p. 101.
54. A. S. Packard, "The study of natural history in college"; Scudder, *The Butterflies,* 1:953; Wayman, *Morse,* pp. 24–25, 226; H. J. Clark, *Mind in Nature;* LeConte, *Autobiography.*
55. Appel, "Jeffries Wyman."
56. [Dana], "On Cephalization"; idem, "The classification of animals"; idem, "On the higher subdivisions"; idem, "On the homologies of the insectean and crustacean types"; idem, "Note on the position of amphibians"; idem, "The classification of animals, no. 2"; idem, "The classification of animals, no. 3." Examples of favorable allusion to Dana's ideas are Packard, "Notes on the family Zygaenidae," p. 15; idem, "Observations on the development and position of the Hymenoptera"; p. 282; idem, *Guide to the Study of Insects,"* p. 8; E. S. Morse [Principle of cephalization applied to the classification of Mollusca].
57. Morse, "A classification of Mollusca based on the 'Principle of Cephalization.' "
58. B. Wilder, "On morphology and teleology"; H. J. Clark, *Mind in Nature.*
59. Herber, *Correspondence between Baird and Agassiz,* p. 63.
60. Lurie, *Louis Agassiz,* pp. 177–78; Ward, *Henry Augustus Ward,* p. 56.
61. L. Agassiz, "The study of nature," cited in A. Tachikawa, "The two sciences and religion in ante-bellum New England," p. 230.
62. Wayman, *Morse,* p. 91.
63. R. W. Dexter, "The 'Salem secession' of Agassiz zoologists," p. 28.
64. A. E. Verrill, Private Journal, 26 May 1859, Harvard University Archives.
65. Wayman, *Morse,* p. 97.
66. Ibid., p. 90.
67. Ibid., pp. 84–85.
68. H. A. Warren, "Survey of the life of Louis Agassiz."
69. A. H. Dupree, *Asa Gray,* chaps. 14 and 15; Lurie, *Louis Agassiz,* pp. 292–301.
70. Verrill [Notes of Private Lectures of L. Agassiz], 25 June [1860], Harvard University Archives.
71. J. A. Allen, *Autobiographical Notes,* p. 9.
72. *Annual Report* for 1860, p. 34.
73. *Annual Report* for 1862, p. 31.
74. Scudder, "Remarks upon the division of the class of insects into orders"; idem, "Inquiry into the zoological relations of . . . fossil neuropterous insects," p. 182.
75. *Annual Report* for 1862, p. 31.

76. Packard, "Observations on the development and position of the Hymenoptera," p. 291.

77. Verrill, Private Journal, 11 September 1860, Harvard University Archives.

78. Ibid., 31 December 1861.

79. Shaler, "Lateral symmetry in Brachiopoda."

80. Morse, "A classification of Mollusca," p. 162.

81. Ibid., p. 163n. This cluster of attention is described from a different point of view by Dexter, "Historical aspects of studies on the Brachiopoda."

82. A. Hyatt, [Remarks on natural selection]; Packard, "A half-century of evolution"; idem, [review of] "Wallace's *Contributions*"; Scudder, *Butterflies,* 1:531–35, 2:950–53; Morse, "The evolution theory."

83. Hyatt, "On the parallelism between the different stages of life"; E. J. Pfeifer, "The genesis of American Neo-Lamarckism."

84. D. S. Jordan, *Days of a Man,* p. 114.

85. L. D. Stephens, *Joseph LeConte,* pp. 161–63.

86. Morse, "What American zoologists have done for evolution," p. 140.

Chapter Two

1. Verrill, Private Journal, 30 May 1862, Harvard University Archives.

2. Wayman, *Morse,* pp. 126–29; Dexter, "The 'Salem secession,'" p. 4.

3. Ibid., p. 127.

4. Ibid.

5. Verrill, Private Journal, 27 July 1860, Harvard University Archives.

6. Wayman, *Morse,* pp. 150–51; Verrill, Private Journal, 13 November 1860, Harvard University Archives.

7. Wayman, *Morse,* p. 170.

8. *Annual Report* for 1862, p. 5.

9. *Annual Report* for 1863, pp. 16–17.

10. A. Agassiz to Lyman, 21 April 1862, M.C.Z. Archives.

11. F. Tuckerman, "Henry James Clark"; Packard, "Memoir of Henry James Clark."

12. H. J. Clark, Diary, 19 May and 28 May 1857, Harvard University Archives.

13. L. Agassiz, *Contributions,* 1: xv–xvi.

14. H. J. Clark, Diary, 19 January 1859, Harvard University Archives.

15. Ibid.

16. E. C. Agassiz, *Agassiz: His Life and Correspondence,* 2:561.

17. H. J. Clark, Diary, 11 June 1859, Harvard University Archives.

18. Ibid.

19. H. J. Clark, *Mind in Nature.*

20. Harvard College Papers, ser. 2, 26:189, Harvard University Archives.

21. H. J. Clark to L. Agassiz, 25 June [1863], Harvard College Papers, ser. 2, 30:162, Harvard University Archives.

22. L. Agassiz, *Contributions,* 3:vi.

23. H. J. Clark to L. Agassiz, 25 June [1863], Harvard College Papers, ser. 2, 30:161, Harvard University Archives.

24. L. Agassiz to T. Hill, 14 May 1863,.Harvard College Papers, ser. 2, 30:120–21, Harvard University Archives.

25. H. J. Clark, "Lucernaria the coenotype of Acalephae."

26. L. Agassiz to T. Hill, 14 May 1863, Harvard College Papers, ser. 2, 30:120–21, Harvard University Archives.

27. Verrill, Private Journal, 12 December 1862, Harvard University Archives.

28. Lurie, *Louis Agassiz,* pp. 110, 154; G. H. O. Volger, *Leben und Leistungen des Naturforschers Karl Schimper,* pp. 6–9; C. Vogt, *Aus meinem Leben,* pp. 188–202.

29. L. Lesquereux to P. Lesley, 21 January 1865, Lesley Papers, American Philosophical Society, Philadelphia.

30. Shaler, *Autobiography,* p. 104.

31. Harvard College Papers, ser. 2, 30:120–21, Harvard University Archives.

32. L. Agassiz *Contributions,* 4:175; H. J. Clark, "Lucernaria the coenotype of Acalephae."

33. Verrill, Private Journal, 13 December 1862, Harvard University Archives.

34. Verrill, "Revision of the polyps."

35. Wayman, *Morse,* p. 97.

36. Herber, *Correspondence between Baird and Agassiz,* p. 158.

37. Lyman to A. Agassiz, 18 December 1862, M.C.Z. Archives.

38. H. J. Clark, Diary, 20 March 1863, Harvard University Archives.

39. A. Agassiz to Lyman, 17 March 1863, M.C.Z. Archives.

40. L. Agassiz to H. J. Clark, 24 March 1863, College Papers, ser. 2, 30:67, Harvard University Archives.

41. H. J. Clark, Diary, 25 March, 28 March, 30 March, 19 April, 23 April, 3 July 1863, Harvard University Archives.

42. H. J. Clark to T. Hill, 14 May 1863, College Papers, ser. 2, 30:119, Harvard University Archives.

43. H. J. Clark, Diary, 26 March 1863, Harvard University Archives.

44. H. J. Clark, Diary, 3–4 April 1863; H. J. Clark to L. Agassiz, 25 June [1863], Harvard College Papers, ser. 2, 30:161, Harvard University Archives.

45. *Annual Report* for 1865, pp. 48–49.

46. Verrill, "Revision of the polyps"; Scudder, "Materials for a monograph of North American Orthoptera"; Packard, "Synopsis of the Bombycidae."

47. *Annual Report* for 1864, p. 11.

48. Dexter, "Salem secession," pp. 7–8.

49. Verrill, Private Journal, 19 December 1863, Harvard University Archives.

50. L. Agassiz to H. Wheatland, 18 December 1863, Letterbooks, 2:354–57, M.C.Z. Archives.

51. L. Agassiz to P. R. Uhler, 6 April 1864, Letterbooks, 2:401–4, M.C.Z. Archives.

52. L. Agassiz to Dana, 15 October 1866, Letterbooks, 4:37–38, M.C.Z. Archives.

53. *Annual Report* for 1863, p. 9.

54. *Annual Report* for 1866, p. 9.

55. Lurie, *Louis Agassiz,* p. 2.

56. Ibid., pp. 316–17.

57. Ibid., p. 316.

Chapter Three

1. L. H. Tharp, *Adventurous Alliance,* p. 166.

2. L. Agassiz and E. C. Agassiz, *Journey in Brazil,* p. 3.

3. Ibid., p. 4.

4. Ibid., p. 8.

5. Ibid., p. 33.

6. Ibid., p. 160.

7. Ibid., pp. 236–37.

8. Ibid., p. 160.

9. Ibid., p. 220.

10. Ibid., p. 183.

11. Quoted in Lurie, *Louis Agassiz,* p. 347.

12. L. Agassiz and E. C. Agassiz, *Journey in Brazil,* p. 195.

13. E. C. Agassiz, *Agassiz: His Life and Correspondence,* 2:640.

14. Lyman, Private Journal, 6 February and 20 February 1866, Massachusetts Historical Society, Boston.

15. E. C. Agassiz, *Agassiz: His Life and Correspondence,* 2:647.

16. L. Agassiz and E. C. Agassiz, *Journey in Brazil,* pp. 8–10.

17. Darwin, *Origin of Species,* p. 347.

18. Ibid., p. 349.

19. Ibid., p. 350, my emphasis.

20. L. Agassiz, J. E. Cabot, et al., *Lake Superior,* pp. 246–48.

21. Ibid., p. 247.

22. L. Agassiz, *Essay on Classification,* pp. 36 and 43.

23. L. Agassiz and E. C. Agassiz, *Journey in Brazil,* pp. 29–34.

24. Ibid., p. 8.

25. F. M. Chapman, "Biographical memoir of J. A. Allen"; Allen, *Autobiographical Notes.*

26. Allen, "Catalogue of the mammals of Massachusetts," p. 149.

27. Ibid., p. 152n.

28. *Annual Report* for 1868, p. 23.

29. L. Agassiz, *Essay on Classification,* pp. 86–87.

30. *Annual Report* for 1869, p. 12.

31. Allen, "Catalogue of the mammals of Massachusetts"; idem, "Notes of an ornithological reconnaissance"; idem, "On the mammals and winter birds of East

Florida; idem, [Geographical variation in mammals and birds]; idem, "On geographical variation in color among North American squirrels."

32. Allen, "Catalogue of the mammals of Massachusetts," p. 214.

33. Allen, "The influence of physical conditions in the genesis of species."

34. R. Rocker, *Pioneers of American Freedom,* pp. 118–20. I am indebted to the New-York Historical Society for access to a copy of the *Radical Review* in its original wrappers, on which Morse is listed.

Chapter Four

1. Wyman to R. Owen, June 1863, Owen Papers, British Museum (Natural History), London.

2. *Annual Report* for 1862, p. 6.

3. *Annual Report* for 1864, p. 12.

4. Hyatt, "The genesis of the Tertiary species of Planorbis"; Allen, discussed above; Morse, "On the tarsus and carpus of birds," idem, "On the identity of the ascending process of the astragalus"; Allen, "The influence of physical conditions in the genesis of species."

5. A. Agassiz to Lyman, 21 April 1862, M.C.Z. Archives.

6. L. Agassiz to J. L. LeConte, 23 March 1864, M.C.Z. Archives.

7. L. Agassiz to Uhler, 6 April 1864, M.C.Z. Archives.

8. A. Mallis, *American Entomologists,* pp. 205–6.

9. A. Agassiz to Lyman, 27 June 1864, Massachusetts Historical Society, Boston.

10. Bickmore and Uhler, "Plan for a Museum of Natural History" and letters [Autumn 1864], Albert S. Bickmore Papers, Archives of American Museum of Natural History; also Bickmore to C. Abbe, November 1864, Cleveland Abbe Papers, Library of Congress. I am indebted to Professor Edward Lurie for calling my attention to the Abbe papers.

11. Uhler to Baird, 15 November 1864, Spencer Fullerton Baird Papers, Smithsonian Institution Archives, Washington, D.C. I am indebted to Edward Lurie for calling my attention to this letter.

12. Minutes of the board of trustees of the Peabody Institute, 3 January 1867, George Peabody Department, Enoch Pratt Free Library, Baltimore. The head of the department, Lyn Hart, kindly supplied me with this information.

13. L. Agassiz to Uhler, 6 April 1864, M.C.Z. Archives.

14. Uhler to O. W. Holmes, 22 January 1867, M.C.Z. Archives.

15. Mallis, *American Entomologists,* pp. 120–21.

16. L. Agassiz to Lyman, 23 July 1867, M.C.Z. Archives.

17. L. Agassiz to Hagen, 3 April and 12 June 1867, M.C.Z. Archives; I am indebted to Mina Brand of the M.C.Z. library for translating these letters for me. See also L. Agassiz, [statement of intention] 23 July 1867, M.C.Z. Archives.

18. Hagen to Henshaw, 13 October 1885, M.C.Z. Archives.

19. Lyman, Private Notebooks, 12 October 1867, Massachusetts Historical Society, Boston.

20. *Annual Report* for 1865, pp. 9–10; for 1866, pp. 4–5; for 1867, pp. 4–5; for 1868, p. 5.

21. Allen to Uhler, 19 December 1867, M.C.Z. Archives.

22. L. Agassiz to Lyman, 21 July 1868, M.C.Z. Archives.

23. L. Agassiz, "Arrangement of a museum"; "Instructions given by Prof. Agassiz [concerning entomology collections]," 1867, Harvard University Archives; *Annual Report* for 1860–61, pp. 11–17; for 1862, p. 14; for 1867, pp. 8, 19–21; for 1868, pp. 27–30; for 1869, pp. 19–21; for 1870, p. 19.

24. *Annual Report* for 1868, p. 28.

25. L. Agassiz to Hagen, 21 January 1868, M.C.Z. Archives.

26. Hagen, "Monograph of the North American Astacidae," p. 16.

27. Ibid., p. [iii].

28. Herber, *Correspondence between Baird and Agassiz;* L. Agassiz, *Lake Superior.*

29. C. Girard, "A revision of the North American Astaci"; Catalogue of Crustacea, Department of Invertebrate Zoology, M.C.Z.

30. Herber, *Correspondence between Baird and Agassiz,* pp. 73–75.

31. E. Lurie, *The Founding;* idem, *Louis Agassiz,* pp. 215–34; *Annual Report* for 1861, pp. 40–49; for 1862, pp. 13 and 15.

32. Herber, *Correspondence between Baird and Agassiz,* p. 63.

33. Ibid., pp. 77–78.

34. L. Agassiz, *Essay on Classification,* p. 178.

35. Herber, *Correspondence between Baird and Agassiz,* p. 134.

36. Winsor, "Louis Agassiz."

37. L. Agassiz and E. C. Agassiz. *Journey in Brazil,* pp. 183–84, 194, 378–80.

38. Lurie, *Louis Agassiz,* pp. 341–44.

39. Girard, "Revision of the North American Astaci."

40. W. F. Erichson, "Untersicht der Arten der Gattung Astacus."

41. L. Agassiz, "Communication."

42. Girard, "Reply to Prof. Agassiz's communication."

43. J. [E.] LeConte, "Descriptions of new species of Astacus from Georgia."

44. Girard, "Revision of the North American Astaci," p. 87.

45. Herber, *Correspondence between Baird and Agassiz,* p. 74.

46. Hagen, "Monograph," p. 12.

47. Ibid., p. 22.

48. Dexter, "Historical aspects of Louis Agassiz's lectures," p. 18.

49. Hagen to Uhler, 29 September 1877, M.C.Z. Archives.

50. Hagen, "Monograph," p. 12.

51. P. Brocchi, "Recherches sur les organes génitaux mâles."

52. David Barr, Royal Ontario Museum, personal communication.

53. H. H. Hobbs, Jr., "The current status of the crayfishes listed by Girard"; idem, *Crayfishes (Astacidae) of North and Middle America.*

54. Walsh to Hagen, 17 October 1864, M.C.Z. Archives.

55. Ibid.

56. J. L. LeConte, "Descriptions of new species of Astacus from Georgia," p. 400.

57. Hagen, "Monograph," p. 18.

58. E. A. Andrews, "Conjugation in the crayfish."

59. J. K. Waage, "Sperm competition."

60. B. D. Walsh, "North American Neuroptera," 2:217. I have substituted the words "male" and "female" for symbols in the original.

61. M. Ridley, *Problems of Evolution,* pp. 102–14.

62. Hagen, "Monograph," p. 38.

63. Ibid., p. 71.

64. Ibid., p. 50.

65. Darwin, *Origin of Species,* pp. 296–97.

66. Ibid., pp. 44–47.

67. Ibid., p. 49.

68. Ibid., p. 485.

69. Ibid., p. 484.

70. Ibid., p. 485.

71. Ibid., p. 484.

72. H. L. Clark, "So-called species and subspecies"; J. A. Allen, "So-called species and subspecies"; Stresemann, *Ornithology from Aristotle to the Present,* p. 247.

73. Winsor, *Starfish, Jellyfish,* pp. 82–86.

74. Owen, *Lectures on Invertebrate Animals.*

75. Hagen, "Monograph," p. 30.

76. Hobbs, "Notes on the Blandingii section."

77. Hobbs, "Adaptations and convergence."

78. B. Silliman, "On the Mammoth Cave of Kentucky."

79. L. Agassiz, "Observations on the blind fish of the Mammoth Cave."

80. Ibid., p. 128.

81. J. Wyman, [Brief account]; idem, "On the eye and the organ of hearing in the blind fishes." Owen contradicted Wyman's result (Dexter, "Sir Richard Owen's interpretation of optic lobes in blind fishes.")

82. Darwin, *Origin of Species,* p. 137.

83. Shaler, *Autobiography,* p. 47; *Annual Report* for 1860, p. 37.

84. F. W. Putnam and Packard, "The Mammoth Cave and its inhabitants"; Packard, "On the crustaceans and insects," pp. 744–61. For Cooke's presence, see Packard, "On the crustaceans and insects," p. 747; for E. D. Cope's presence, see his "Wyandotte Cave," p. 406.

85. A thorough survey of the history of this species is given in Hobbs and T. C. Barr, Jr., *Origins and Affinities of the Troglobitic Crayfishes . . . Orconectes.*

86. Hagen, "Monograph," p. 55–56.

87. Packard, pp. 750–51.

88. Cope, "Wyandotte Cave." Hobbs and Barr show that it was probably a nearby cave rather than the Wyandotte Cave itself.

89. Hobbs and Barr, *Origins and Affinities . . . Orconectes,* p. 36.
90. Cope, "Wyandotte Cave," p. 410.
91. Hagen, "Blind crayfish."
92. W. Faxon, "A revision of the Astacidae," p. 43.
93. Hagen, "Blind crayfish," p. 494–95.
94. Cope and Packard, "The fauna of the Nickajack Cave," p. 881.
95. Hobbs and Barr, "The origins and affinities of the troglobitic crayfishes . . . *Cambarus*."
96. Cope and Packard, "The fauna of the Nickajack Cave," p. 879.
97. Packard, "The effect of cave life on animals," p. 393.
98. Darwin, *Origin of Species,* p. 139.
99. Ibid., p. 420.
100. Ibid., p. 427.
101. Packard to Faxon, 25 February 1883, M.C.Z. Archives.
102. Packard, "The cave-fauna of North America," p. 123.
103. Faxon, "Revision of the Astacidae," p. 42.
104. Ibid., pp. 45–46; L. B. Holthius, "On the status of two allegedly European crayfishes."
105. Faxon, "Revision of the Astacidae," p. 46.
106. Faxon, "Notes on the crayfishes in the United States National Museum."
107. Hagen, "Monograph," p. 74.
108. Winsor, "The impact of Darwinism upon the Linnaean enterprise, with special reference to the work of T. H. Huxley."
109. Ibid.
110. T. H. Huxley, *An Introduction to the Study of Zoology,* p. 319.
111. T. H. Huxley, "On the classification and the distribution of the crayfishes."
112. T. H. Huxley, *An Introduction to the Study of Zoology,* p. 346.
113. S. Garman, "Cave animals from southwestern Missouri."
114. *Annual Report* for 1882–83, p. 24.
115. Huxley to S. F. Baird, 28 June and 8 August [1882], "Exposition Records of the Smithsonian Institution and the United States National Museum, 1875–1916," Smithsonian Institution Archives, kindly located for me by staff of the archives.

Chapter Five

1. L. Agassiz to Lyman, 4 May 1866, M.C.Z. Archives.
2. A. E. Günther, "General history of the department of zoology," p. 2.
3. J. E. Gray, "On museums, their use and improvement."
4. W. H. Flower, "Modern museums."
5. Günther "General history of the department of zoology," p. 2
6. Owen, "On a national museum of natural history."
7. Smithsonian Institution, *Annual Report of the Board of Regents of the Smithsonian Institution* for 1856, p. 43.

8. L. V. Coleman, *The Museum in America,* 2:249.
9. *Annual Report* for 1877–78, p. 5.
10. A. R. Wallace, "American museums: The Museum of Comparative Zoology, Harvard University."
11. S. Henshaw, in *Annual Report* for 1906–7, p. 3; A. B. Meyer, "Studies of the museums and kindred institutions of New York City, Albany, Buffalo, and Chicago," p. 325; Coleman, *Museum in America,* 2:249.
12. *Annual Report* for 1864, p. 13.
13. Ibid.
14. L. Agassiz, [speech to visiting committee]; idem, "Arrangement of a museum."
15. *Annual Report* for 1863, p. 36 (italics mine).
16. L. Agassiz, "Des Musées d'Histoire naturelle."
17. E. C. Agassiz, *Louis Agassiz,* 2:562.
18. *Annual Report* for 1867, p. 4.
19. *Annual Report* for 1869, p. 8.
20. *Annual Report* for 1869, p. 11.
21. *Annual Report* for 1867, p. 4.
22. *Annual Report* for 1868, p. 4.
23. *Annual Report* for 1865, p. 12.
24. *Annual Report* for 1868, p. 30.
25. W. M. Whitehill, *The East India Marine Society,* p. 66.
26. L. Agassiz to H. A. Ward, 13 December 1870, University of Rochester Archives.
27. Ibid., 22 November 1870.
28. Ibid., 10 July 1871.
29. *Annual Report* for 1872, p. 7.
30. *Annual Report* for 1872, p. 5.
31. *Annual Report* for 1871, pp. 21–22.
32. J. M. Kennedy, "Philanthropy and science."
33. A. Agassiz to Lyman, 6 August 1870, M.C.Z. Archives.
34. "Dedication of the Museum of the Society of Natural History," *Boston Daily Advertiser,* 3 June 1864; Bouvé, "Historical sketch of the Boston Society of Natural History."
35. Hyatt, [Annual Report of the Custodian], 1871.
36. *Annual Report* for 1872, p. 6.
37. A. Agassiz to Lyman, 21 February 1862, M.C.Z. Archives.
38. Lyman to A. Agassiz, n.d. [bAg 584.10.1(31–40)], M.C.Z. Archives.
39. College Papers, Harvard University Archives, ser. 2, 26:189.
40. *Annual Report* for 1860, pp. 15–17.
41. A. Agassiz to Lyman, [March 1866], Lyman Family Papers, Massachusetts Historical Society, Boston.
42. Lyman to A. Agassiz, 27 August 1872, M.C.Z. Archives.
43. G. R. Agassiz, *Letters and Recollections,* pp. 93–94.

44. Eliot to Lyman, 3 March 1873, Lyman Family Papers, Massachusetts Historical Society.

45. A. Agassiz to T. H. Huxley, 14 July 1874, Imperial College Archives, London.

46. A. Agassiz to T. H. Huxley, 8 November 1882, and 20 March 1883, Imperial College Archives, London.

47. *Annual Report* for 1861, pp. 16–17.

48. *Annual Report* for 1866, p. 6.

49. G. R. Agassiz, *Letters and Recollections,* pp. 24–25.

50. A. Agassiz to Lyman, [March 1866], Lyman Family Papers, Massachusetts Historical Society.

51. Lyman to A. Agassiz, 12 March 1866, M.C.Z. Archives.

52. A. Agassiz to Lyman, [March 1866], Lyman Family Papers, Massachusetts Historical Society.

53. Tharp, *Adventurous Alliance;* J. W. Dean, "Descendants of the Rev. Daniel Rogers."

54. A. Forbes and J. W. Greene, *The Rich Men of Massachusetts.*

55. W. B. Gates, Jr., *Michigan Copper and Boston Dollars,* p. 2.

56. L. Agassiz, "Geological relations of the various copper deposits of Lake Superior."

57. G. R. Agassiz, *Letters and Recollections,* pp. 57–85; letters between A. Agassiz and Lyman, M.C.Z. Archives; Lyman, Private Journal, Massachusetts Historical Society.

58. Lyman, Private Journal (3 and 25 March, 4 April, 15 and 16 July, 10 September 1867); G. P. Shaw to Lyman, 8 April 1967; Lyman to Howland Shaw, 4 March 1867; Francis George Shaw to Lyman, 13 March 1867), Massachusetts Historical Society, Boston.

59. Gates, *Michigan Copper and Boston Dollars,* pp. 45–46.

60. L. Lankton, *Cradle to Grave.*

61. A. Agassiz to Lyman, 26 April 1872, M.C.Z. Archives.

62. A. Agassiz, "Biographical memoir of Louis François de Pourtalès"; *Dictionnaire Historique & Bibliographique de la Suisse.*

63. L. Agassiz to S. D. Schlesinger, 6 November 1872, M.C.Z. Archives.

64. Lyman, Private Journal, 23 December 1873, Massachusetts Historical Society, Boston.

65. G. R. Agassiz, *Letters and Recollections,* p. 127.

66. Lyman to Putnam, 23 July 1874, Harvard University Archives.

67. A. Agassiz, "The Anderson School," *Boston Daily Advertiser,* 16 October 1875.

68. *Annual Report* for 1875, p. 42.

69. G. B. Emerson, [Address on Louis Agassiz], p. 158.

70. *Annual Report* for 1868, pp. 6–7.

71. L. Agassiz to Uhler, 6 April 1864, M.C.Z. Archives.

72. W. Lawrence, *Life of Amos A. Lawrence,* p. 160.

73. *Annual Report* for 1866, p. 17.
74. Lyman, "Recollections of Agassiz," p. 227.
75. *Annual Report* for 1873, p. 8.
76. *Annual Report* for 1877–78, p. 5.
77. *Annual Report* for 1875, pp. 12–13.
78. Ibid., pp. 6–7.
79. Ward, *Henry Augustus;* S. G. Kohlstedt, "Henry A. Ward"; catalogue of Ward's Scientific Establishment, n.d., University of Rochester Archives.
80. A. Agassiz to H. A. Ward, 18 April 1884, University of Rochester Archives.
81. *Annual Report* for 1893–94, p. 12.
82. A. Agassiz to Eliot, 16 May 1882, Eliot Papers, Harvard University Archives.
83. A. Agassiz to T. H. Huxley, 8 November 1882, Imperial College, London.
84. A. Agassiz to Eliot, 8 August 1884, Eliot Papers, Harvard University Archives.

Chapter Six

1. Winsor, *Starfish, Jellyfish,* pp. 142–68.
2. G. R. Agassiz, *Letters and Recollections,* p. 97.
3. A. Agassiz to Lyman, 26 January 1870, M.C.Z. Archives.
4. Lurie, *Louis Agassiz,* pp. 111–13.
5. G. R. Agassiz, *Letters and Recollections,* pp. 12–13.
6. Lurie, *Louis Agassiz,* p. 383; G. R. Agassiz, *Letters and Recollections,* pp. 122, 128; L. Agassiz, "Evolution and the permanence of type"; A. Agassiz, "Revision of the Echini."
7. A. Agassiz, "Revision of the Echini," pp. 753–54.
8. G. R. Agassiz, *Letters and Recollections,* pp. 119–20.
9. A. Agassiz, "Revision of the Echini," p. 16.
10. Ibid., p. 17.
11. Ibid., pp. 17–18.
12. A. Agassiz, "Paleontological and embryological development," pp. 408–9.
13. Ibid., p. 411.
14. Ibid., p. 412.
15. Ibid., p. 413.
16. Ibid., p. 413.
17. I. Todhunter, *Algebra for the Use of Colleges and Schools.*
18. A. Agassiz, *Report of the Echinoidea . . . Challenger,* pp. 18–24.
19. A. Agassiz, "Paleontological and embryological development," p. 395.
20. Ibid., p. 414.
21. Ibid., p. 396.
22. Emil Du Bois-Reymond, "The limits of our knowledge of nature," p. 32.
23. A. Agassiz, "Paleontological and embryological development," p. 414.

24. [J. Michels], [on the Boston AAAS meeting].
25. Ibid.
26. C[ope], "Editors' Table."
27. Ibid., p. 728.
28. Ibid., p. 727.
29. H. W. MacIntosh, "The echinoids of the 'Challenger,' " p. 41.
30. G. R. Agassiz, *Letters and Recollections,* pp. 162–63.
31. Ibid., pp. 163–64.
32. F. A. Bather, "Biological classification."
33. W. M. Fitch, "On the problem of discovering the most parsimonious tree"; R. L. Graham and L. R. Foulds, "Unlikelihood that minimal phylogenies for a realistic biological study can be constructed in reasonable computation time." Literature in this area could be approached from E. O. Wiley, *Phylogenetics.*

Chapter Seven

1. *Annual Report* for 1871, p. 8.
2. Shaler, *Autobiography,* p. 250.
3. Ibid., p. 227.
4. Ibid., p. 249.
5. Ibid., p. 248.
6. H. James, *Charles W. Eliot,* 1:184–235; H. Hawkins, *Between Harvard and America.*
7. James, *Charles W. Eliot,* 1:194.
8. Eliot, *The Man and His Beliefs,* 1:26.
9. Lyman to Eliot, 3 August 1870, Harvard University Archives.
10. Ibid.
11. *Annual Report* for 1871, p. 29.
12. *Annual Report* for 1872, p. 29.
13. Wilder, "Louis Agassiz, teacher," p. 604.
14. *Annual Report* for 1872, p. 29.
15. *Annual Report* for 1870, p. 2.
16. Ibid.
17. Both letters are excerpted in Shaler, *Autobiography,* p. 251.
18. D. N. Livingstone, *Nathaniel Southgate Shaler,* p. 254.
19. Shaler, "Notice concerning the Summer School." December 1898, Eliot Papers, Harvard University Archives; *Annual Report* for 1872, pp. 30–31; Livingstone, *Nathaniel Southgate Shaler,* p. 255.
20. A. Agassiz to Lyman, 5 August 1872, M.C.Z. Archives.
21. Lyman to A. Agassiz, 27 August 1872, M.C.Z. Archives.
22. *Annual Report* for 1876, p. 9.
23. *Annual Report* for 1877–78, pp. 11–12; for 1878–79, p. 21; for 1879–80, pp. 6–7; for 1880–81, p. 5.
24. *Annual Report* for 1875, p. 9.
25. *Annual Report* for 1876, pp. 6–8.

26. E. L. Mark, "[Talk given to] Shop Club," 13 January 1898, Harvard University Archives.
27. *Annual Report* for 1879–80, pp. 14–17.
28. Ibid., p. 15.
29. *Annual Report* for 1875, pp. 10–12.
30. P. R. Cutright, *Theodore Roosevelt,* p. 127.
31. Ibid.
32. Mark, "Zoology, 1847–1921," p. 386.
33. A. Agassiz to Eliot, 25 May 1886, Eliot Papers, Harvard University Archives.
34. Parker, *The World Expands,* p. 52.
35. Ibid., p. 65.
36. Parker, *Mark Anniversary Volume.*
37. Jordan, *Days of a Man,* 1:118.
38. C. Eigenmann, "On the egg membranes and micropyle of some osseous fishes."
39. Hagen to C. F. Dunbar, February 1880, M.C.Z. Archives.
40. R. T. Jackson, "Phylogeny of the Echini," p. 9; C. E. Beecher, *Studies in Evolution.*
41. A. G. Mayer, "Alpheus Hyatt," p. 137.
42. Barbour to L. Stejneger, 27 February 1905, Smithsonian Institution Archives, Washington.
43. Barbour to Stejneger, 29 May 1911, Smithsonian Institution Archives, Washington.
44. K. Sterling, *Last of the Naturalists.*
45. Ibid., p. 200.
46. Merriam, "Roosevelt," p. 182.
47. Merriam, "Biology in our colleges," pp. 352–53.
48. Harvard University, "Report of the Committee on Zoölogy," p. 196.
49. J. Christian Bay, "A plea for a fair valuation."
50. J. P. Campbell, "Biological teaching in the colleges of the United States," idem, "Editorial"; C. MacMillan, "Open letters"; idem, "On the emergence of a sham biology in America"; idem, "On methods of defending the existence of a sham biology in America"; F. H. Herrick, "On the teaching of biology"; J. Wood, "What is biology?"; H. F. Nachtrieb, "Sham biology"; H. T. Fernald, "Teaching of biology."
51. E. Mendelsohn, "The emergence of science as a profession in nineteenth-century Europe," p. 4. An introduction to the literature on professionalization of American science is given by Kohlstedt, "Institutional history."
52. A. Agassiz to H. P. Walcott, 16 September 1894, private collection of Dr. Freddy Homburger, Cambridge, Massachusetts.
53. A. Agassiz to Eliot, 25 May 1899, Eliot Papers, Harvard University Archives.
54. *Annual Report* for 1897–98, p. 7.

Chapter Eight

1. *Annual Report* for 1875, p. 44.
2. Ibid.
3. *Annual Report* for 1875, pp. 5–6.
4. A. Agassiz to Eliot, 17 September 1884, Harvard University Archives.
5. *Annual Report* for 1875, p. 13.
6. Ibid.
7. W. H. Dall, *Spencer Fullerton Baird,* pp. 388, 401, 424–26; D. Allard "The Fishing Commission Laboratory."
8. Lyman, Private Notebooks, [August or September] 1871, Massachusetts Historical Society, Boston.
9. A. Agassiz to Mark, 2 January 1894, M.C.Z. Archives.
10. *Annual Report* for 1890–91, p. 3; for 1891–92, p. 11.
11. *Annual Report* for 1891–92, pp. 11–12; for 1894–95, pp. 6–7.
12. *Annual Report* for 1891–92, pp. 12–13, 18; for 1894–95, pp. 4–7.
13. A. Agassiz to E. H. Clark, 2 May 1898, privately owned.
14. *Annual Report* for 1881–82, p. 5; for 1882–83, p. 7; for 1883–84, p. 4; for 1884–85, p. 10; for 1885–86, p. 4; for 1886–87, pp. 4–5.
15. Fewkes to A. Agassiz, 1 August 1888, M.C.Z. Archives.
16. A. Agassiz to Carpenter, 13 March 1889, M.C.Z. Archives.
17. Fewkes, "A preliminary notice of a stalked bryozoan"; idem, "Contribution to Passamaquoddy folk-lore," p. 258n.
18. A. Agassiz to Fewkes, 9 June 1889, M.C.Z. Archives.
19. Fewkes, "An aid to the collector of the Coelenterata and Echinodermata of New England." The reference librarian of the James Duncan Phillips Library of the Essex Institute kindly checked for me the minutes of the Publication Committee of the Essex Institute.
20. J. F. Hunt to A. Agassiz, 12 and 22 September, and 1 October 1891, 8 March 1892; Fewkes to A. Agassiz, 21 and 22 September 1891; A. Agassiz to Fewkes, 10 and 24 September 1891, M.C.Z. Archives.
21. A. Agassiz to Eliot, 4 September 1891, Harvard University Archives.
22. A. Agassiz to Fewkes, 6 October 1891, M.C.Z. Archives.
23. Fewkes to A. Agassiz, 22 September 1891, M.C.Z. Archives.
24. A. Agassiz to Eliot, 4 November 1891, Harvard University Archives.
25. G. R. Agassiz, *Letters and Recollections,* p. 208; *Annual Report* for 1882–83, p. 7; for 1883–84, p. 4.
26. A. Agassiz to Whitman, 13 July 1892, Letterbooks, vol. 2, M.C.Z. Archives.
27. A. Agassiz to Eliot, August 1906, Eliot Papers, Harvard University Archives.
28. A. Agassiz to C. B. Davenport, 21 May 1899, American Philosophical Society, Philadelphia.

29. Mark to Davenport, 4 January 1900, American Philosophical Society, Philadelphia.

30. A. Agassiz, "Testimony before the Joint Commission . . . William B. Allison, Chairman."

Chapter Nine

1. Gates, *Michigan Copper.*

2. *Annual Report* for 1883–84, p. 6.

3. *Annual Report* for 1883–84, p. 9.

4. List of persons engaged at M.C.Z., Eliot Papers, Harvard University Archives.

5. O. W. Holmes, 10 January 1874, Eliot Papers, Harvard University Archives.

6. *Annual Report* for 1883–84, p. 9.

7. *Annual Report* for 1883–84, p. 6.

8. *Annual Report* for 1883–84, p. 8.

9. Allen, "Autobiographical notes," p. 33.

10. *Annual Report* for 1891–92, p. 6.

11. *Annual Report* for 1901–2, p. 5.

12. Jordan, *The Days of a Man,* 1:325.

13. E. H. Clark [for A. Agassiz], 13 September 1887, M.C.Z. Letterbooks, 10:221r, M.C.Z. Archives.

14. Carl H. Eigenmann to his mother, 11 November 1887, Lilly Library, Indiana University, Bloomington, Indiana.

15. Barbour, *Naturalist at Large,* pp. 136–37; Garman to A. Agassiz, August 1882, M.C.Z. Archives.

16. C. L. Hubbs, "History of ichthyology," p. 56.

17. R. S. Eigenmann to her mother, 23 August 1887, Eigenmann MSS, Lilly Library, Indiana University, Bloomington, Indiana.

18. C. H. Eigenmann and R. S. Eigenmann, "Revision of the edentulous genera of Curimatinae," p. 409; idem, "Catalogue of the fresh-water fishes of South America," p. 12.

19. Eigenmann and Eigenmann, "Revision of the edentulous genera of Curimatinae," p. 409.

20. Eigenmann and Eigenmann, "A revision of the South American Nematognathi," p. 3.

21. Hagen to P. P. Calvert, [postmarked 5 May 1890]; Henshaw to Calvert, 5 December 1890, 2 March, 11 July, and 4 November 1891, 17 May 1893, Academy of Natural Sciences, Philadelphia; *Annual Report* for 1890–91, pp. 4–5.

22. *Annual Report* for 1890–91, p. 8.

23. *Annual Report* for 1891–92, p. 6; for 1892–93, p. 6; for 1894–95, p. 8.

24. C. H. Eigenmann to Davenport, 3 January 1895, Davenport Papers, American Philosophical Society, Philadelphia.

25. C. H. Eigenmann to A. Agassiz, 9 December 1889, M.C.Z. Archives.
26. C. H. Eigenmann to Davenport, 3 January 1895, Davenport Papers, American Philosophical Society, Philadelphia.
27. Hagen to A. Agassiz, 23 November 1883, M.C.Z. Archives.
28. *Annual Report* for 1897–98, pp. 43–44.
29. G. R. Agassiz, *Letters and Recollections,* p. 89.
30. A. Agassiz, [single printed sheet] 1 August 1898, Smithsonian Institution Archives, RU 189, assistant secretary in charge of the USNM, Incoming Correspondence, 1860–1908, Box 2, folder 3.
31. Woodworth to A. Agassiz, 30 June 1900, M.C.Z. Archives.
32. Harvard College, Class of 1888, *Secretary's Report* no. 7, June 1913, p. 165.
33. A. Agassiz to E. H. Clark, 3 January 1902. Privately owned by the heirs of Alexander Agassiz.
34. G. L. Goodale to C. W. Eliot, 18 July 1904, Harvard University Archives.
35. H. P. Walcott to C. W. Eliot, 12 July 1904, Harvard University Archives.
36. Woodworth to A. Agassiz, 3 July 1905, M.C.Z. Archives.
37. T. Barbour to H. L. Clark, 24 December 1935, M.C.Z. Archives.
38. C. H. Eigenmann to the sons of Nathaniel Thayer, 16 July 1906, M.C.Z. Archives.
39. C. H. Eigenmann to Henshaw, 26 July 1907, M.C.Z. Archives.
40. C. H. Eigenmann to Henshaw, 7 March 1911, M.C.Z. Archives.
41. Marginalia on letter of Eigenmann to John E. Thayer, 7 March 1911, M.C.Z. Archives.
42. C. B. Davenport, *Inheritance in Poultry.*
43. C. H. Eigenmann to Davenport, 7 April 1907, Davenport Papers, American Philosophical Society, Philadelphia.
44. There is an extensive correspondence between the Eigenmanns and Henshaw in the M.C.Z. Archives; this quote is from a letter of 17 November 1907.
45. C. H. Eigenmann to Davenport, 17 May 1908, Davenport Papers, American Philosophical Society, Philadelphia.
46. C. H. Eigenmann, "Adaptation," p. 193.
47. C. H. Eigenmann, "American Characidae," p. 296.
48. Ibid., pp. 199–200.
49. C. H. Eigenmann, "Adaptation," p. 196.
50. Ibid., p. 208.
51. C. H. Eigenmann, "American Characidae," p. 46.
52. Ibid., p. 47.
53. Ibid.
54. Ibid., pp. 47–48.
55. Ibid., p. 38.
56. G. S. Myers, "Amoenitates biologicae: The influence of Louis Agassiz on the ichthyology of Brazil," p. 132.

Chapter Ten

1. H. L. Clark to Henshaw, 20 May 1905, M.C.Z. Archives.

2. H. L. Clark to Henshaw, 8 June, 26 August, 2 September, and 11 September 1905; H. L. Clark to A. Agassiz, 4 September 1905, M.C.Z. Archives.

3. H. L. Clark to A. Agassiz, 10 October 1901, 21 April 1902, 15 and 30 October 1903; A. Agassiz to H. L. Clark, June 1903, M.C.Z. Archives.

4. A. Agassiz, "Revision of the Echini," p. 662, also published as "The homologies of pedicellariae," *American Naturalist* 7 (1873), 398–406.

5. Ibid., pp. 667–70.

6. Ibid., p. 670.

7. A. Agassiz to H. L. Clark, June 1903, M.C.Z. Archives.

8. H. L. Clark to A. Agassiz, 4 September 1905, M.C.Z. Archives.

9. T. Mortensen to H. L. Clark, 12 January 1906, M.C.Z. Archives.

10. Günther, *A Century of Zoology*, p. 411.

11. Mortensen to H. L. Clark, 28 November 1911, M.C.Z. Archives.

12. G. R. Agassiz, *Letters and Recollections*, pp. 30–31.

13. H. L. Clark to A. Agassiz, 13 February 1904, M.C.Z. Archives.

14. A. Agassiz, "Reports on an exploration of the West Coasts of Mexico," p. x.

15. Mortensen to H. L. Clark, 27 October 1915, M.C.Z. Archives.

16. Günther, *A Century of Zoology*, pp. 406–14.

17. H. J. van Cleave, "An index to the opinions rendered."

18. Ludwig Döderlein, "Die Echinoiden der deutschen Tiefsee-Expedition"; I. C. H. de Meijere, "*Echinoidea der Siboga-Expedition.*"

19. A. Agassiz, "Revision of the Echini," p. 423.

20. H. L. Clark and A. Agassiz, "Hawaiian and other Pacific Echini . . . The Cidaridea," p. 10.

21. Mortensen to H. L. Clark, 26 February 1910, M.C.Z. Archives.

22. H. L. Clark to A. Agassiz, 21 September 1908, M.C.Z. Archives.

23. A. M. D'yakonov, *Echinoidea*, pp. 93–94.

24. H. L. Clark, [review of Jackson, "Phylogeny of the Echini"].

25. S. J. Gould, *Ontogeny and Phylogeny.*

26. H. L Clark, "The classification of the regular Echini."

27. T. Mortensen to H. L. Clark, 19 March 1912, M.C.Z. Archives.

28. Ibid.

29. Mortensen to H. L. Clark, 11 May 1915, M.C.Z. Archives.

30. Mortensen to H. L. Clark, 27 October 1915, M.C.Z. Archives.

31. E. Deichmann and F. A. Chace, Jr., "Obituary of Hubert Lyman Clark."

Chapter Eleven

1. H. B. Bigelow, "Thomas Barbour"; T. Barbour, *Naturalist at Large;* idem, *A Naturalist's Scrapbook.*

2. T. Barbour, *Naturalist at Large*, p. 22.

3. T. Barbour to E. W. Nelson, 1 November 1932, Barbour Papers, Harvard University Archives.

4. T. Barbour, *A Naturalist's Scrapbook,* p. 21.

5. T. Barbour, "Notes on a zoological collecting trip to Dutch New Guinea"; idem, "Further notes on Dutch New Guinea"; idem, "Notes on Burma."

6. Darwin, *Origin of Species,* p. 407.

7. Wallace, *The Malay Archipelago* 1:10.

8. Ibid. p. 110.

9. T. Barbour, *Naturalist at Large,* p. 41.

10. A. Agassiz, *Three Cruises of the "Blake."*

11. *Annual Report* for 1878–79, pp. 3, 15.

12. R. T. Hill, "Notes on the geology of the island of Cuba"; idem, "The geology and physical geography of Jamaica."

13. C. Schuchert, *Historical Geology of the Antillean-Caribbean Region,* p. 2.

14. T. Barbour and R. Barbour, *Letters Written While on a Collecting Trip,* p. 216.

15. A. Agassiz to Henshaw, 30 May and 23 October 1906, M.C.Z. Archives; G. R. Agassiz, *Letters and Recollections,* p. 436.

16. *Proceedings of the 7th International Zoölogical Congress,* pp. 855–69.

17. Ibid., p. 857.

18. T. Barbour to Stejneger, 27 February 1905, Smithsonian Institution Archives, Washington.

19. T. Barbour to A. Agassiz, 22 August n.d.; M.C.Z. Archives; T. Barbour to R. F. Scharff, Harvard University Archives.

20. T. Barbour, "Notes on the herpetology of Jamaica"; idem, "A contribution to the zoögeography of the East Indian Islands"; idem, "A contribution to the zoögeography of the West Indies."

21. T. Barbour, "A contribution to the zoögeography of the West Indies," p. 203.

22. G. Baur, "On the origin of the Galapagos Islands"; idem, "New observations on the origin of the Galapagos Islands"; H. A. Pilsbry, "The genesis of mid-Pacific faunas."

23. P. N. van Kampen, "The zoogeography of the East Indian Archipelago."

24. T. Barbour, "Notes on the herpetology of Jamaica," p. 277.

25. Matthew, "Climate and evolution," p. 207.

26. T. Barbour and G. K. Noble, "Ameiva," p. 423.

27. Ibid., p. 419.

28. Ibid., p. 424.

29. Ibid., pp. 420–21.

30. Ibid., p. 421.

31. T. Barbour, "Some remarks upon Matthew's 'Climate and evolution.' "

32. Matthew, "Supplementary note."

33. Matthew to T. Barbour, 18 and 26 November 1915, Barbour Papers, Harvard University Archives.

34. "It must be about ten years ago that Dr. Walcott . . . spoke of you . . . ," W. M. Davis to T. Barbour, 25 December 1927. Barbour Papers, Harvard University Archives.

35. T. Barbour to A. L. Lowell, 24 December [1926], Lowell Papers, Harvard University Archives.

36. G. R. Agassiz to Lowell, 14 June 1926; Lowell to A. Agassiz, 15 June 1926; Lowell to Henshaw, 25 June 1927, Lowell Papers, Harvard University Archives.

37. A. S. Romer, "Thomas Barbour," p. 229.

38. M. A. Evans and H. E. Evans, *William Morton Wheeler.*

39. T. Barbour, *Naturalist at Large,* pp. 299–310 .

40. P. J. Darlington, Jr., "The origin of the fauna of the Greater Antilles."

41. F. M. C[arpenter], "Remarks for the memorial service [of P. J. Darlington]," M.C.Z. Archives.

42. Darlington, *Zoogeography.*

43. Mayr, "The birds of Timor and Sumba," p. 618.

44. G. G. Simpson, *Concession to the Improbable,* pp. 33–34, 271–72.

45. L. Croizat, *Panbiogeography* 1:xi–xii.

46. Ibid., p. 636n.

47. "Symposium: World perspectives in biogeography"; D. E. Rosen, "Vicariant patterns and historical explanation in biogeography."

48. Rosen, "A vicariance model of Caribbean biogeography"; S. B. Hedges, "Caribbean biogeography"; J. C. Briggs, "Freshwater fishes and biogeography"; M. Rauchenberger, "Historical biogeography of poeciliid fishes."

49. F. J. Sulloway, "Darwin's conversion."

50. L. Huxley, *Life and Letters of Sir Joseph Dalton Hooker,* 2:299.

51. Ibid., pp. 237–39, 416–17.

52. T. Barbour to George H. Lyman, 8 October 1934, Harvard University Archives.

Concluding Remarks

1. Günther, *A Century of Zoology;* W. T. Stearn, *The Natural History Museum,* pp. 138–42.

2. G. B. Goode, "The museums of the future," p. 436.

3. Darwin, *Autobiography,* p. 24.

4. C. Groeben, ed., *Charles Darwin, 1809–1882, Anton Dohrn, 1840–1909: Correspondence,* p. 67.

5. Mayr, "The role of systematics in biology," p. 417.

6. P. J. Pauly, "The appearance of academic biology."

7. C. Limoges, "The development of the Muséum d'Histoire Naturelle," pp. 211–240, 234–235.

8. L. Agassiz to Uhler, 6 April 1864, M.C.Z. Archives.

Bibliography

Agassiz, Alexander. "Louis François de Pourtalès: 1824–1880." *Biographical Memoirs of the National Academy of Sciences* 5 (1905): 78–89.

———. "The homologies of pedicellariae." *American Naturalist* 7 (1873): 398–406.

———. "Paleontological and embryological development." *Proceedings of the American Association for the Advancement of Science* 29 (1880): 389–414. Also *Science* 1 (1880): 142–49; and *American Journal of Science* 20 (1880): 294–302, 375–89.

———. *Report of the Echinoidea, Dredged by H. M. S. Challenger during the Years 1873–1876*. Challenger Expedition. Report on the Scientific Results of the Voyage of H. M. S. *Challenger*. Vol. 3, no. 1, ed. Sir Charles Wyville Thomson. London, 1881.

———. "Reports on an exploration of the West Coasts of Mexico, Central and South America, and off the Galapagos Islands, in charge of Alexander Agassiz, by the U.S. Fish Commission steamer 'Albatross,' during 1891, Lieut. Commander Z. L. Tanner, U.S.N., commanding: 32 The Panamic Deep Sea Echini." *Memoirs of the Museum of Comparative Zoology* 31 (1904): 1–243.

———. "Revision of the Echini." *Illustrated Catalogue of the Museum of Comparative Zoology at Harvard College*, 7 (1872–74).

———. "Testimony before the Joint Commission to consider the present organizations of the Signal Service, Geological Survey, Coast and Geodetic Survey . . . William B. Allison, Chairman." U.S. Congress, Senate. 49th Cong., 1st sess., 1886. Misc. Doc. 82, pp. 1014–15.

———. "Three cruises of the United States Coast and Geodetic Survey steamer 'Blake' in the Gulf of Mexico, in the Caribbean Sea, and along the Atlantic Coast of the United States, from 1877 to 1880." *Bulletin of the Museum of Comparative Zoology* 14 and 15 (1888). Also published as *A Contribution to American Thalassography: Three Cruises of the . . . "Blake" . . .* 2 vols. Boston: Houghton Mifflin, 1888.

Agassiz, Alexander, and Charles Otis Whitman. "The development of osseous fishes. I. The pelagic stages of young fishes." *Memoirs of the Museum of Comparative Zoology* 14 (1885): 1–56.

———. "The development of osseous fishes. II. The preembryonic stages of development. Part first. The history of the egg from fertilization to cleavage." *Memoirs of the Museum of Comparative Zoology* 14 (1889): 1–40.

Agassiz, Elizabeth Cary. *Louis Agassiz: His Life and Correspondence.* 2 vols. Boston, 1886.

Agassiz, Elizabeth Cary, and Alexander Agassiz. *Seaside Studies in Natural History: Marine Animals of Massachusetts Bay. Radiates.* Boston, 1865.

Agassiz, George Russell. *The Founding of the Calumet & Hecla Mine: 1866–1916.* Privately printed, 1916.

———. *Letters and Recollections of Alexander Agassiz.* Boston: Houghton Mifflin, 1913.

Agassiz, Louis. "Arrangement of a museum." *Proceedings of the Boston Society of Natural History* 7 (1859): 191–92.

———. "Communication." *Proceedings of the Academy of Natural Sciences of Philadelphia* 6 (1853): 375.

———. *Contributions to the Natural History of the United States of America.* 4 vols. Boston, 1857–62.

———. "Des Musées d'histoire naturelle" [lecture delivered 11 July 1860], *Bibliothèque Universelle et Revue Suisse* 47 (1862): 527–40.

———. *Essay on Classification.* Edited by Edward Lurie. 1859. Reprint. Cambridge, Mass.: Harvard University Press, 1962.

———. "Evolution and the permanence of type." *Atlantic Monthly* 33 (1874): 92–101.

———. "Geological relations of the various copper deposits of Lake Superior." In *Lake Superior: . . .*, edited by Louis Agassiz et al., pp. 427–28. Boston, 1850.

———. "Observations on the blind fish of the Mammoth Cave." *American Journal of Science* 11, ser. 2 (1851): 127–28.

———. "On the Origin of Species." *American Journal of Science* 30 (1860): 151. Also in his *Contributions* 3: 97.

———. [Speech to visiting committee]. *American Journal of Science* 27 (1859): 297–99.

———. "The study of nature." *Boston Weekly Courier* 8 (26 March 1859).

Agassiz, Louis, and Elizabeth Cary Agassiz. *A Journey in Brazil.* Boston: Ticknor & Fields, 1868. Reprint. New York: Praeger, 1969.

Agassiz, Louis, J. Elliot Cabot, et al. *Lake Superior: Its Physical Character, Vegetation, and Animals, Compared with Those of Other and Similar Regions.* Boston, 1850.

Allard, Dean C. "The Fish Commission laboratory and its influence on the founding of the Marine Biological Laboratory." *Journal of the History of Biology* 23 (1990): 251–70.

Allen, Joel Asaph. *Autobiographical Notes and Bibliography.* New York: American Museum of Natural History, 1916.

———. "Catalogue of the mammals of Massachusetts: with a critical revision of the species." *Bulletin of the Museum of Comparative Zoology* 1 (1869): 143–252.

———. "On geographical variation in color among North American squirrels with a list of the species and varieties of the American Sciuridae occurring north of Mexico." *Proceedings of the Boston Society of Natural History* 16 (1874): 276–94.

———. [Geographical variation in mammals and birds.] *Proceedings of the Boston Society of Natural History* 15 (1872): 156–59, 212–19.

———. "The influence of physical conditions in the genesis of species." *Radical Review* 1 (1877): 108–40. Reprinted in *Smithsonian Institution Annual Report for 1905,* pp. 375–402. Washington, D.C.: 1906.

———. "On the mammals and winter birds of East Florida, with an examination of certain assumed specific characters in birds, and a sketch of the bird-fauna of Eastern North America." *Bulletin of the Museum of Comparative Zoology* 2 (1871): 161–450.

———. "Notes of an ornithological reconnaissance of portions of Kansas, Colorado, Wyoming, and Utah." *Bulletin of the Museum of Comparative Zoology* 2 (1871): 113–83.

———. "'So-called species and subspecies.'" *Science* 16 (1902): 383–86.

Andrews, Ethan A. "Conjugation in the crayfish, Cambarus affinis." *Journal of Experimental Zoology* 9 (1910): 235–64.

Annual Report. From 1861 through 1877, these were titled "Annual Report of the Trustees of the Museum of Comparative Zoölogy . . ." and were documents of the Massachusetts Senate, submitted and published in January or February to cover the preceding calendar year. From 1878 they were titled "Annual Report of the curator [or director, after 1911] of the Museum of Comparative Zoölogy . . ." and were submitted to the president and fellows of Harvard in the autumn, covering the preceding academic year.

Anonymous. "[Review of] *Vestiges of the Natural History of Creation.*" *American Journal of Science* 48 (1845): 395.

Appel, Toby A. *The Cuvier-Geoffroy Debate: French Biology in the Decades before Darwin.* New York: Oxford University Press, 1987.

———. "Jeffries Wyman, philosophical anatomy, and the scientific reception of Darwin in America." *Journal of the History of Biology* 21 (1988): 69–94.

Baird, Spencer Fullerton, Thomas Mayo Brewer, and Robert Ridgway. "The water birds of North America." *Memoirs of the Museum of Comparative Zoology* 12–13 (1884).

Barbour, Thomas. "A contribution to the zoögeography of the East Indian Islands." *Memoirs of the Museum of Comparative Zoology* 44 (1912): 5–168.

———. "A contribution to the zoögeography of the West Indies, with especial reference to amphibians and reptiles." *Memoirs of the Museum of Comparative Zoology* 44 (1914): 209–359.

———. "Further notes on Dutch New Guinea." *National Geographic Magazine* 19 (1908): 528–45.

———. *Naturalist at Large*. Boston: Little, Brown & Co., 1943.

———. *A Naturalist's Scrapbook*. Cambridge, Mass.: Harvard University Press, 1946.

———. "Notes on Burma." *National Geographic Magazine* 20 (1909): 841–66.

———. "Notes on the herpetology of Jamaica." *Bulletin of the Museum of Comparative Zoology* 52 (1910): 273–301.

———. "Notes on a zoological collecting trip to Dutch New Guinea." *National Geographic Magazine* 19 (1908): 469–81.

———. "Some remarks upon Matthew's 'Climate and evolution.'" *Annals of the New York Academy of Science* 27 (1916): 1–10.

Barbour, Thomas, and Glover M. Allen. *Narrative of a Trip to the Bahamas*. Cambridge, Mass.: Privately printed, 1904.

Barbour, Thomas, and Rosamond Barbour. *Letters Written While on a Collecting Trip in the East Indies*. Paterson, N.J.: Privately printed, 1913.

Barbour, Thomas, and G. K. Noble. "A revision of the lizards of the genus Ameiva." *Bulletin of the Museum of Comparative Zoology* 59 (1915): 417–79.

Bateson, William. "Heredity and variation in modern lights." In *Darwin and Modern Science,* ed. A. C. Seward, pp. 85–101. Cambridge: Cambridge University Press, 1909.

Bather, Francis A. "Biological classification, past and future." *Quarterly Journal of the Geological Society of London* 83 (1927): lx–civ.

Baur, George. "New observations on the origin of the Galapagos Islands, with remarks on the geological age of the Pacific Ocean." *American Naturalist* 31 (1897): 661–80, 864–96.

———. "On the origin of the Galapagos Islands." *American Naturalist* 25 (1891): 217–29, 307–26.

Bay, J. Christian. "A plea for a fair valuation of experimental physiology in biological courses." *Science* (14 July 1893): 21–22.

Bechtel, William. "The nature of scientific integration." In *Integrating Scientific Disciplines,* edited by William Bechtel, pp. 3–52. Boston: Martinus Nijhoff, 1986.

Beecher, Charles Emerson. *Studies in Evolution*. New York, 1901.

Benson, Keith R. "From museum research to laboratory research: the transformation of natural history into academic biology." In *The American Development of Biology,* edited by Ron Rainger, Keith R. Benson, and Jane Maienschein, pp. 49–83. Philadelphia: University of Pennsylvania Press, 1988.

———. "Laboratories on the New England shore: the 'somewhat different direction' of American marine biology." *New England Quarterly* 61 (1988): 55–78.

———. "William Keith Brooks (1848–1908): a case study in morphology and the development of American biology." Ph.D. diss., Oregon State University, 1979.

Bigelow, Henry Bryant. "Thomas Barbour." *Biographical Memoirs of the National Academy of Sciences* 27 (1952): 13–45.

Bouvé, Thomas T. "Historical sketch of the Boston Society of Natural History;

with a notice of the Linnaean Society, which preceded it," 3–250. *Anniversary Memoirs of the Boston Society of Natural History,* 1880.

Briggs, John C. "Freshwater fishes and biogeography of Central America and the Antilles." *Systematic Zoology* 33 (1984): 428–35.

Brocchi, Paul-Louis-Antoine. "Recherches sur les organes génitaux mâles des Crustacés décapodes." *Annales des Sciences Naturelles* 2 (1875): 27–28, 119–20.

Campbell, John P. "Biological teaching in the colleges of the United States." *U.S. Bureau of Education Circular of Information,* no. 9. Washington, D.C., 1891.

———. "Editorial." *Botanical Gazette* 17 (1892): 260–62.

Castelnau, F. de. *Animaux nouveaux ou rares recueillis pendant l'expédition dans les parties Centrales de l'Amerique du Sud, de Rio de Janeiro à Lima, et de Lima au Para.* Paris, 1855.

Chapman, Frank M. "Joel Asaph Allen." *Biographical Memoirs of the National Academy of Sciences* 21 (1941): 1–20.

Clark, Henry James. "Lucernaria, the coenotype of Acalephae." *Proceedings of the Boston Society of Natural History* 9 (1862): 47–54.

———. *Mind in Nature: or, The Origin of Life, and the Mode of Development of Animals.* New York: D. Appleton and Co., 1865.

———. *A Claim for Scientific Property.* Privately printed, n.p., n.d. [1863].

Clark, Hubert Lyman. "Autonomy in Linckia." *Zoologischer Anzeiger* 42 (1913): 156–59.

———. "Bermuda echinoderms. A report on observations and collections made in 1899." *Proceedings of the Boston Society of Natural History* 29 (1901): 339–45.

———. "Cidaridae." *Bulletin of the Museum of Comparative Zoology* 51 (1908): 165–230.

———. "The classification of a regular Echini." *Zoologischer Anzeiger* 45 (1915): 172–72.

———. *The Echinoderm Fauna of Australia, Its Composition and Origin.* Washington, D.C.: Carnegie Institution, 1947.

———. "The echinoderms of Porto Rico." *Bulletin of the United States Fish Commission [for 1882]* 2 (1883): 231–63.

———. "Echinoderms from Puget Sound: observations made on the echinoderms collected by the parties from Columbia University, in Puget Sound in 1896 and 1897." *Proceedings of the Boston Society of Natural History* 29 (1901): 323–37.

———. "The echinoderms of the Woods Hole region." *Bulletin of the Bureau of Fisheries* 24 (1904): 545–76.

———. "Fauna of New England, 4. List of the Echinodermata." *Occasional Papers of the Boston Society of Natural History* 7, no. 4 (1905).

———. "Further notes on the echinoderms of Bermuda." *Annals of the New York Academy of Science* 12 (1899): 117–38.

———. "How I became an ornithologist and how I fell from grace." *Bulletin of the Massachusetts Audubon Society* 30 (1947): 321–25.

———. "The limits of difference in specific and subspecific distinctions." *Annual Report of the Michigan Academy of Science* 5 (1903): 216–18.
———. "Notes on the echinoderms of Bermuda." *Annals of the New York Academy of Science* 11 (1898): 407–13.
———. "Notes on some North Pacific Holothurians." *Zoologischer Anzeiger* 25 (1902): 562–64.
———. "Papers from the Hopkins Stanford Galapagos Expedition, 1898–1899." *Proceedings of the Washington Academy of Science* 4 (1902): 521–31.
———. "The purpose and some principles of systematic zoology." *Popular Science Monthly* 79 (1911): 261–71.
———. "The quest of the vital force." *Amherst Graduates' Quarterly* (1913): 294–301.
———. [Review of "Phylogeny of Echini" by R. T. Jackson]. *Science* 35 (1912): 986–93.
———. "Samuel Garman." *Dictionary of American Biography.* Vol. 4, p. 154. New York: Chas. Scribner's Sons, 1964.
———. "So-called species and subspecies." *Science* 16 (1902): 229–31.
———. "Synapta vivipara: a contribution to the morphology of echinoderms." *Memoirs of the Boston Society of Natural History* 5 (1898): 53–88.
———. "The synaptas of the New England Coast." *Bulletin of the United States Fish Commission* 19 (1899): 21–31.
Clark, Hubert Lyman, and Alexander Agassiz. "Hawaiian and other Pacific Echini: based upon collections made by the U.S. Fish Commission steamer 'Albatross' in 1902, Commander Chauncey Thomas, U.S.N., commanding. The Cidaridae." *Memoirs of the Museum of Comparative Zoology* 34 (1907): 1–42.
———. "Hawaiian and other Pacific Echini. . . . The Echinothuridae." *Memoirs of the Museum of Comparative Zoology* 34 (1909): 133–204.
———. "Hawaiian and other Pacific Echini: the Salenidae, Arbaciadae, Aspidodiadematidae, and Diadematidae." *Memoirs of the Museum of Comparative Zoology* 34 (1908): 47–132.
———. "Preliminary report on the Echini collected, in 1902, among the Hawaiian Islands by the U.S. Fish Commission steamer 'Albatross' in charge of Commander Chauncey Thomas, U.S.N., commanding." *Bulletin of the Museum of Comparative Zoology* 50 (1907): 231–59.
Coleman, Lawrence Vail. *The Museum in America: A Critical Study.* 3 vols. Washington, D.C.: American Association of Museums, 1939.
Coleman, William. "The cognitive basis of the discipline: Claude Bernard on physiology." *Isis* 76 (1985): 49–70.
Cooper, Lane. *Louis Agassiz as a Teacher.* 1917. Revised edition. New York: Comstock, 1945.
Cope, Edward Drinker. "Editors' table." *American Naturalist* 14 (1880): 726–28.
———. "On the Origin of Genera." [1868] In *The Origin of the Fittest: Essays on Evolution.* New York: D. Appleton, 1887.
———. "On the Wyandotte Cave and its fauna." *American Naturalist* 6 (1872): 406–22.

Cope, Edward Drinker, and A. S. Packard. "The fauna of the Nickajack Cave." *American Naturalist* 15 (1881): 877–82.

Croizat, Léon. *Panbiogeography.* 3 vols. Caracas, Venezuela: Privately printed, 1958.

Cutright, Paul Russell. *Theodore Roosevelt: The Making of a Conservationist.* Chicago: University of Illinois Press, 1985.

Dall, William Healey. *Spencer Fullerton Baird: A Biography, Including Selections from His Correspondence with Audubon, Agassiz, Dana, and Others.* Philadelphia: J. B. Lippincott Co., 1915.

Dana, James Dwight. "Agassiz's contributions to the natural history of the United States." *American Journal of Science* 25 (1858): 321–41.

———. "On cephalization, and on megastenes and microsthenes, in classification." *American Journal of Science* 36 (1863): 1–10.

———. "The classification of animals based on the principle of cephalization." *American Journal of Science* 36 (1863): 321–52.

———. "The classification of animals based on the principle of cephalization, no. 2, classification of insects." *American Journal of Science* 37 (1864): 10–33.

———. "The classification of animals based on the principle of cephalization, no. 3, classification of herbivores." *American Journal of Science* 37 (1864): 157–83.

———. "On the higher subdivisions in the classification of mammals." *American Journal of Science* 35 (1863): 65–71.

———. "On the homologies of the insectean and crustacean types." *American Journal of Science* 36 (1863): 233–35.

———. "Note on the position of amphibians among the classes of vertebrates." *American Journal of Science* 37 (1864): 184–86.

Darlington, Philip Jackson, Jr. "The origin of the fauna of the Greater Antilles, with discussion of dispersal of animals over water and through the air." *Quarterly Review of Biology* 13 (1938): 274–300.

———. *Zoogeography: The Geographical Distribution of Animals.* New York, 1957.

Darwin, Charles Robert. *Charles Darwin: His Life Told in an Autobiographical Chapter and in a Selected Series of His Published Letters.* New York: D. Appleton and Co., 1892. Reprinted as *The Autobiography of Charles Darwin and Selected Letters.* Edited by Francis Darwin. New York: Dover Publications, 1958.

———. *On the Origin of Species.* London, 1859. Facsimile reprint. Cambridge: Harvard University Press, 1964.

Daudin, Henri. *Cuvier et Lamarck: Les Classes zoologiques et l'idée de série animale (1790–1830).* 2 vols. Paris: Félix Alcan, 1926.

———. *De Linné à Jussieu: Méthodes de classification et idée de série en botanique et en zoologie (1740–1790).* Paris: Félix Alcan, 1926.

Davenport, C. B. *Inheritance in Poultry.* Washington, D.C.: Carnegie Institution of Washington, 1906.

Dean, John Ward. "Descendants of the Rev. Daniel Rogers of Littleton, Mass." *New England Historical and Genealogical Register* 39 (1885): 225–26.

"Dedication of the museum of the Society of Natural History." *Boston Daily Advertiser* (3 June 1864).

Deichmann, Elisabeth, and Fenner A. Chace, Jr. "Obituary of Hubert Lyman Clark, 1870–1947." *Science* 106 (1947): 611–12.

de Meijere, I. C. H. "Echinoidea der Siboga-Expedition." *Siboga-Expeditie 1899–1900,* vol. 43, book 14. Leiden: E. J. Brill, 1904.

Dexter, Ralph W. "An early defense of Darwinism." *American Naturalist* 93 (1959): 138–39.

———. "Historical aspects of F. W. Putnam's systematic studies on fishes." *Journal of the History of Biology* 3 (1970): 131–48.

———. "Historical aspects of Louis Agassiz's lectures on the nature of the species." *Bios* 48 (1977): 12–19.

———. "Historical aspects of studies on the Brachiopoda by E. S. Morse." *Systematic Zoology* 15 (1966): 241–44.

———. "The impact of evolutionary theories on the Salem group of Agassiz zoologists (Morse, Hyatt, Packard, Putnam)." *Essex Institute Historical Collections* 115 (1979): 144–71.

———. "The 'Salem secession' of Agassiz zoologists." *Essex Institute Historical Collections* 101 (1965): 27–39.

———. "Sir Richard Owen's interpretation of optic lobes in blind fishes." *American Naturalist* 100 (1966): 271–72.

Dick, Myvanwy M. "Stations of the Thayer Expedition to Brazil 1865–1866." *Breviora,* no. 444 (1977):1–37.

Dictionnaire Historique & Bibliographique de la Suisse. Edited by Victor Attinger, Marcel Godet, and Heinrich Turler. Neuchâtel, 1930.

Döderlein, Ludwig. *Die Echinoiden der deutschen Tiefsee-Expedition.* Wissenschaftliche Ergebnisse der deutschen Tiefsee-Expedition auf dem Dampfer "Valdivia" 1898–1899. Vol. 5, ed. Carl Chun. Jena, 1906.

Du Bois-Reymond, Emil. "The limits of our knowledge of nature." *Popular Science Monthly* 5 (1874): 17–32.

Dupree, A. Hunter. *Asa Gray.* Cambridge, Mass.: Harvard University Press, 1959.

D'yakonov [Diakonov], A. M. *Echinoidea* (Fauna of Russia and adjacent countries, edited by N. V. Nasonov, vol. 1, no. 1). Petrograd, 1923. Trans. Y. Salkind. Jerusalem: Israel Program for Scientific Translations, 1969.

Eigenmann, Carl H. "Adaptation." In *Fifty Years of Darwinism: Modern Aspects of Evolution, Centennial Addresses in Honor of Charles Darwin before the American Association for the Advancement of Science, Baltimore, Friday, January 1, 1909,* pp. 182–208. New York, 1909.

———. "The American Characidae." *Memoirs of the Museum of Comparative Zoology* 43 (1917–1927).

———. "On the egg membranes and micropyle of some osseous fishes." *Bulletin of the Museum of Comparative Zoology* 19 (1890): 129–54.

Eigenmann, Carl H., and Rosa Smith Eigenmann. "A catalogue of the fresh-water fishes of South America." *Proceedings of the United States National Museum* 14 (1891): 1–81.

———. "A revision of the edentulous genera of Curimatinae." *Annals of the New York Academy of Science* 4 (1889): 409–40.

———. "A revision of the South American Nematognathi or catfishes." *Occasional Papers of the California Academy of Science* 1 (1890): 1–508.

Eliot, Charles W. *The Man and His Beliefs*. Edited by William Allan Neilson. 2 vols. New York: Harper and Bros., 1926.

Emerson, George B. [Address on Louis Agassiz]. *Anniversary Memoirs of the Boston Society of Natural History* 1880: 155–64.

Erichson, Wilhelm Ferdinand. "Untersicht der Arten der Gattung Astacus." *Archiv für Naturgeschichte* 12 (1846): 86–103.

Evans, Mary Alice, and Howard Ensign Evans. *William Morton Wheeler, Biologist.* Cambridge, Mass.: Harvard University Press, 1970.

Faxon, Walter. "Notes on the crayfishes in the United States National Museum and the Museum of Comparative Zoology with descriptions of new species and subspecies to which is appended a catalogue of the known species and subspecies." *Memoirs of the Museum of Comparative Zoology* 40 (1914): 347–427.

———. "Observations on the Astacidae in the United States National Museum and in the Museum of Comparative Zoology, with descriptions of new species." *Proceedings of the United States National Museum* 20 (1898): 643–94.

———. "A revision of the Astacidae." *Memoirs of the Museum of Comparative Zoology* 10, no. 4 (1885): 1–186.

———. "On the so-called dimorphism in the genus Cambarus." *American Journal of Science* 27, ser. 3 (1884): 42–43.

Fernald, H. T. "Teaching of biology." *Science* 21 (1893): 347–48.

Fewkes, Jesse Walter. "An aid to the collector of the Coelenterata and Echinodermata of New England." *Bulletin of the Essex Institute* 23 (1891): 1–92.

———. "A contribution to Passamaquoddy folk-lore." *Journal of American Folk-Lore* 3 (1890): 257–80.

———. "A preliminary notice of a stalked bryozoan (Ascorhiza occidentalis)." *Annals and Magazine of Natural History* 3 (1889): 1–6.

Fitch, Walter M. "On the problem of discovering the most parsimonious tree." *American Naturalist* 111 (1977): 223–57.

Fitzpatrick, J. F., Jr. "The Propinquus group of the Crawfish Genus *Orconectes* (Decapoda: Astacidae)." *Ohio Journal of Science* 67 (1967): 129–72.

Flower, William Henry. "Modern museums (Presidential address to the Museums Association, London, 3 July 1893)." In *Essays on Museums and Other Subjects Connected with Natural History*, pp. 30–53. London, 1898.

Forbes, Abner, and J. W. Greene. *The Rich Men of Massachusetts: Containing a Statement of the Reputed Wealth of about Fifteen Hundred Persons with Brief Sketches of More Than One Thousand Characters*. Boston, 1851.

Frankel, Henry. "Alfred Wegener and the specialists." *Centaurus* 20 (1976): 305–24.

Garman, Samuel. "Cave animals from southwestern Missouri." *Bulletin of the Museum of Comparative Zoology* 17 (1889): 225–39.

———. "On the lateral canal system of the Selacia and Holocephala." *Bulletin of the Museum of Comparative Zoology* 17 (1888): 57–121.

Gates, William B., Jr. *Michigan Copper and Boston Dollars: An Economic History of the Michigan Copper Mining Industry.* Cambridge, Mass.: Harvard University Press, 1951.

Gerstfeldt, Georg. "Ueber Flusskrebse Europa's." *Mémoires présentés à l'Académie Impériale des Sciences de St. Pétersbourg* 9 (1859): 551–89.

Ghiselin, Michael. *Triumph of the Darwinian Method.* Berkeley: University of California Press, 1969.

Girard, Charles. "Reply to Prof. Agassiz's communication." *Proceedings of the Academy of Natural Sciences of Philadelphia* 6 (1853): 380–81.

———. "A revision of the North American Astaci, with observations on their habits and geographical distribution." *Proceedings of the Academy of Natural Sciences of Philadelphia* 6 (1852): 87–91.

Goode, George Brown. "The museums of the future." *Annual Report of the United States National Museum* for 1888–89, pp. 427–45. Washington, D.C., 1891.

Gould, Stephen Jay. *Ontogeny and Phylogeny.* Cambridge, Mass.: Harvard University Press, 1977.

Graham, R. L., and L. R. Foulds. "Unlikelihood that minimal phylogenies for a realistic biological study can be constructed in reasonable computation time." *Mathematical Biosciences* 60 (1982): 133–42.

Gray, James Edward. "On museums, their use and improvement, and on the acclimatization of animals." *Report of the British Association for the Advancement of Science* (1865): 75–86. Also in *Annals and Magazine of Natural History* 14 (1864): 283–97.

Groeben, Christiane, ed. *Charles Darwin, 1809–1882, Anton Dohrn, 1840–1909: Correspondence.* Naples: Macchiaroli, 1982.

Günther, Albert E. *A Century of Zoology at the British Museum through the Lives of Two Keepers, 1815–1914.* Folkestone: William Dawson, 1975.

———. "General history of the department of zoology from 1856 to 1895." *The History of the Collections Contained in the Natural History Departments of the British Museum.* Vol. 2, appendix. London: Printed by Order of the Trustees of the British Museum, 1912.

Haeckel, Ernst. *Ziele und Wege der heutigen Entwicklungsgeschichte.* Jena, 1875.

Hagen, Hermann. "The blind crayfish." *American Naturalist* 6 (1872): 494–95.

———. "Monograph of the North American Astacidae." *Illustrated Catalogue of the Museum of Comparative Zoology* 3 (1870). Also published as *Memoirs of the Museum of Comparative Zoology* 2 (1871).

Harlan, Richard. "Description of a new species of the Genus Astacus." *Transactions of the American Philosophical Society* 3 (1830): 465.

Harvard University. "Doctors of Philosophy and Doctors of Science who have received their degree in course from Harvard University 1873–1926 with the titles of their theses." *Official Register of Harvard University,* vol. 23, no. 39. Cambridge, Mass.: Harvard University Press, 1926.

———. "Report of the Committee on Zoölogy." Reports of the Visiting Committees to the Board of Overseers, no. 40, 1893, pp. 195–201. Harvard University Archives.

Hawkins, Hugh. *Between Harvard and America: The Educational Leadership of Charles W. Eliot.* New York: Oxford University Press, 1972.

Hedges, S. Blair. "Caribbean biogeography: implications of recent plate tectonic studies." *Systematic Zoology* 3 (1982): 518–22.

Henshaw, Samuel. "Hermann August Hagen." *Daedalus* 29 (1894): 419–23.

Herber, Elmer Charles, ed. *Correspondence between Spencer Fullerton Baird and Louis Agassiz.* Washington, D.C.: Smithsonian Institution, 1963.

Herrick, Francis H. "On the teaching of biology." *Science* 21 (1893): 220–21.

Hill, Robert Thomas. "The geology and physical geography of Jamaica: study of the type of Antillean development. Based on surveys made for A. Agassiz." *Bulletin of the Museum of Comparative Zoology* 34 (1899): 1–226.

———. "Notes on the geology of the island of Cuba, based upon a reconnaissance made for Alexander Agassiz." *Bulletin of the Museum of Comparative Zoology* 16 (1895): 243–88.

Hobbs, Horton H., Jr. "Adaptations and convergence in North American crayfishes." In *Freshwater Crayfish,* edited by J. W. Avault, pp. 541–51.

———. *Crayfishes (Astacidae) of North and Middle America.* Water Pollution Control Research Series 18050 ELDO5/72. U.S. Environmental Protection Agency, 1972.

———. "The current status of the crayfishes listed by Girard (1852) in his 'A revision of the North American Astaci . . .' (Decapoda, Astacidae)." *Crustaceana* 12 (1967): 124–32.

———. "Notes on the affinities of the members of the Blandingii section of the crayfish genus *Procambarus* (Decapoda, Astacidae)." *Tulane Studies in Zoology* 9 (1962): 273–93.

Hobbs, Horton H., Jr., and Thomas C. Barr, Jr. "The origins and affinities of the troglobitic crayfishes of North America (Decapoda, Astacidae). I. The genus *Cambarus.*" *American Midland Naturalist* 64 (1960): 12–33.

———. *Origins and Affinities of the Troglobitic Crayfishes of North America (Decapoda: Astacidae) II. Genus Orconectes.* Smithsonian Contributions to Zoology, no. 105. 1972.

Hodge, Michael J. S. "Darwin, species, and the theory of natural selection." In *Histoire du concept d'espèce dans les sciences de la vie,* pp. 227–52. Paris: Singer-Polignac, 1987.

Holthuis, L. B. "On the status of two allegedly European crayfishes, *Cambarus typhlobius* Joseph, 1880, and *Austropotamobius pallipes bispinosus* Karaman, 1961 (Decapoda, Astacidae)." *Crustaceana* 7 (1964): 42–48.

Hough, Walter. "Jesse Walter Fewkes." *Biographical Memoirs of the National Academy of Sciences* 15 (1934): 258–83.

Howe, Mark A. De Wolfe. *Memories of a Hostess*. Boston: Atlantic Monthly Press, 1923.

Hubbs, Carl L. "History of ichthyology in the United States after 1850." *Copeia* (1964): 42–60.

Hull, David. *Science as a Process*. Chicago: University of Chicago Press, 1988.

Hume, David. *Dialogues Concerning Natural Religion*. London, 1779.

Huxley, Leonard. *Life and Letters of Sir Joseph Dalton Hooker.* 2 vols. London, 1918.

Huxley, Thomas Henry. "On the classification and the distribution of the crayfishes." *Proceedings of the Zoological Society of London* (1878): 752–88.

———. *An Introduction to the Study of Zoology, Illustrated by the Crayfish*. New York, 1888.

———. "A lobster: or, The study of zoology." In *Discourses Biological and Geological,* pp. 196–228. New York, 1897.

———. *Scientific Memoirs*. Edited by Edwin Ray Lankester and Michael Foster. 4 vols. London: Macmillan, 1898–1902.

Hyatt, Alpheus. [Annual report of the custodian.] *Proceedings of the Boston Society of Natural History* 14 (1871): 207–33.

———. "The genesis of the Tertiary species of Planorbis at Steinheim." *Anniversary Memoirs of the Boston Society of Natural History,* 1880.

———. "On the parallelism between the different stages of life in the individual and those in the entire group of the molluscous order Tetrabranchiata." [1866.] *Memoirs of the Boston Society of Natural History* 1 (1869): 193–209.

———. [Remarks on natural selection.] *Proceedings of the Boston Society of Natural History* 14 (1872): 146–48.

Jackson, Robert Tracy. "Phylogeny of the Echini, with a revision of Palaeozoic species." *Memoirs of the Boston Society of Natural History* 7 (1912).

James, Henry. *Charles W. Eliot: President of Harvard University 1869–1909.* 2 vols. Boston: Houghton Mifflin, 1930.

Jordan, David Starr. *The Days of a Man: Being Memories of a Naturalist, Teacher and Minor Prophet of Democracy.* 2 vols. Yonkers-on-Hudson, N.Y.: World Book Co., 1922.

Kähsbauer, Paul. "Intendant Dr. Franz Steindachner, sein Leben und Werk." *Annalen des Naturhistorisches Museums in Wien* 63 (1959):1–30.

Kennedy, John M. "Philanthropy and science in New York City: The American Museum of Natural History, 1868–1968." Ph.D. diss., Yale University, 1968.

Kohlstedt, Sally Gregory. "Henry A. Ward." *Journal of the Society for the Bibliography of Natural History* 9 (1980): 647–61.

———. "Institutional history." *Osiris* 1 (1985): 17–36.

———. "Museums on campus: a tradition of inquiry and teaching." In *The American Development of Biology,* edited by R. Rainger, K. R. Benson, J. Maienschein, pp. 15–47. Philadelphia: University of Pennsylvania Press, 1988.

———. "The nineteenth-century amateur tradition: the case of the Boston Society of Natural History." *Boston Studies in the Philosophy of Science* 33 (1976): 173–90.

Lankton, Larry. *Cradle to Grave: Life, Work and Death at the Lake Superior Copper Mines.* Oxford: Oxford University Press, forthcoming.

Lawrence, William. *Life of Amos A. Lawrence: With Extracts from His Diary and Correspondence.* Boston: Houghton Mifflin, 1888.

LeConte, John [Eatton]. "Descriptions of new species of Astacus from Georgia." *Proceedings of the Academy of Natural Sciences of Philadelphia* 7 (1855): 400–402.

LeConte, Joseph. *Autobiography.* Edited by William Dallam Armes. New York, 1903.

Limoges, Camille. "The development of the Muséum d'Histoire Naturelle of Paris, 1800–1914." In *The Organization of Science and Technology in France 1808–1914,* edited by Robert Fox and George Weisz. Cambridge: Cambridge University Press, 1980.

Livingstone, David N. *Nathaniel Southgate Shaler and the Culture of American Science.* Tuscaloosa: University of Alabama Press, 1987.

Lurie, Edward. *The Founding of the Museum of Comparative Zoology.* Privately printed for the centennial of the M.C.Z., 1959.

———. *Louis Agassiz: A Life in Science.* Chicago: University of Chicago Press, 1960.

Lyman, Theodore. "Recollections of Agassiz." *Atlantic Monthly* 33 (1874): 221–29.

Lyon, John. "The 'initial discourse' to Buffon's *Histoire naturelle:* the first complete English translation." *Journal of the History of Biology* 9 (1976): 133–81.

MacIntosh, H. W. "The echinoids of the 'Challenger.'" *Nature* 25 (10 November 1881): 41–2.

MacMillan, Conway. "On the emergence of a sham biology in America." *Science* 21 (1893): 184–86.

———. "On methods of defending the existence of a sham biology in America." *Science* 21 (1893): 289–91.

———. "Open letters: Dr. J. P. Campbell's 'biological instruction.'" *Botanical Gazette* 17 (1892): 301–2.

Mallis, Arnold. *American Entomologists.* New Brunswick, N.J.: Rutgers University Press, 1971.

Marcou, Jules. *Life, Letters, and Works of Agassiz.* 2 vols. New York, 1896.

Mark, Edward Laurens. "Zoology, 1847–1921." In *The Development of Harvard University since the Inauguration of President Eliot: 1869–1929,* edited by Samuel Eliot Morison, pp. 378–93. Cambridge, Mass.: Harvard University Press, 1930.

Matthew, William Diller. "Climate and evolution." *Annals of the New York Academy of Science* 24 (1915): 171–318.

———. "Supplementary note." *Annals of the New York Academy of Science* 27 (1916): 11–15.

Mayer, Alfred Goldsborough. "Alpheus Hyatt, 1838–1902." *Popular Science Monthly* 78 (1911): 129–46.

Mayr, Ernst. "The birds of Timor and Sumba." *Bulletin of the American Museum of Natural History* 83 (1944): 127–94, excerpted as "Land bridges and dispersal faculties" in his *Evolution and the Diversity of Life,* pp. 618–25.

———. *Evolution and the Diversity of Life: Selected Essays.* Cambridge, Mass.: Harvard University Press, Belknap Press, 1976.

———. *The Growth of Biological Thought.* Cambridge, Mass.: Harvard University Press, 1982.

———. *The Museum of Comparative Zoology and Its Role in the Harvard Community.* Cambridge, Mass.: Museum of Comparative Zoology, 1969.

———. "The role of systematics in biology." [1969.] In his *Evolution and the Diversity of Life,* pp. 416–24.

Mayr, Ernst, and Richard Goodwin. *Biological Material: Part I: Preserved Material and Museum Collection.* National Academy of Science–National Research Council, Publication no. 399, n.d.

Mendelsohn, Everett. "The emergence of science as a profession in nineteenth-century Europe." In *The Management of Scientists,* edited by Karl B. Hill. Boston: Beacon Hill Press, 1964.

Merriam, C. Hart. "Biology in our colleges: a plea for a broader and more liberal biology." *Science* 21 (1893): 352–55.

———. "Roosevelt the naturalist." *Science* 75 (1932): 181–83.

Meyer, A. B. "Studies of the museums and kindred institutions of New York City, Albany, Buffalo, and Chicago, with notes on some European institutions." *Report of the United States National Museum for 1903,* pp. 311–608. Washington, D.C.: Smithsonian Institution, 1905.

[Michels, John]. [On the Boston AAAS meeting]. *Science* 1 (1880): 141.

Morison, Samuel Eliot, ed. *The Development of Harvard University since the Inauguration of President Eliot: 1869–1929.* Cambridge, Mass.: Harvard University Press, 1930.

Morse, Edward Sylvester. "A classification of Mollusca based on the 'principle of cephalization.'" *Proceedings of the Essex Institute* 4 (1865): 162–80.

———. "The evolution theory." *Canadian Naturalist* 25 (1874): 153–55.

———. "On the identity of the ascending process of the astragalus in birds with the intermedium." *Anniversary Memoirs of the Boston Society of Natural History* (1880): 3–10.

———. [Principle of cephalization applied to the classification of Mollusca.] [1867.] *Proceedings of the Boston Society of Natural History* 11 (1868): 287.

———. "On the tarsus and carpus of birds." [1872.] *Annals of the Lyceum of Natural History of New York* 10 (1873): 141–58.

———. "Variability of species." *Northern Monthly* 1 (1864): 267–70.

———. "What American zoologists have done for evolution." *Proceedings of the American Association for the Advancement of Science* 25B (1876): 137–76.

Mortensen, Theodor. “Echinoidea.” *The Danish Ingolf-Expedition*, vol. 4. Copenhagen, 1903–7.

———. *A Monograph of the Echinoidea*. Copenhagen: C. A. Reitzel, 1928–52.

Myers, George S. “Amoenitates biologicae: the influence of Louis Agassiz on the ichthyology of Brazil.” *Revista Brasileira de Biologia* 3 (1943): 127–33.

Nachtrieb, Henry F. “Sham biology.” *Science* 26, no. 538 (20 May 1893): 287–89.

Nininger, H. H. “Zoology and the college curriculum.” *Scientific Monthly* 16 (1923): 66–72.

Ortmann, Arnold E. “The crawfishes of the state of Pennsylvania.” *Memoirs of the Carnegie Museum of Pittsburgh* 2 (1906): 343–523.

Owen, Richard. *Lectures on the Comparative Anatomy and Physiology of the Invertebrate Animals*. London, 1843.

———. “On a national museum of natural history.” *The Athenaeum* (27 July 1861): 118–20, 153–55, 187–89.

———. “Presidential address.” *Report of the British Association for the Advancement of Science* (1859): xlix–cx.

Packard, Alpheus Spring, Jr. “The cave-fauna of North America, with remarks on the anatomy of the brain and origin of the blind species.” *Memoirs of the National Academy of Sciences* 4 (1888): 3–156.

———. “On the crustaceans and insects.” *American Naturalist* 5 (1871): 744–61.

———. “The effect of cave life on animals, and its bearing on the evolution theory.” *Popular Science Monthly* 36 (1890): 389–97.

———. *Guide to the Study of Insects*. Salem, Mass.: Naturalist’s Book Agency, 1868.

———. “A half-century of evolution, with special reference to the effects of geological changes on animal life.” *Proceedings of the American Association for the Advancement of Science* 47 (1898): 311–56.

———. “Henry James Clark.” *Biographical Memoirs of the National Academy of Sciences* 1 (1877): 317–28.

———. “Notes on the family Zygaenidae.” *Communications of the Essex Institute* 4 (1864): 7–47.

———. “Observations on the development and position of the Hymenoptera, with notes on the morphology of insects.” *Proceedings of the Boston Society of Natural History* 10 (1866): 279–95.

———. [Review of] “Wallace’s *Contributions to the Theory of Natural Selection*.” *American Naturalist* 4 (1871): 419–22.

———. “The study of natural history in college.” *University Quarterly* (1861): 319–25.

———. “Synopsis of the Bombycidae of the United States.” *Proceedings of the Entomological Society of Philadelphia* 3 (1864): 97–130, 331–96.

———. “On synthetic types in insects.” *Journal of the Boston Society of Natural History* 7 (1863): 590–603.

Parker, George Howard, ed. *Mark Anniversary Volume*. New York: Henry Holt and Co., 1903.

———. *The World Expands: Recollections of a Zoologist*. Cambridge, Mass.: Harvard University Press, 1946.

Pauly, Philip J. "The appearance of academic biology in late nineteenth-century America." *Journal of the History of Biology* 17 (1984): 369–97.

———. "Summer resort and scientific discipline: Woods Hole and the structure of American biology, 1882–1925." In *The American Development of Biology*, edited by Ronald Rainger, Keith Rodney Benson, Jane Maienschein, pp. 121–50. Philadelphia: University of Pennsylvania Press, 1988.

Pearl, Raymond. "Trends of modern biology." *Science* 56 (1922): 581–92.

Peters, James L. "Outram Bangs, 1863–1932." *Auk* 50 (1933): 264–74.

Pfeifer, Edward J. "The genesis of American Neo-Lamarckism." *Isis* 56 (1965): 156–67.

Pilsbry, Henry Augustus. "The genesis of mid-Pacific faunas." *Proceedings of the Academy of Natural Sciences of Philadelphia* (1900): 568–81.

Powell, Baden. *Essays on the Spirit of the Inductive Philosophy, the Unity of the Worlds, and the Philosophy of Creation*. London, 1855.

Proceedings of the 7th International Zoölogical Congress. (Boston, 19–24 August 1907.) Cambridge: Cambridge University Press, 1912.

Putnam, Frederic W., and A. S. Packard. "The Mammoth Cave and its inhabitants." *American Naturalist* 5 (1871): 739–44.

Rauchenberger, Mary. "Historical biogeography of poeciliid fishes in the Caribbean." *Systematic Zoology* 37 (1988): 356–65.

Ridley, Mark. *Evolution and Classification: The Reformation of Cladism*. London: Longman, 1986.

———. *The Problems of Evolution*. Oxford: Oxford University Press, 1985.

Rieppel, Olivier. "Louis Agassiz (1807–1873) and the reality of natural groups." *Biology and Philosophy* 3 (1988): 29–47.

Rocker, Rudolf. *Pioneers of American Freedom*. Los Angeles: Rocker Publications Committee, 1949.

Romer, Alfred Sherwood. "George Howard Parker." *Biographical Memoirs of the National Academy of Sciences* 39 (1967): 358–90.

———. "Thomas Barbour." *Systematic Zoology* 13 (1964): 227–35.

Rosen, Donn E. "A vicariance model of Caribbean biogeography." *Systematic Zoology* 24 (1975): 431–64.

———. "Vicariant patterns and historical explanation in biogeography." *Systematic Zoology* 27 (1978): 159–88.

Russell, Edward Stuart. *Form and Function*. London: J. Murray, 1916. Reprint. Chicago: University of Chicago Press, 1982.

Scholes, Robert. "Is there a fish in this text?" *College English* 46 (1984): 653–64.

Schuchert, Charles. *Historical Geology of the Antillean-Caribbean Region*. 1935. Facsimile reprint. New York and London: Hafner Publishing Co., 1968.

Scudder, Samuel. *The Butterflies of the Eastern United States and Canada*. 3 vols. Cambridge, 1889.

———. "Inquiry into the zoological relations of . . . fossil neuropterous insects . . . " *Memoirs of the Boston Society of Natural History* 1 (1866): 173–92.

———. "In the laboratory with Agassiz." *Every Saturday* (4 April 1874), pp. 369–70.

———. "Materials for a monograph of North American Orthoptera." *Boston Journal of Natural History* 7 (1863): 409–80.

———. [Remarks upon the division of the class of insects into orders.] *Proceedings of the Boston Society of Natural History* 9 (1862): 69.

Shaler, Nathaniel Southgate. *Autobiography, with a Supplementary Memoir by His Wife*. Boston: Houghton Mifflin, 1909.

———. "Lateral symmetry in Brachiopoda." *Proceedings of the Boston Society of Natural History* 8 (1862): 274–79.

Sharp, Dallas Lore. "Turtle eggs for Agassiz." *Atlantic Monthly* 150 (1932): 537–45.

Silliman, Benjamin. "On the Mammoth Cave of Kentucky." *American Journal of Science* 11 (1851): 336.

Simpson, George Gaylord. *Concession to the Improbable: An Unconventional Autobiography*. New Haven: Yale University Press, 1978.

Smithsonian Institution. *Annual Report of the Board of Regents* for 1856. (U.S. Congress, House. 34th Cong., 3d sess. Misc. Doc. 55.) Washington, D.C.

Sonneborn, T. M. "Herbert Spencer Jennings." *Biographical Memoirs of the National Academy of Sciences* 47 (1975): 143–223.

Stearn, William Thomas. *The Natural History Museum at South Kensington: A History of the British Museum (Natural History) 1753–1980*. London: Heinemann, 1981.

Steindachner, Franz. "Die Süsswasserfische des südöstlichen Brasilien." *Sitzungberichte der Mathematisch—Naturwissenschaftlichen Klasse der kaiserlichen Akademie der Wissenschaften* 70 (1875): 499–538.

Stejneger, Leonhard. "Carl H. Eigenmann." *Biographical Memoirs of the National Academy of Sciences* 19 (1938): 305–36.

Stephens, Lester D. *Joseph LeConte: Gentle Prophet of Evolution*. Baton Rouge: Louisiana State University Press, 1982.

Sterling, Keir. *Last of the Naturalists: The Career of C. Hart Merriam*. New York: Arno Press, 1977.

Stott, P. "History of biogeography." In *Themes in Biogeography*, edited by J. A. Taylor, pp. 1–24. London: Croom Helm, 1984.

Stresemann, Erwin. *Ornithology from Aristotle to the Present*. Cambridge, Mass.: Harvard University Press, 1975.

Strickland, Hugh. "On the structural relations of organized beings." *Philosophical Magazine* 28 (1846): 354–64.

liscovering the natural system in zoology and bota- *Natural History* 6 (1841): 184–94.

Sulloway, Frank. "Darwin's conversion: The *Beagle* voyage and its aftermath." *Journal of the History of Biology* 15 (1982): 327–98.

"Symposium: World perspectives in biogeography." *Systematic Zoology* 24 (1975).

Tachikawa, Akira. "The two sciences and religion in antebellum New England: the founding of the Museum of Comparative Zoology and the Massachusetts Institute of Technology." Ph.D. diss., University of Wisconsin, 1978.

Tharp, Louise Hall, *Adventurous Alliance: The Story of the Agassiz Family of Boston.* Boston: Little, Brown and Co., 1959.

Todhunter, Isaac. *Algebra for the Use of Colleges and Schools with Numerous Examples.* London: MacMillan & Co., 1879.

Tuckerman, Frederick. "Henry James Clark: Teacher and investigator." *Science* 35 (1912): 725–30.

van Cleave, Harley J. "An index to the opinions rendered by the International Commission on Zoological Nomenclature." *American Midland Naturalist* 80 (1943): 223–40.

van Kampen, P. N. "The zoogeography of the East Indian Archipelago." Translated from the Dutch by T. Barbour. *American Naturalist* 45 (1911): 537–60.

Verrill, Addison Emory. "Revision of the polyps of the Eastern coast of the United States." *Memoirs of the Boston Society of Natural History* 1 (1864): 1–45.

Veysey, Laurence R. *The Emergence of the American University.* Chicago: University of Chicago Press, 1965.

Vogt, Carl. *Aus meinem Leben: Erinnerungen und Rückblicke.* Stuttgart, 1896.

Volger, G. H. O. *Leben und Leistungen des Naturforschers Karl Schimper.* Frankfurt am Main, 1889.

Waage, Jonathan K. "Sperm competition and the evolution of odonate mating systems." In *Sperm Competition and the Evolution of Animal Mating Systems,* ed. Robert L. Smith, 251–90. Orlando, Fla.: Academic Press, 1984.

Wallace, Alfred Russel. "American museums: the Museum of Comparative Zoology, Harvard University." *Fortnightly Review* 42 (1887): 347–59.

———. *The Geographical Distribution of Animals.* 2 vols. London, 1876.

———. *Island Life: or, The Phenomena and Causes of Insular Faunas and Floras.* 2 vols. London, 1880.

———. *The Malay Archipelago.* 2 vols. London, 1869.

Walsh, Benjamin Dann. "On certain entomological speculations of the New England school of naturalists." *Proceedings of the Entomological Society of Philadelphia* 3 (1864): 207–49.

———. "North American Neuroptera." *Proceedings of the Entomological Society of Philadelphia* 2 (1863): 167–272.

———. "Notes and descriptions of about twenty new species." *Proceedings of the Entomological Society of Philadelphia* 2 (1863): 217.

Ward, Roswell. *Henry Augustus Ward: Museum Builder to America.* Rochester, N.Y.: Rochester Historical Society Publications, 1948.

Warren, Helen Ann. "Survey of the life of Louis Agassiz: the centenary of the glacial theory, prepared under the direction of Dr. Henry Fairfield Osborn." *Scientific Monthly* 27 (1928): 355–66.

Wayman, Dorothy G. *Edward Sylvester Morse: A Biography.* Cambridge, Mass.: Harvard University Press, 1942.

Wheeler, William Morton. "The Bussey Institution." In *The Development of Harvard University since the Inauguration of President Eliot: 1869–1929,* edited by Samuel Eliot Morison, pp. 508–17. Cambridge, Mass.: Harvard University Press, 1930.

Whitehill, Walter Muir. *The East India Marine Society and the Peabody Museum of Salem: A Sesquicentennial History.* Salem, Mass.: Peabody Museum, 1949.

Whitman, Charles Otis. "A marine biological observatory." *Popular Science Monthly* 42 (1893): 459–71.

———. "A marine observatory, the prime need of American biology." *Atlantic Monthly* 71 (1893): 808–15.

Wilder, Burt. "Louis Agassiz, teacher." *Harvard Graduates' Magazine* (June 1907): 605–6.

———. "On morphology and teleology, especially in the limbs of Mammalia." *Memoirs of the Boston Society of Natural History* 1 (1866): 46–80.

Wiley, E. O. *Phylogenetics: The Theory and Practice of Phylogenetic Systematics.* New York: J. Wiley and Sons, 1981.

Williams, T. *Fiji and the Fijians.* London, 1858.

Winsor, Mary P. "The development of Linnaean insect classification." *Taxon* 25 (1976): 57–67.

———. "A historical investigation of the siphonophores." *Proceedings of the Royal Society of Edinburgh* 73 (1972): 315–23.

———. "The impact of Darwinism on the Linnaean enterprise, with special reference to the work of T. H. Huxley." In *Contemporary Perspectives on Carl von Linné,* edited by J. M. Weinstock, pp. 55–84. Lanham, Md.: University Press of America, 1985.

———. "Louis Agassiz and the species question." *Studies in History of Biology* 3 (1979): 89–117.

———. *Starfish, Jellyfish, and the Order of Life.* New Haven: Yale University Press, 1976.

Wood, James. "What is biology?" *Science* 21 (1893): 275–76.

Wyman, Jeffries. [Brief account of dissections of some of the blind animals from the Mammoth Cave]. *Proceedings of the Boston Society of Natural History* 3 (1850): 349.

———. "On the eye and the organ of hearing in the blind fishes (*Ambylopsis spelaeus,* Dekay) of the Mammoth Cave." *American Journal of Science* 17 (1854): 258–61.

Index